Springer Series in Optical Sciences

Volume 185

For further volumes:
http://www.springer.com/series/624

Springer Series in Optical Sciences

The Springer Series in Optical Sciences, under the leadership of Editor-in-Chief William T. Rhodes, Georgia Institute of Technology, USA, provides an expanding selection of research monographs in all major areas of optics: lasers and quantum optics, ultrafast phenomena, optical spectroscopy techniques, optoelectronics, quantum information, information optics, applied laser technology, industrial applications, and other topics of contemporary interest.
With this broad coverage of topics, the series is of use to all research scientists and engineers who need up-to-date reference books.
The editors encourage prospective authors to correspond with them in advance of submitting a manuscript. Submission of manuscripts should be made to the Editor-in-Chief or one of the Editors. See also www.springer.com/series/624

Minoru Yamada

Theory of Semiconductor Lasers

From Basis of Quantum Electronics to Analyses of the Mode Competition Phenomena and Noise

Springer

Minoru Yamada
Faculty of Electrical and Computer
Engineering
Kanazawa University
Kanazawa
Japan

ISSN 0342-4111 ISSN 1556-1534 (electronic)
ISBN 978-4-431-54888-1 ISBN 978-4-431-54889-8 (eBook)
DOI 10.1007/978-4-431-54889-8
Springer Tokyo Heidelberg New York Dordrecht London

Library of Congress Control Number: 2014935278

Printed on acid-free paper

Springer is part of Springer Science+Business Media (www.springer.com)

Preface

"Laser" is a coined word, formed by taking the initials of words in the phrase "Light Amplification by Stimulated Emission of Radiation." The laser was invented in 1960 and was a ruby laser in which optical light was emitted and amplified in a ruby crystal. The semiconductor laser was invented in 1962.

Semiconductor lasers are being used as principal devices in the technical fields of optoelectronics, such as the optical fiber communication system, the optical disc system, and the optical precise measuring system. Under certain operating situations, however, semiconductor lasers often show undesired performance characteristics such as instability of lasing mode and generation of extra noise. Understanding of these complex characteristics is not easy to achieve, because semiconductor lasers have specific physics that do not play a part in other electrical and optical devices.

This book is written for students, engineers, and researchers who wish to study a unified theoretical account of the operating principles of semiconductor lasers based on theoretical physics rather than on phenomenological treatments. The above-mentioned complex operating characteristics are logically analyzed in this book, which consists of four major sections.

In Part I, "Fundamental Theory," descriptions from quantum electronics focusing specifically on the semiconductor laser are provided from the broader perspective of quantum mechanics. The basic subject matter required to construct a standard semiconductor laser theory is treated here. The author explains the theoretical developments of semiconductor laser theory in a concise but self-contained manner, relying on the Appendices to provide help in understanding related principles and theorems.

The required background material for readers includes classical electromagnetic theory, quantum mechanics, and semiconductor physics. Familiarity with the content of these fields is assumed. Readers wishing to skip the derivation of lasing gain and go directly to a study of the mechanism of laser instability and noise can jump to Chap. 9 in Part II and omit the fundamental theory in Part I.

Although laser quality is evaluated with a view toward achieving a lower threshold current level, higher efficiency, and higher optical power, properties of the lasing modes, and noise problems are also subjects required in the evaluation of laser quality. Criteria for these properties require an understanding of nonlinear phenomena and other complex physics in semiconductor lasers.

One example, namely, whether we can achieve single-mode operation, is analyzed from the perspective of nonlinear gain and discussion of mode competition phenomena. Another example, which is a serious issue for semiconductor lasers, is so-called feedback noise, induced by reflections of output light from connecting optical devices such as optical lenses, optical fibers, or optical discs. The mechanism for feedback noise generation and the technology to suppress it are explained in terms of nonlinear properties and quantum mechanical fluctuations. Theoretical treatments of nonlinear properties, mode competition phenomena, and noise are given in Part II, "Advanced Theory for Mode Competition and Noise."

The standard structures used in semiconductor lasers are the double heterojunction, the stripe structure, and the Fabry–Perot cavity made of cleaved facets. Besides these standard structures, many other structures have been developed to achieve more superior operating characteristics and wider applications.

In Part III, "Structures for Superior Characteristics in the Semiconductor Lasers," the quantum well structure, distributed feedback and mode-selective lasers, and surface-emitting lasers are introduced.

The Appendices, comprising the fourth major section, present related principles and theorems to facilitate understanding of the basis of the theories covered in the book, making it unnecessary to draw upon other sources to achieve that understanding.

Kanazawa, Japan Minoru Yamada

Contents

Part II Advanced Theory for Mode Competition and Noise

Part I
Fundamental Theory

Chapter 1
Overview of the Operating Principles of Lasers

In the beginning of this book, operating mechanisms of the semiconductor lasers are briefly reviewed in this chapter, based on physical understanding without introducing detailed mathematical derivations.

1.1 Optical Emission

The emission of light from any material results from the vibrating motions of charged particles, specifically electrons, within atoms constituting the material or the bulk material itself. For example, consider a hydrogen-like atomic system with the charge on the electron being negative and the nucleus positive, as sketched in Fig. 1.1a. The electron rotates about the nucleus along a spherical trajectory forming an orbit centered on the nucleus. When the electron is located either on the inner orbit a or the outer orbit b, the center of the electron's orbit coincides with the center of the nucleus. Under such circumstances, the electric field produced by the superposition of the electric fields from the orbiting electron and nucleus vanishes outside the atom. However, if the electron orbit spreads over both a and b yielding a distorted orbit, the center of the electron orbit is displaced from the center of the nucleus. With the two charge centers displaced, the electron and nucleus form an electric *dipole*, as illustrated in Fig. 1.1b.

We denote the electron energy in the inner and outer orbits by W_a and W_b, respectively. The respective angular frequencies of the electron in these orbits are then $\omega_a = W_a / \hbar$ and $\omega_b = W_b / \hbar$, where $\hbar$ is the *reduced Planck constant*, i.e., *Planck's constant* h divided by 2π. The displaced electron orbit thus rotates around the nucleus with angular frequency

$$\omega_{ba} = \frac{W_b - W_a}{\hbar}. \tag{1.1}$$

M. Yamada, *Theory of Semiconductor Lasers*,
Springer Series in Optical Sciences 185,
DOI: 10.1007/978-4-431-54889-8_1,

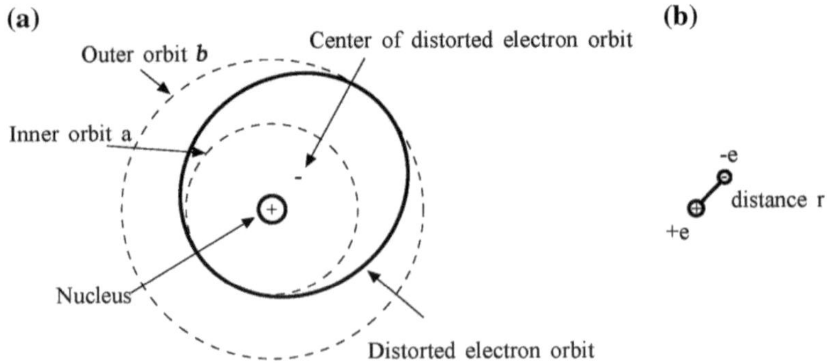

Fig. 1.1 Formation of a dipole in a hydrogen-like atom. **a** Distribution of electron orbit. **b** Image of dipole

Fig. 1.2 Energy levels

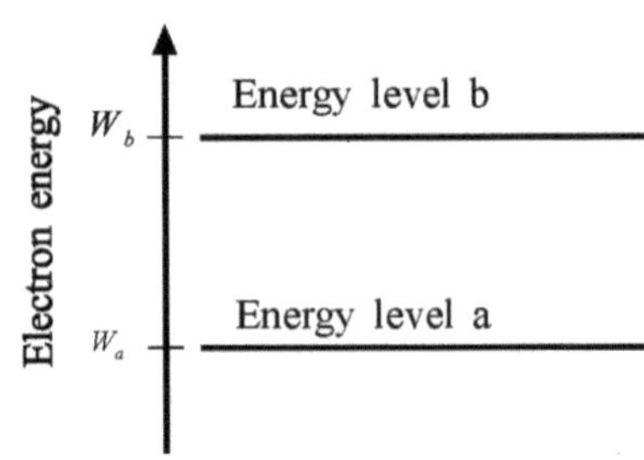

This rotating dipole results in either the emission or absorption of *light*, which is typically an *electromagnetic wave* in the visible range and hence often referred to as an *optical wave*.

If the electron, initially located in the outer orbit b, makes a transition to the inner orbit a, this dipole transition results in an *optical emission* with angular frequency $\omega \approx \omega_{ba}$, where ω is the angular frequency of the emitted light. In transiting, the electron loses energy.

In contrast, an electron, initially located in the inner orbit a, is excited to the outer orbit b; the dipole transition results in an *optical absorption* with angular frequency $\omega \approx \omega_{ba}$. The electron then gains energy in the transition.

The above-described dynamics can also be represented as an electron transition between *two energy levels*, as illustrated in Fig. 1.2. To each level is ascribed a probability of finding the electron, f_a for the lower energy level a and f_b for the upper energy level b. The electron transition from the upper level to the lower level is proportional to the occupation probability f_b in the upper level b and the vacancy probability $(1 - f_a)$ in the lower level a of the electron. The electron transition must also be proportional to the intensity or energy of the existing optical wave surrounding the electron, because the dipole is induced by the electric field of the optical wave.

The energy W_{opt} of the optical wave, to be derived more rigorously in Chap. 2, is

$$W_{opt} = \left(S + \frac{1}{2}\right)\hbar\omega, \tag{1.2}$$

where S is the *photon number*. The optical energy includes the term $\hbar\omega/2$ even if the photon number S is zero. This term, called the *zero-point energy*, represents an energy that can never be absorbed. In essence, the subtraction of the energy of one photon from the zero-point energy would give a photon number of $1/2 - 1 = -1/2$, a negative value of the photon number, which is not physically allowed. Moreover, because a detector needs to absorb at least one photon to measure the energy of the optical wave, we cannot directly measure the zero-point energy.

However, the zero-point energy plays the same role in optical emission as any other photon energy. Because increases in photon number occur only by integer values, the optical emission probability is proportional to $(S + 1)$ and not $(S + 1/2)$. Hence, the *electron transition probability* for an electron undergoing a transition from the higher to the lower energy level by optical emission is

$$P_{b \to a} \propto f_b(1 - f_a)(S + 1). \tag{1.3}$$

The inverse transition from the lower level a to the upper level b occurs through optical absorption. Because the zero-point energy is never absorbed, the electron transition probability is

$$P_{a \to b} \propto f_a(1 - f_b)S. \tag{1.4}$$

The net probability for the electron transition from the upper level to the lower level is then the difference

$$P = P_{b \to a} - P_{a \to b} \propto (f_b - f_a)S + f_b(1 - f_a). \tag{1.5}$$

If $f_b < f_a$, the term $(f_b - f_a)S$ is negative corresponding to optical absorption; if $f_b > f_a$, the term $(f_b - f_a)S$ is positive and we have *stimulated emission*. Under normal circumstances at *thermal equilibrium*, the condition $f_b < f_a$ must hold; hence, the condition $f_b > f_a$ is called *population inversion*. It is this condition under which lasers need to operate. The last term $f_b(1 - f_a)$ is always positive and represents *spontaneous emission*.

We need to keep in mind that an optical wave is always present even if the photon number S is zero because of the presence of the zero-point energy $\hbar\omega/2$, and distortions of the electron orbit in the atom can always be induced even if there is no measurable optical wave.

Optical emission and absorption in *semiconductor materials* are explained with the aid of a rather advanced model. An *electron* in a semiconductor material is not

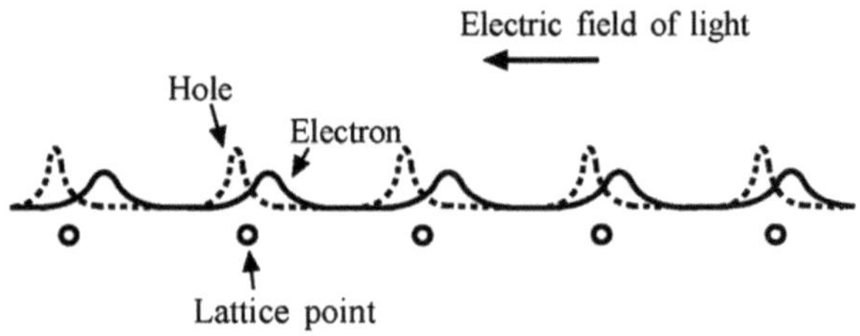

Fig. 1.3 Formation of a dipole in a semiconductor material

bound to a single atom but spreads itself over a large number of atoms. The electron orbits are classified under various continuous energy levels which make up the *band structure* of the material. The distributed electron wave has a precise form, one that is periodically repeated about all *lattice points*, as illustrated in Fig. 1.3.

A *hole* wave located on the opposite side of the lattice point to the electron wave forms as the result of the electric field of the optical wave. The pair of an electron and a hole constitutes the *dipole.*

In a semiconductor material, the upper and the lower levels as sketched in Fig. 1.2 should be counted among the energy levels in the *conduction* and *valence bands,* respectively. Both occupation probabilities f_b and f_a are defined for the electron. The vacancy probability $(1 - f_a)$ in the lower level corresponds to the existence of the hole in the valence band. Then, the electron transition probability is proportional to $f_b(1 - f_a)$, also called the *recombination probability* of the electron and hole.

1.2 Coherent Light

When a hydrogen-like atom or an electron-hole pair in a semiconductor material emits light through an electron transition from upper level b to lower level a, a temporal variation of the electric field E of the emitted light arises as in Fig. 1.4a. As the emission has to diminish after the electron transition, the temporal variation corresponds to a damped harmonic oscillation with decay time τ_s. The frequency spectrum of this emitted light is obtained using *Fourier analysis* of the temporal variation. The transform of the spectrum takes a Gaussian form, as shown in Fig. 1.4b, where its *half-width* is $1/\tau_s$.

Continuous optical emission from any material is given by a series of emissions from a large number of atoms or electron-hole pairs. For a series of emissions, two mechanisms are possible. One is that each atom or electron-hole pair emits light independently at different times, as illustrated in Fig. 1.5a. Because atoms or electron-hole pairs weakly interact with each other, the energy difference W_{ba} should differ slightly for each atom or electron-hole pair. Each light wave must then have a different angular frequency ω_i, where i denotes the emitting atom or electron-hole pair. In summing up the phases of each contributing light wave, the overall phase becomes random. The total optical power of the summed light wave is

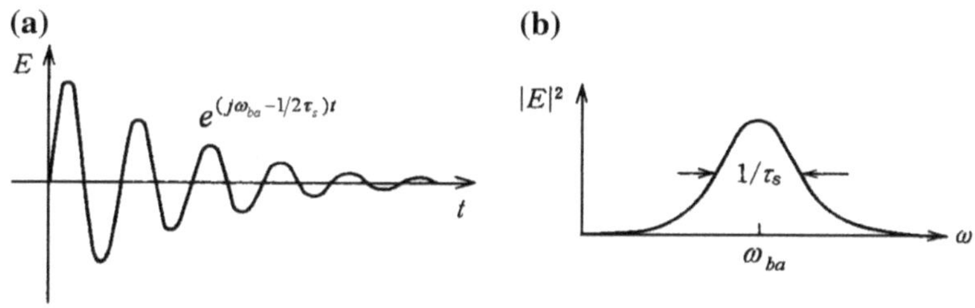

Fig. 1.4 Optical emission from an atom or an electron-hole pair. **a** Temporal variation. **b** Frequency spectrum

given by simply summing the optical power of each contributing light wave. The frequency spectrum for the series emission is approximately given as $1/\tau_s + \Delta\omega_i$, where $\Delta\omega_i$ ranges over the angular frequencies arising from the mutual interactions among the atoms or the electron-hole pairs. Because of its random nature, we call the light output from this type of series of emissions *incoherent light*. Light seen in nature and that emitted from conventional lamps are examples of incoherent light.

Another mechanism for a series of emissions is that a new emission starts by adopting both the *vibrational phase* and *frequency* of the preceding emitted optical wave, as illustrated in Fig. 1.5b. In summing up each light contribution, the overall light wave retains the same amplitude and vibrational frequency; that is, it is monochromatic. The overall spectrum corresponds to a line spectrum, because there is no damping. Light output from this type of series emissions is called *coherent light*.

In general, *coherency* refers to the capability of light waves to interfere, and is classified as either temporal or spatial. The above discussion concerns temporal coherency. Temporal coherent light does not exist in nature and is generated artificially only within *lasers*.

Another aspect of laser light is the superior ease with which in may be focused onto a very small area yielding high power densities. As is generally known, using a convex lens we can form an inverse real image of an illuminated lamp, as illustrated in Fig. 1.6a. This effect means that light emitted from different parts of the thermally-excited filament are focused onto different spatial points, resulting in a copy of the full shape of the lamp. In contrast, for a laser, the light emitted from different positions within the laser material is focused by the convex lens onto an almost identical point, as illustrated in Fig. 1.6b. As a result the optical power density can become high.

1.3 Fundamental Structure of Laser

A schematic diagram illustrating the fundamental elements of a laser is given in Fig. 1.7. The laser consists of three parts: the first is the *laser material* that produces the optical emission and optical amplification; the second is the *equipment*

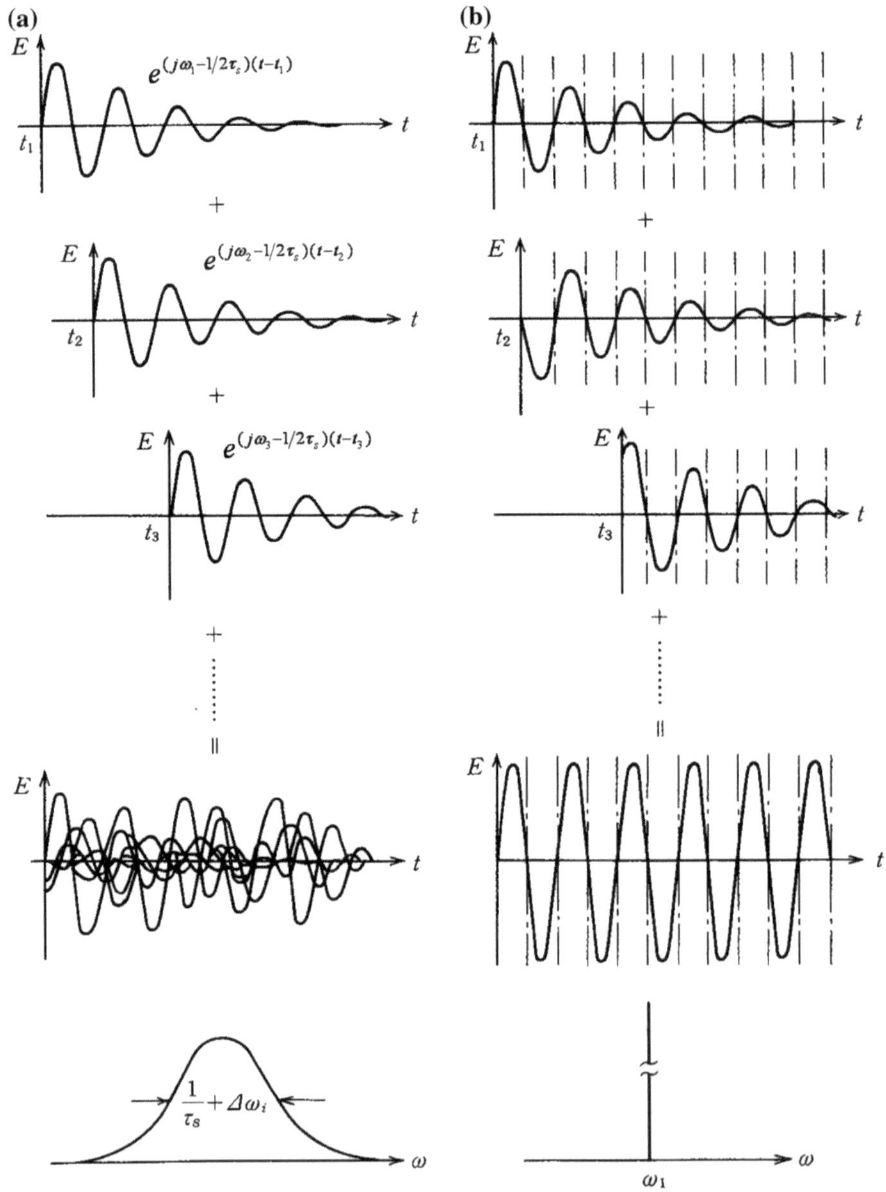

Fig. 1.5 Optical emission from an ensemble of atoms or electron-hole pairs. **a** Incoherent light. **b** Coherent light

supplying the energy to excite the electrons in the laser material; and the third is the *cavity* comprising two *mirrors* that confine the optical wave.

The optical wave is generated in the laser material by spontaneous emission and is amplified in the material aided by stimulated emission. The amplified optical

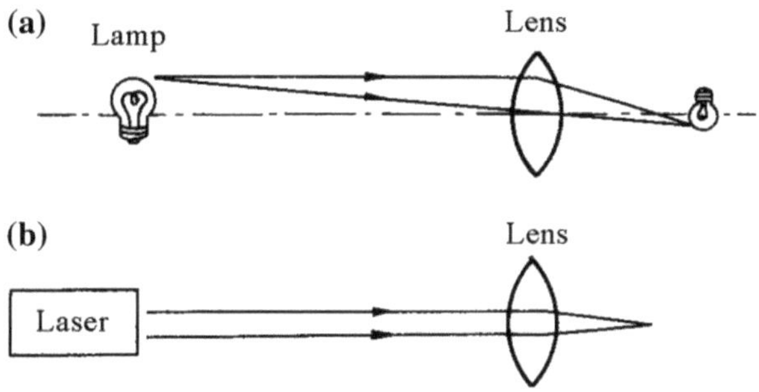

Fig. 1.6 Differences in focusing between conventional light and laser light. **a** Conventional light. **b** Laser light

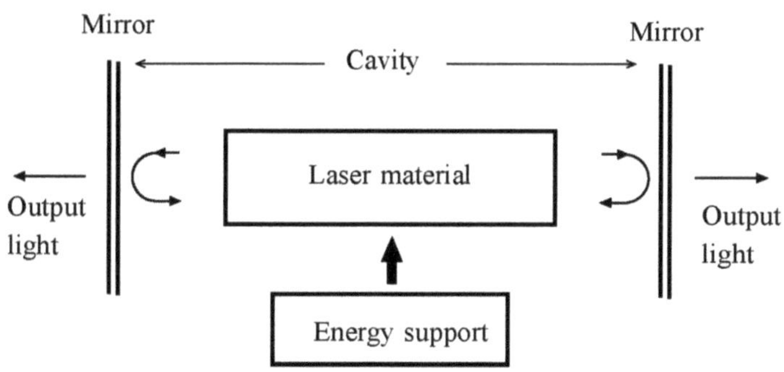

Fig. 1.7 Fundamental construction of a laser

wave is reflected by one mirror, returning to the laser material amplified again, and reflected by an opposing mirror. The optical wave undergoes repeated amplification and reflection in the cavity. The mirrors are not 100 % reflective; a portion of the wave passes through the mirrors and escapes as *output light*. The condition to maintain lasing is to preserve both the amplitude and vibrational phase of the optical wave after each cycle in the cavity with those at the start. Because this operating condition is the same as the oscillation condition in conventional electronics, we call the lasing operation the *laser oscillation*.

Depending on the material used, lasers are classified into four types: *gas lasers*, *liquid lasers*, *solid lasers* (or solid state laser) and *semiconductor lasers*. Gas lasers make use of the excited energy levels of atoms or molecules in a gas. The excitation energy is supplied from gas discharges by applying a high electric voltage. He-Ne, He-Cd, Ar, and CO_2 are typical gas media in such lasers. Liquid lasers use the energy levels of atoms or molecules dissolved in a *liquid*. Dye lasers are a common type of liquid laser, where the lasing material is typically a dye dissolved in ethanol. Similarly, solid lasers use the energy levels of atoms embedded in a *host material*. For example, in a Nd-YAG laser, an yttrium-aluminum-garnet

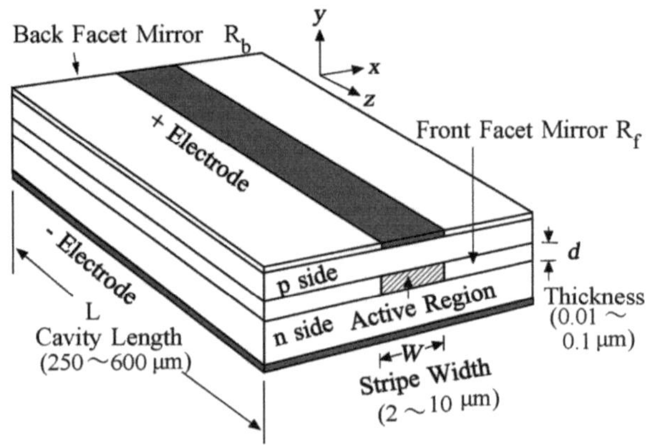

Fig. 1.8 Typical structure of semiconductor laser

(YAG) crystal is the host material holding the lasing material composed of Nd atoms. The excitation energies in both liquid and solid lasers are supplied through optical irradiation.

Semiconductor lasers have a number of different features from the other three laser types. The first concerns the energy levels used. Semiconductor lasers use the band energy levels of a semiconductor material, whereas other lasers use the energy levels of isolated atoms or molecules. Semiconductor lasers utilizes a *current injection* based on *p-n* (or *p-i-n*) junctions (to be described below) to provide the energy supply, whereas other lasers operate through gas discharge or optical irradiation.

1.4 Typical Structure of Semiconductor Laser

A typical structure of a semiconductor laser is illustrated in Fig. 1.8. The laser is constructed with *p-i-n* semiconductor *heterojunctions*; here, *p* and *n* denote materials with positive and negative charge carriers, and *i* signifies an intrinsic material. The *active region* is the main region of the laser, in which the optical wave is generated and amplified. Electrons and holes are injected into the active region from the *n* and *p* side electrodes, respectively. The optical wave propagates along the active region in the *longitudinal z* direction. In our setup, the *x-y* plane is the *transverse* cross-section to the longitudinal direction. The active region works as the *core* of a *waveguide*. The opposing ends of the active region are cleaved surfaces of the crystal and are used as *facet mirrors*. The optical light is reflected back by these facet mirrors and forms a *standing wave* in the longitudinal direction. Lasing is achieved, aided by the waveguide and the facet mirrors as well as wave amplification in the active region.

The cavity structure with simple two facet mirrors is sometimes called *Fabry-Perot type cavity*.

Because the reflectivity R_f of the front facet is less than 100 %, a portion of the optical wave escapes through the front facet and constitutes the optical output of

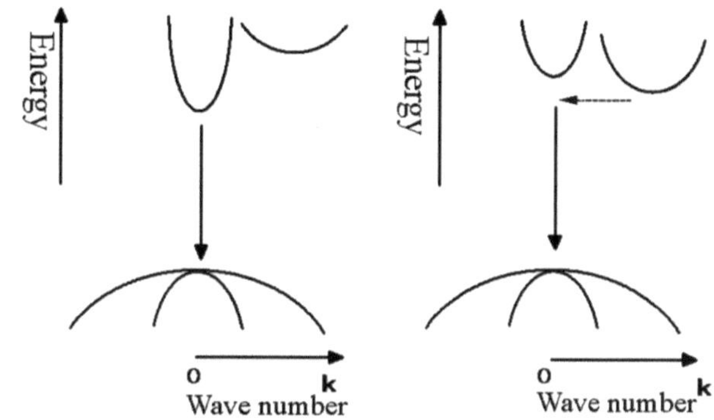

Fig. 1.9 Direct transition and indirect transition

the laser. The intensity profile of the output optical light is determined by the geometrical design of the active region and surrounding *cladding regions*.

The operating current of a semiconductor laser is called the *injection current*, because the current is used to inject electrons and holes into the active region.

1.5 Material

The energy levels of the semiconductor material form the band structure and are characterized by the *wave number* $\boldsymbol{k}$ of the electrons. The $\boldsymbol{k}$-dependence of the energy is approximately parabolic as depicted in Fig. 1.9.

Based on the band structure, semiconductor materials are classified into two groups, *direct transition materials* and *indirect transition materials*. The energy levels of the first group have at position $\boldsymbol{k} = 0$ a lowest point in the conduction band and a highest point in the valence band. The second group, in contrast, have these points in the two bands located at different wavenumbers. The electrons fill the conduction band around the lowest point, whereas the holes occupy the valence band around the highest point. When the electron transits from the conduction band to the valence band, the wave number of the electron has to coincide with that of the hole because of *momentum conservation*. For direct transition materials this is straightforward.

For indirect transition materials, the electron must change its wave number either by emitting a phonon or through some other mechanism, to effect a transition to the valence band. Because this transition is achieved via a two-step process, the transition probability is low. Therefore, indirect transition materials are not suitable as key materials for the active region.

Typical direct transition materials include GaN, GaAs, and InP; indirect transition materials include Si, Ge, and AlAs. The composition of *complex compound materials*, such as $Al_xGa_{1\text{-}x}As$ and $Ga_xIn_{1\text{-}x}As_yP_{1\text{-}y}$, can be controlled so that the materials operate as either a direct or an indirect transition materials.

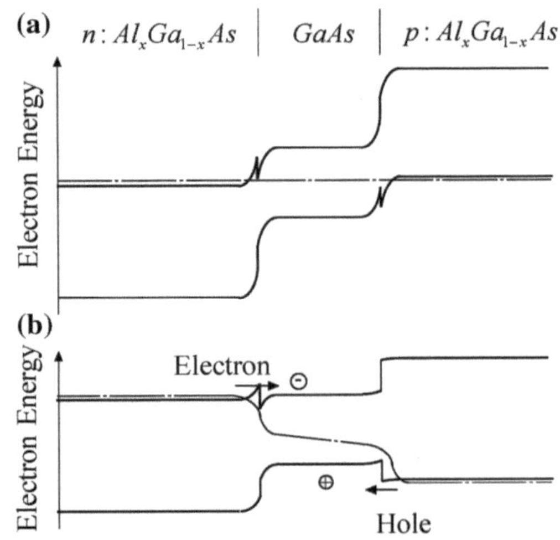

Fig. 1.10 Electron and hole injection using a double hetero (DH) junction. **a** Thermal equilibrium. **b** Voltage applied

1.6 Injection Mechanism

Population inversion of a semiconductor laser is achieved with *p-i-n junctions* with a band structure, illustrated in Fig. 1.10. The center region is the active region, which is made of a direct transition material such as GaAs. In most cases, the material of the active region does not have any impurity dopants. The adjoining regions are *n*-type and *p*-type materials that have larger *band gaps* than the active region. If the junction is constructed from different materials such as GaAs and $Al_xGa_{1-x}As$, the structure is called a *heterojunction*. Furthermore, a structure which has two junction planes, as drawn in Fig. 1.10, is called a *double hetero (DH) junction* [1, 2].

When a voltage is applied from an external electric source, the energy levels of the *n*-type side are raised and those of the *p*-type side are lowered. Electrons and holes are then injected into the active region from the *n*-type and *p*-type side, respectively. Because the band gaps of the adjoining regions are higher than that in the active region, the injected electrons and holes are blocked from escaping to the opposite side by the potential walls, and are confined to the active region. To avoid unnecessary recombination of the electrons and holes, the adjoining regions to the active region must be composed of indirect transition materials.

1.7 Output Light

When we apply an electric source to the *p-i-n* junction from the electrode in the *forward direction*, an electric current is injected into the active region causing spontaneous and stimulated emissions to occur. The variation of the optical output with injection current *I* is illustrated in Fig. 1.11. When the injection current is

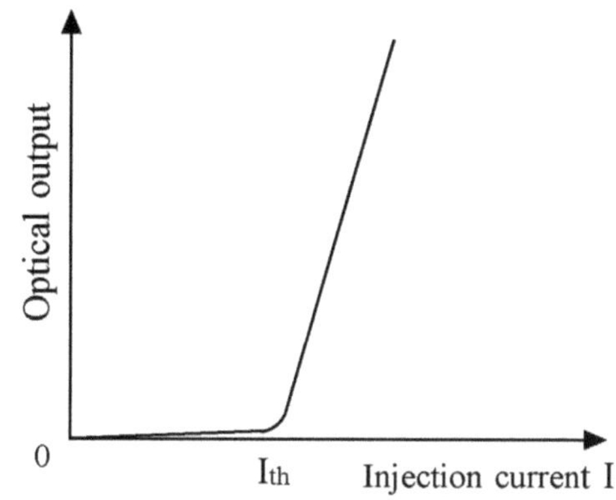

Fig. 1.11 Optical output versus injection current

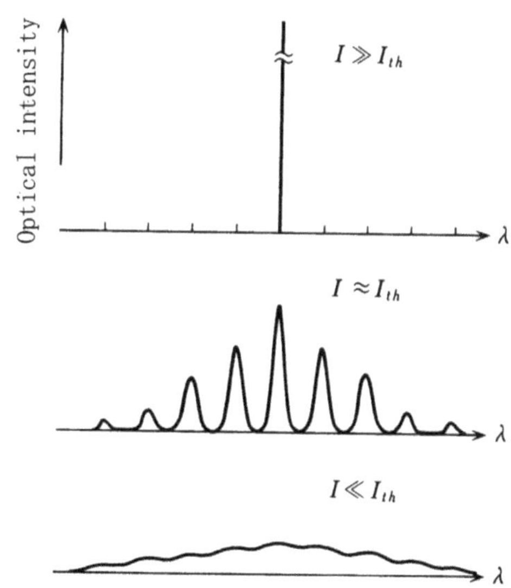

Fig. 1.12 Variation of optical spectrum around the threshold current

low, the laser produces only a small output arising from the spontaneous emissions. By increasing the injection current, the spontaneous emissions are amplified by the stimulated emissions. The optical output rapidly rises from a specific current level called the *threshold current* I_{th}, increasing almost linearly with injection current I.

The variation of the *optical spectrum* of the output light is illustrated in Fig. 1.12. When the injection current is much lower than the threshold current, $I \ll I_{th}$, the output spectrum is broad because of the nature of spontaneous emission, but exhibits periodic modulation in the intensity due to resonances in the laser cavity. When the laser current nears the threshold, $I \approx I_{th}$, the spectrum shows well-separated *multiple peaks*. With further increases in the injection current, the laser shows either a *single-mode* spectrum as shown in Figs. 1.12 and 1.13a, or a *multimode* spectrum, as illustrated in Fig. 1.13b.

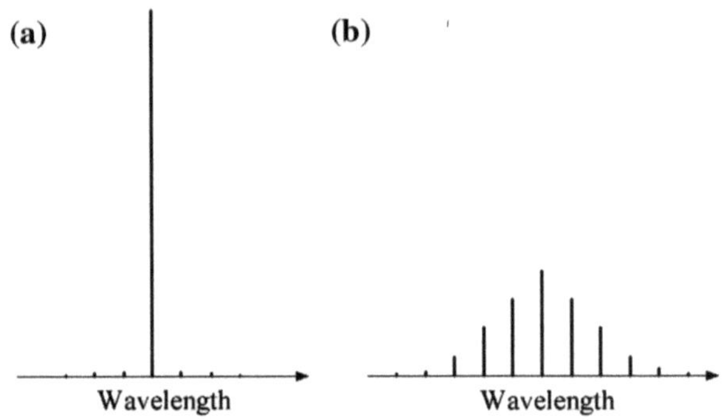

Fig. 1.13 Spectrum of optical output in lasing operation. **a** Single mode operation. **b** Multimode operation

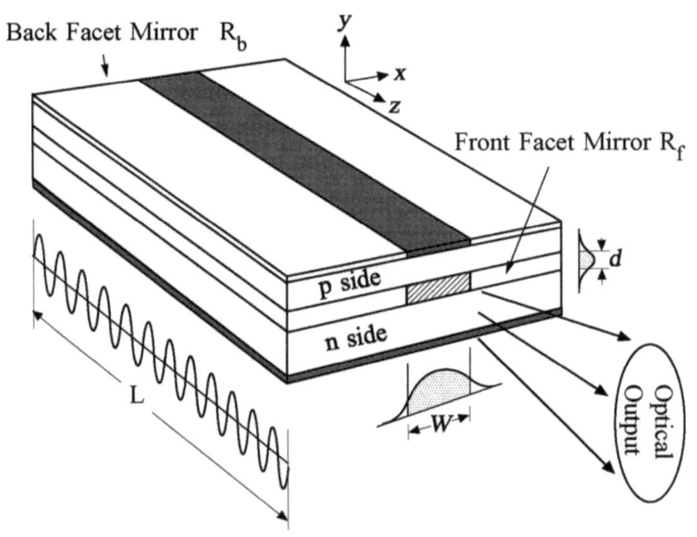

Fig. 1.14 Field distribution of the optical wave

1.8 Lasing Mode

The properties of the optical wave are characterized according to *mode*. The lasing mode is classified by the three-dimensional spatial distribution of the optical wave and the direction of the electric polarization. An optical wave distributed along its propagation direction is called a *longitudinal mode*. In Fig. 1.14, the longitudinal direction is by convention taken to be the z-axis. Waves distributed in the two lateral directions to the propagation direction, the x- and y-axes in Fig. 1.14, are called *transverse modes*. The transverse modes are conventionally taken to be along the width direction x and the thickness direction y of the active region. The heterojunction of the active region is designed to achieve a low threshold current and a suitable optical wave profile along the thickness direction. The field distribution along the width direction is controlled using an artificially designed structure called the *stripe structure*.

As the field distribution along the thickness direction is usually sharper than that in the width direction, the output optical wave spreads more widely along the thickness direction than the width direction. Additionally, the direction of polarization is mainly characterized by the field distribution along the thickness direction. The main component of the *transverse electric (TE) mode* is E_x, and that of the *transverse magnetic (TM) mode* is E_y.

The spectrum of the output optical wave directly corresponds to the longitudinal mode. There are two kinds of spectrum. One corresponds to the *single-mode* (or *single-longitudinal-mode*) operation and the other is the *multimode* (or *multi-longitudinal-mode*) operation as illustrated in Fig. 1.13. Control of the longitudinal mode is not easy and relates to rather complicated physics. One purpose of this book is to provide a theoretical understanding into how the lasing mode can be controlled.

Bibliography

1. H. Kressel, J K. Butler, *Semiconductor Lasers and Heterojunction LEDs* (Academic Press, New York, 1977)
2. H.C. Casey Jr., M.B. Panish, *Heterostructure Lasers (Part A: Fundamental Principles)* (Academic Press, New York, 1978)
3. D. Marcuse, *Principles of Quantum Electronics*. (Academic Press Incorporated, New York, 1970)
4. A. Yariv, *Quantum Electronics*, 3rd edn. (Wiley, New York, 1989). ISBN: 978-0-471-60997-1
5. D.F. Walls, G.J. Milburn, *Quantum Optics*, 2nd edn. (Springer, Berlin, 2008). ISBN: 978-3-540-28574-8
6. M. Sargent III, M.O Scully, W.E. Lamb Jr., *Laser Physics* (Perseus Books Group, New York, 1978)
7. A.E. Siegman, *Lasers* (University Science Book, Mill Valley, 1986). ISBN: 0-935702-11-3
8. M. Ohtsu, *Highly Coherent Semiconductor Laser* (Artech House, Boston, 1992)
9. W.W. Chow, S.W. Koch, M. Sargent, *Semiconductor-Laser Physics* (Springer-Verlag, Berlin, 1994)
10. Y. Suematsu, A.R. Adams (eds.), *Handbook of Semiconductor Lasers and Photonic Integrated Circuits* (Chapmann & Hall, London, 1994)
11. T. Numai, *Fundamentals of Semiconductor Lasers* (Springer, New York, 2004). Springer Series in Optical Sciences
12. T. Suhara, *Semiconductor Laser Fundamentals* (Marcel Dekker Inc., New York, 2004)
13. D. Kane, A. Shore, *Unlocking Dynamical Diversity: Optical Feedback Effects on Semiconductor Lasers* (Wiley, Chichester, 2005)
14. J. Otsubo, *Semiconductor Lasers: Stability, Instability and Chaos*, (Springer, Berlin, 2006). Springer Series in Optical Sciences
15. Y. Yamamoto, *Coherence, Amplification, and Quantum Effects in Semiconductor Lasers*, (Wiley-Interscience, New York, 2007)
16. The Landolt-Börnstein Database, *Advanced Materials and Technologies, Laser Physics and Applications, Laser Systems. Part 3,* (Springer, Heidelberg, 2011)
17. J.J. Coleman, A.C. Bryce, *Advances in Semiconductor Lasers* (Academic Press, San Diego, 2012)

Chapter 2
The Photon

Theoretical analyses toward building up a semiconductor laser theory are start from this chapter. This book is aimed not require to read other books for understanding basis of the theory. When the readers wish to refer other books to compare our treatments with them, bibliographies are listed in the end of the last chapter.

2.1 Analytical Approaches

Laser operation is based on the interaction between an optical wave and the semiconducting material, or more specifically between an electromagnetic (EM) wave and electron-hole pairs in the semiconductor as depicted in Fig. 2.1. As already discussed in Sect. 1.1, these interactions are classified according to whether they give rise to optical absorption, stimulated emission, or spontaneous emission. The accompanying noise is understood as a summed effect of fluctuations occurring from these interactions.

Nonetheless, the analytical method of any branch of physics is grounded in *classical mechanics* and *quantum mechanics*. The analytical approaches to these interactions are classified into three types as listed in Table 2.1. The first, called *full quantization*, treats both the optical wave and the material using quantum mechanics. The second, called *semi-classical*, continues to treat the material quantum mechanically but treats the optical wave using classical mechanics. The third, called *fully classical*, treats both optical wave and material using classical mechanics.

If we use classical mechanics to analyze the electron dynamics in the material, we can describe optical absorption by the material, but cannot explain stimulated emission. Spontaneous emission can be theoretically described only using a full quantization approach. The term *photon* is given to the massless particle representing a quantum of light following the *quantization* of the optical wave. However, results using quantized optical fields do not always match those from conventional treatments in electronics originally founded on classical mechanics,

M. Yamada, *Theory of Semiconductor Lasers*,
Springer Series in Optical Sciences 185,
DOI: 10.1007/978-4-431-54889-8_2, © Springer Japan 2014

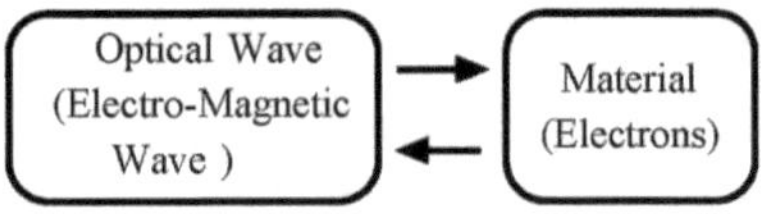

Fig. 2.1 Interaction between optical wave and material

Table 2.1 Relationships between the analytical approach and phenomenological treatment

Analytical approach	Mechanics		Phenomena			
	Optical wave	Material	Optical absorption	Stimulated emission	Spontaneous emission	Noise
Full quantization	Quantum	Quantum	Theoretical	Theoretical	Theoretical	Theoretical
Semi-classical	Classical	Quantum	Theoretical	Theoretical	Phenomeno-logical	Phenomeno-logical
Fully classical	Classical	Classical	Theoretical	Invalid	Phenomeno-logical	Phenomeno-logical

such as EM theory, circuit theory, control theory, and signal processing theory. The classical mechanics require us simultaneous representation of the energy and the vibrating phase. From a quantum mechanical viewpoint, there exists an *uncertainty* between the energy and vibrational phase of an optical wave as will be shown in later. The word of *temporal coherency* originates from classical mechanics not from the quantum mechanics. Therefore, in many cases, we apply a semi-classical approach in which the optical wave is analyzed using the classical EM equations of Maxwell. Spontaneous emission and noise are introduced via phenomenological equations aided by results obtained from quantizing the optical wave.

2.2 Principles of Classical Mechanics

Many people find quantum mechanics is challenging to understand and to envisage. This stems from the different logical framework of quantum mechanics from that of Newtonian mechanics. Classical and quantum mechanics are compared in Table 2.2. There are two types of mathematical equations used to describe dynamics. One is a theory based on differential equations corresponding to *causal relationships*. The other is a theory with integral equations corresponding to a *variational principle*. Even in classical mechanics, Newtonian and analytical mechanics are founded on completely different logical frameworks. Let us review these frames of classical mechanics, before moving onto quantum mechanics.

In *Newtonian mechanics*, it is postulated that the rate of change of momentum is given by a force; force is the "cause" resulting in a momentum change, the

Table 2.2 Logical frameworks and classification of mechanics

	Theory with differential equation (causal relationship)	Theory with integral equation (variational principle)	Reproducibility
Classical	Newtonian mechanics	Analytical mechanics	Assumed
Quantum	?	Quantum mechanics	Uncertain

"effect". This postulate can be described mathematically using a differential equation

$$\frac{dp}{dt} = F, \tag{2.1}$$

where p is the momentum of a material object and F the force applied to that object. In Newtonian mechanics, the relation $p = m\ dx/dt$ *is presupposed*, where x and m are the position and mass of the object, respectively. The Maxwell EM equations are also described in classical theory using differential equations.

In *classical analytical mechanics*, the relation between x and p is supposed *to be unknown initially*. The postulate of this mechanics is that the relation between x and $\dot{x}$ is determined by the *principle of least action*, i.e., the minimization of the value of the action J defined as

$$J = \int_{t_1}^{t_2} L(x, \dot{x}, t)dt, \tag{2.2}$$

with the *Lagrangian*

$$L(x, \dot{x}, t) = T(x, \dot{x}, t) - U(x, t), \tag{2.3}$$

where $T(x, \dot{x}, t)$ is the kinetic energy, $U(x, t)$ the potential energy, and

$$\dot{x} = \frac{\partial x}{\partial t} \tag{2.4}$$

is the velocity. In classical analytical mechanics, the relation between x and $\dot{x}$ is also unknown initially. To find this relation through (2.2) is the fundamental objective of classical analytical mechanics. The momentum p is defined as

$$p = \frac{\partial L(x, \dot{x}, t)}{\partial \dot{x}}, \tag{2.5}$$

With p called the *conjugate* of x. The *classical Hamiltonian* $H(x, p, t)$ is defined using $\dot{x}$, p, and $L(x, \dot{x}, t)$ as

$$H(x, p, t) = \dot{x}p - L(x, \dot{x}, t) = T(x, \dot{x}, t) + U(x, t). \tag{2.6}$$

The Hamiltonian determines the total energy of the mechanical system.

The Lagrangian $L(x, \dot{x}, t)$ and Hamiltonian $H(x,\ p,\ t)$ are called generating functions. Physical quantities such as x and p are obtained by differentiating the classical Hamiltonian:

$$\dot{p} = -\frac{\partial H(x, p, t)}{\partial x}, \tag{2.7}$$

$$\dot{x} = \frac{\partial H(x, p, t)}{\partial p}. \tag{2.8}$$

Note that in classical analytical mechanics the physical quantities x and p have been introduced as independent parameters. The relation $p = m\ dx/dt$ is obtained as *a derived result of classical analytical mechanics.*

As given in the above discussion, the basic postulate and the logical framework of classical analytical mechanics differ completely from those of Newtonian mechanics.

Here, we find a situation in (2.2) that is strange. The lower limit t_1 of the integration corresponds to the starting time of the dynamics which is in the *past,* whereas the upper limit t_2 corresponds to the finishing time of the dynamics which must be in the *future.* Can we know the future? The answer to this question in classical mechanics is "Yes". The common assumption in classical mechanics is that any dynamics is *reproducible* for temporal variation. Hence, both Newtonian mechanics and classical analytical mechanics are consistent and compatible with each other. However, we know through personal experiences that nature and human life are never reproducible.

Because the fundamental equation (2.1) in Newtonian mechanics is given for the instantaneous time *now*, everyone can readily accept its outcomes. However, everyone might not accept the outcomes from classical analytical mechanics because (2.2) requires us to suppose that our *future* is reproducible.

Even in Newtonian mechanics, we can predict our future by integrating (2.1) into the future, if that is what we wish to know. The key concept underlying classical mechanics is the assumption of *reproducibility*.

Here, we present an example illustrating the logical difficulty in classical mechanics without the assumption of *reproducibility.* We consider an experiment to measure position and velocity of an electron moving in a vacuum using metal slits, as depicted in Fig. 2.2. When an electron passes a slit set at position x, the induced electric current in the slit rises to a peak and falls way. We can then fix a time t when the electron passes this slit, as sketched in Fig. 2.2a. Here, we also suppose that the sensitivity and resolution of the measuring system are sufficient enough to register sharp measurements.

Meanwhile, the instantaneous velocity v of the moving electron as specified within Newtonian mechanics is

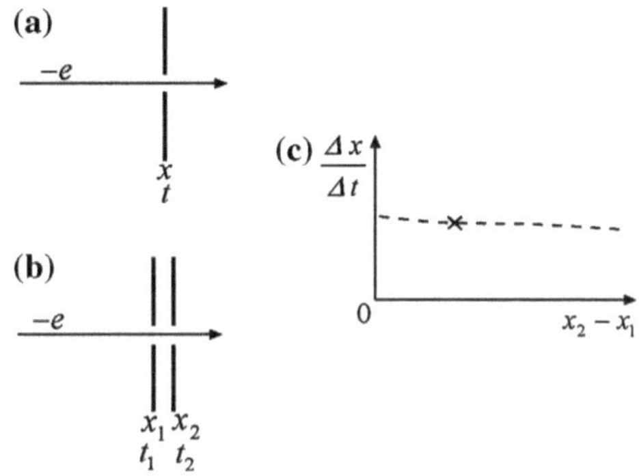

Fig. 2.2 Example to show the relationship between reproducibility and uncertainty. **a** Position measurment. **b** Velocity measurment. **c** Variation of measured data

$$v = \lim_{\Delta t \to 0} \frac{\Delta x}{\Delta t}. \tag{2.9}$$

Thus, we need to prepare two slits, as shown in Fig. 2.2b, to measure the difference in position and the interval of time

$$\Delta x = x_2 - x_1, \tag{2.10}$$

$$\Delta t = t_2 - t_1. \tag{2.11}$$

We can then plot point $(x_2 - x_1)/(t_2 - t_1)$ on the graph of $\Delta x/\Delta t$ versus $x_2 - x_1$, as illustrated in Fig. 2.2c. Another requirement in determining the velocity is that we need to repeat the measurement to be able to obtain the extrapolated point $\lim_{\Delta t \to 0} \Delta x/\Delta t = \lim_{\Delta x \to 0} \Delta x/\Delta t$ by reducing the distance $x_2 - x_1$ between the two slits, as shown in Fig. 2.2c. Here, we pose the following questions:

Is the experiment reproducible? If we repeat the experiment, can we get exactly the same result? Is there any proof that the extrapolated point $\lim_{\Delta t \to 0} \Delta x/\Delta t = \lim_{\Delta x \to 0} \Delta x/\Delta t$ converges to a unique point?

Most people might respond to these questions using a statistical treatment, for example, to repeat the experiment many times and to take an averaged value. However, statistical treatments are only supported on the *assumption of reproducibility*.

If we suppose that we can do the experiment only once and repeating the experiment is not permitted, what sorts of logical differences will be generated. The aim of the experiment is to measure the position and the velocity simultaneously.

We need to use two slits, because the velocity cannot be measured by one slit only. We might set the slit distance $x_2 - x_1$ as small as possible to obtain $(x_2 - x_1)/(t_2 - t_1)$. However, this value is not exactly the value $v \equiv \lim_{\Delta t \to 0} \Delta x/\Delta t = \lim_{\Delta x \to 0} \Delta x/\Delta t$, because $t_2 - t_1$ is still nonzero albeit small. The next question is whether we can ascertain the position x of the electron exactly when the electron has velocity v. We can only say that the electron is present in the space between x_1 and x_2 during the time interval $t_1 - t_2$. Thus, we cannot measure precisely the position x and velocity v simultaneously from only a single measurement. An accurate measurement of the position requires an instantaneous time t, while in principle the measurement of velocity requires two instantaneous times t_1 and t_2. Such incompatible requirements are at the root of *uncertainty* in quantum mechanics as we shall explain in the following sections.

The reader should bear in mind that classical mechanics is established under the assumption of reproducibility.

2.3 Principles of Quantum Mechanics

The basic philosophy of quantum mechanics is different from that of classical mechanics. Quantum mechanics does not suppose reproducibility for the temporal variation of physical quantities but permits uncertainty in the dynamics. The rules and postulates of quantum mechanics are summarized in four axioms:

(1) Any *physical quantity* is expressed as an expectation value of an *operator* A and *a state* $|\Psi\rangle$ as $\langle A\rangle = \langle\Psi|A|\Psi\rangle$, where $\langle A\rangle$ is said to be the *expectation value* of the physical quantity.
(2) The temporal variation of any physical quantity is *not* given directly by the dynamics of the *physical quantity* itself, but is given by the dynamics of the quantum mechanical *operator* or *state.*
(3) The temporal variation of the quantum mechanical operator or state is determined from one of the following equations:

- the *Schrödinger equation* for arbitrary state $|\Psi\rangle$,

$$j\hbar\frac{d|\Psi\rangle}{dt} = H\,|\Psi\rangle, \tag{2.12}$$

where j is the imaginary unit and H is the *quantum mechanical Hamiltonian.*
- the *Heisenberg equation* for arbitrary operator A,

$$\frac{dA}{dt} = \frac{1}{j\hbar}[A,H] = \frac{1}{j\hbar}(AH - HA), \tag{2.13}$$

where [,] denotes the *commutator* of the two arguments.
- the *commutation relation* for *canonical conjugate operators*,

$$[q,p] = qp - pq = j\hbar. \tag{2.14}$$

The applications of these three equations are different in mathematics. However, their intrinsic characteristics are the same. We are able to develop quantum mechanical properties by applying any one of these equations.

(4) Any observable physical quantity is represented by a *real number*. Introducing the imaginary unit into quantum mechanics aids the handling of the mathematics more simply. The *imaginary unit j* (sometimes denoted *i*) merely indicates $\pi/2$ preceding a *phase variation* for periodically vibrating phenomena.

Let us recall some of the mathematical structure of quantum mechanics:

We introduce $\langle \Psi |$ and $| \Psi \rangle$, called the *bra vector* and *ket vector* respectively, to denote the states of a quantum mechanical system. For each operator A acting on the ket vector, $A|\Psi\rangle$, a *conjugate operation* A^+ is defined that acts on the bra vector, $\langle \Psi | A^+$, where the symbol $^+$ is read as *dagger*. When the relation $A^+ = A$ holds, this operator is called a *Hermite operator* or *Hermitian*. The expectation value $\langle A \rangle = \langle A^+ \rangle$ of a Hermite operator is always a *real number*.

2.4 Procedure for Quantization

Because any dynamic behavior can be derived from the quantum mechanical Hamiltonian H, quantum mechanics is categorized as a *theory with an integral equation*, as stated in Table 2.2. Even though the Schrödinger equation and the Heisenberg equation are given in the form of differential equations, these equations do not imply a causal relationship. These equations describe the dynamics of states and operators, and not of physical quantities. Indeed, theories based on causal relations, such as Newton's dynamical equation and Maxwell's equations, are easy to understand. Unfortunately, there is no direct theory to treat the quantum mechanical properties based on causal relationships. As we shall show in Chaps. 5 and 6, the density matrix equation and the rate equations are methods to enable causal relations to be treated by taking into account the quantum mechanical properties in the form of differential equations. However, these are *skillfully modified equations* to cover the weaknesses of quantum mechanics by allowing one to step out of the exact quantum mechanical framework. Therefore, the column of "Theory with differential equation" and "Quantum" in Table 2.2 is noted with " ? " mark.

A way of applying quantum mechanics is to trace and convert categories given in Table 2.2 using the following four steps:

1. Formulate classical differential equations, and analyze them.
2. Calculate the energy associated with the classical Hamiltonian as an integral equation.

3. Regard the dynamic variables as quantum mechanical operators as well as the energy as the quantum mechanical Hamiltonian.
4. Analyze the quantum mechanical features using one of the equations (2.12)–(2.14).

These four steps embody the notion of quantizing a general classical system. Let us now quantize the optical wave by following these steps.

2.5 Classical Form of Maxwell's Equations

The first step towards quantization is to formulate classical differential equations and analyze them. Optical waves in the semiconductor laser propagate along the active region and are reflected back by a facet mirror at either end, thereby forming a standing wave in the laser cavity, as sketched in Fig. 1.14. In this chapter, we assume that reflectivities R_f and R_b of the two facet mirrors are 100 %, and there is neither gain nor loss in the laser cavity. That is, the entire optical wave is confined within the laser cavity in a steady state.

For EM waves without sources, Maxwell's equations are given as

$$\nabla \times \boldsymbol{H} = \varepsilon \frac{\partial \boldsymbol{E}}{\partial t}, \tag{2.15}$$

$$\nabla \times \boldsymbol{E} = -\mu \frac{\partial \boldsymbol{H}}{\partial t}, \tag{2.16}$$

$$\nabla \cdot \boldsymbol{E} = 0, \tag{2.17}$$

$$\nabla \cdot \boldsymbol{H} = 0, \tag{2.18}$$

where $\boldsymbol{E}$ is the *electric field* vector, $\boldsymbol{H}$ the *magnetic field* vector, ε the dielectric constant, and μ the magnetic permeability. Here, we introduce a *vector potential* $\boldsymbol{A}$ subject to the *Coulomb gauge condition*

$$\nabla \cdot \boldsymbol{A} = 0. \tag{2.19}$$

Both $\boldsymbol{E}$ and $\boldsymbol{H}$ are derived from the vector potential:

$$\boldsymbol{E} = -\frac{\partial \boldsymbol{A}}{\partial t}, \tag{2.20}$$

$$\mu \boldsymbol{H} = \nabla \times \boldsymbol{A}. \tag{2.21}$$

Equations (2.20) and (2.21) automatically satisfy (2.17) and (2.18), respectively. By substituting (2.20) and (2.21) into (2.15) and (2.16) and using the well-known formula from vector calculus

$$\nabla \times (\nabla \times \boldsymbol{A}) = \nabla(\nabla \cdot \boldsymbol{A}) - \nabla^2 \boldsymbol{A}, \tag{2.22}$$

we obtain the *wave equation* for the vector potential,

$$\nabla^2 \boldsymbol{A} = \varepsilon\mu \frac{\partial^2 \boldsymbol{A}}{\partial t^2}. \tag{2.23}$$

We propose trial solutions of (2.23) of the form

$$\boldsymbol{A} = \sqrt{\frac{1}{\varepsilon}} q(t) \boldsymbol{\Phi}(x, y, z), \tag{2.24}$$

where $\boldsymbol{\Phi}(x, y, z)$ is a *field distribution function* describing the standing wave in the laser cavity. The distribution function actually consists of three spatial components given by the vector

$$\boldsymbol{\Phi}(x, y, z) = \boldsymbol{e}_{\mathbf{x}} \boldsymbol{\Phi}_x(x, y, z) + \boldsymbol{e}_y \boldsymbol{\Phi}_y(x, y, z) + \boldsymbol{e}_z \boldsymbol{\Phi}_z(x, y, z), \tag{2.25}$$

where $\boldsymbol{e}_x$, $\boldsymbol{e}_y$, and $\boldsymbol{e}_z$ are independent unit vectors indicating the spatial directions. The distribution function is normalized to *unity* over the whole volume of the cavity, that is,

$$\iiint_{cav} |\boldsymbol{\Phi}(x, y, z)|^2 dxdydz = 1. \tag{2.26}$$

The variable $q(t)$ in (2.24) is the *temporal-varying amplitude* of the vector potential.

By substituting (2.24) into (2.23), the terms involving the spatial distribution and the temporal variation can be separated into two equations:

$$\nabla^2 \boldsymbol{\Phi}(x, y, z) = -\mu\varepsilon\omega^2 \boldsymbol{\Phi}(x, y, z), \tag{2.27}$$

$$\frac{\partial^2 q(t)}{\partial t^2} = -\omega^2 q(t). \tag{2.28}$$

Here, we should keep in mind that the parameter ω is defined as an *eigenvalue* that connects these two equations. The mathematical solution to (2.28) is

$$q(t) = ce^{j\omega t} + c^* e^{-j\omega t}, \tag{2.29}$$

where c is an amplitude given with a complex number. The second term on the right-hand side of (2.29) is the complex conjugate of the first term, because $q(t)$ itself should be a real number.

Any observable physical quantity should be a real number both in classical and quantum mechanics. Representations with complex numbers such as in (2.29) stem from the mathematical benefits of using imaginary numbers in the treatment of periodically vibrating phenomena. We find that the eigenvalue ω can be interpreted as an angular frequency of the temporal variation.

The second step towards quantization is to calculate the energy corresponding to the classical Hamiltonian. The stored energy W of the optical wave in the laser cavity as obtained from (2.20), (2.21), and (2.24) is

$$W = \frac{1}{2} \iiint_{cav} \left(\varepsilon \boldsymbol{E}^2 + \mu \boldsymbol{H}^2 \right) dxdydz = \frac{1}{2} \left(\dot{q}^2 + \omega^2 q^2 \right), \tag{2.30}$$

where $\dot{q}$ is the first derivative of q with respect to time t, that is, $\dot{q} = \partial q / \partial t$ in classical mechanics. Equation (2.30) describes the stored energy of the optical wave and can be expressed in the same form as for a harmonic oscillator. If we suppose $q(t) = q(0)\cos(\omega t)$, which gives $\dot{q}(t) = -\omega q(0)\sin(\omega t)$, the time-averaged value of the energy $\bar{W} = \omega^2 q(0)^2/2$ is obtained. In deriving (2.30), we used the formula

$$\iiint (\nabla \times \boldsymbol{\Phi})^2 dxdydz = \iiint \boldsymbol{\Phi} \{\nabla \times (\nabla \times \boldsymbol{\Phi})\} dxdydz, \tag{2.31}$$

together with (2.19), (2.22), (2.26), and (2.27).

2.6 Obtaining Quantum Mechanical Operators from Classical Physical Quantities

The third step towards quantization is to regard the dynamic variables as quantum mechanical operators. In particular, energy becomes the quantum mechanical Hamiltonian. As mentioned previously, the notion of a time derivative for any physical quantity is not *allowed* in quantum mechanics. We suppose that $\dot{q}$ must correspond to the operator p which must satisfy (2.14) in relation to its *canonical conjugate* operator q:

$$q \rightarrow q, \tag{2.32}$$

$$\dot{q} \rightarrow p. \tag{2.33}$$

Operators q and p must be Hermitian because the expectation values of these operator should be real numbers. The stored energy W associated with the optical wave corresponds to the quantum mechanical Hamiltonian H

$$W \to H. \tag{2.34}$$

Substitution of (2.32) and (2.33) into (2.30) yields the *quantum mechanical Hamiltonian* in the form

$$H = \frac{1}{2}\left(p^2 + \omega^2 q^2\right). \tag{2.35}$$

2.7 Quantization of the Optical Wave

The final step to quantization is to analyze the quantum mechanical behavior by applying one of (2.12)–(2.14). Here, we introduce two new operators a and a^+ defined using q and p:

$$a = \sqrt{\frac{1}{2\hbar\omega}}(\omega q + jp), \tag{2.36}$$

$$a^+ = \sqrt{\frac{1}{2\hbar\omega}}(\omega q - jp). \tag{2.37}$$

Although q and p are Hermitian, the new operators a and a^+ are not. Inverting the above relations, q and p can be expressed in terms of a and a^+:

$$q = \sqrt{\frac{2\hbar}{\omega}}\frac{a + a^+}{2}, \tag{2.38}$$

$$p = \sqrt{2\hbar\omega}\,\frac{a - a^+}{2j}. \tag{2.39}$$

By substituting (2.38) and (2.39) into (2.14), the commutation relation for a and a^+ is found to be

$$\left[a, a^+\right] = aa^+ - a^+a = 1. \tag{2.40}$$

By substituting (2.38), (2.39), and (2.40) into (2.35), the Hamiltonian is rewritten with a and a^+ in the form

$$H = \hbar\omega\left(a^+a + \frac{1}{2}\right). \tag{2.41}$$

Based on a more detailed analysis of the operators a and a^+, which is given in Appendix 1, it is known that the operator a acts on the states $|s\rangle$ to reduce the *number s* by one and a^+ acts to increase this number by one:

$$a|s\rangle = \sqrt{s}|s-1\rangle, \tag{2.42}$$

$$a^+|s\rangle = \sqrt{s+1}|s+1\rangle, \tag{2.43}$$

where $|s\rangle$ is an eigenstate of operator a^+a corresponding to eigenvalue s, a positive *integer*. For this reason a is called an *annihilation operator* and a^+ a *creation operator*. Derivations of (2.42) and (2.43) are rather tedious and are given in Appendix 1. From (2.42) and (2.43), we verify the relation

$$a^+a|s\rangle = s|s\rangle. \tag{2.44}$$

Hence, the *eigenequation* of the Hamiltonian is written as

$$H|s\rangle = \hbar\omega\left(s+\frac{1}{2}\right)|s\rangle. \tag{2.45}$$

This states that the eigenvalue W of the *stored energy* of the optical wave in the laser cavity is given by

$$W = \left(s+\frac{1}{2}\right)\hbar\omega. \tag{2.46}$$

Because the energy is characterized by a nonnegative integer s, s is called the *photon number*. The constant $\hbar\omega/2$ remains even if $s = 0$, and is called the *zero-point energy*. Spontaneous emission is induced by this zero-point energy as already explained in Chap. 1 and is proved in Appendix 5.

2.8 Remark on the Photon

It is worthwhile to recapitulate the manner in which the photon number was obtained. The vector potential $\boldsymbol{A}$ of the optical wave is represented by the amplitude $q(t)$ and spatial distribution $\boldsymbol{\Phi}(x, y, z)$. Quantum mechanics is applied only to the *temporal-varying terms*. In particular, the stored energy or Hamiltonian of the optical wave in a laser cavity is represented by such terms. The photon number is derived via the eigenvalue for the stored energy in the whole cavity. Meanwhile, the spatial distribution of the optical wave $\boldsymbol{\Phi}(x, y, z)$ is defined over the entire volume of the laser cavity and is characterized by the eigenfunctions determined by (2.27). That means (2.27) is common to both classical mechanics and quantum mechanics, because the spatial distribution is time-independent.

Because the photon refers to a particle-like object, most people imagine that it is a spatially localized particle even in free space, and the output light emitted from a laser is like a volley from a machine gun. *This image is wrong.* When we obtain *discrete eigenvalues* for any physical quantity, we can associate this basic quantum unit to a physical characteristic of *a particle* for the respective phenomenon. The spatial distribution of the photon needs to be determined subject to the spatial boundary conditions as performed for any classical wave.

2.9 The Uncertainty Principle

One important feature of quantum mechanics is the uncertainty principle, which is given in the following theorem.

Theorem

Given the commutation relation $[A,B] = jC$, *the relation* $\Delta A \cdot \Delta B \geq \frac{1}{2}|\langle\Psi|C|\Psi\rangle|$ *then holds.*

Here, A and B are Hermite operators and C can be a real number or another Hermite operator. ΔA and ΔB are the uncertainty values corresponding to the respective operators and are defined as the "root mean square" of the expectation value:

$$\Delta A = \sqrt{\left\langle\Psi|(A-\bar{A})^2|\Psi\right\rangle}, \tag{2.47}$$

$$\text{with } \bar{A} = <\Psi|A|\Psi>, \tag{2.48}$$

$$\Delta B = \sqrt{<\Psi|(B-\bar{B})^2|\Psi>}, \tag{2.49}$$

$$\text{with } \bar{B} = \langle\Psi|B|\Psi\rangle. \tag{2.50}$$

Proof of this theorem is presented in Appendix 2. Here, we apply this theorem to several examples.

The first example concerns the *relation between the electric and magnetic fields.* The electric field $\boldsymbol{E}$ is given in classical mechanics by the time derivative of the vector potential as stated in (2.20) and (2.24). The time derivative of the amplitude q should be replaced with another operator p as supposed in (2.33). Then, the electric field $\boldsymbol{E}$ in quantum mechanics is written as

$$\boldsymbol{E} = -\sqrt{\frac{1}{\varepsilon}}p\,\boldsymbol{\Phi}(x,y,z). \tag{2.51}$$

The magnetic field $\boldsymbol{H}$ is given by the rotation (or curl) of the vector potential as formulated in (2.21). This operation is not a temporal variation but a spatial distribution. Hence, the magnetic field $\boldsymbol{H}$ in quantum mechanics is written as

$$\boldsymbol{H} = \frac{1}{\mu\sqrt{\varepsilon}} q\nabla \times \boldsymbol{\Phi}(x, y, z). \tag{2.52}$$

Therefore, the commutation relation between the electric and magnetic fields is

$$[\boldsymbol{E}, \boldsymbol{H}] = -\frac{1}{\mu\varepsilon}\boldsymbol{\Phi} \cdot \nabla \times \boldsymbol{\Phi}(pq - qp) = \frac{j\hbar}{\mu\varepsilon}\boldsymbol{\Phi} \cdot \nabla \times \boldsymbol{\Phi}. \tag{2.53}$$

By applying the above theorem to (2.53), we obtain the uncertainty relation

$$\Delta\boldsymbol{E} \cdot \Delta\boldsymbol{H} \geq \frac{\hbar}{2\mu\varepsilon}\boldsymbol{\Phi}(x, y, z) \cdot \nabla \times \boldsymbol{\Phi}(x, y, z). \tag{2.54}$$

Because the right-hand side is nonzero, this uncertainty relation implies that a simultaneous representation for both electric and magnetic fields is not allowed in quantum mechanics, although a simultaneous representation is essential in the classical mechanics. Therefore, the quantum mechanical treatment of the optical wave does not match with its conventional treatment in electronics which is based on classical mechanics.

The second example concerns the *relation between energy and time*. The Heisenberg equation has been given in (2.13). If we regard the operator A in (2.13) as time t, the commutation relation on the right-hand side is

$$\frac{1}{j\hbar}[A, H] = \frac{1}{j\hbar}[t, H]. \tag{2.55}$$

The left-hand side of (2.13) is the time derivative of t, giving the simple relation

$$\frac{dt}{dt} = 1. \tag{2.56}$$

With the expectation value of the Hamiltonian denoted by W, we obtain the uncertainty relation

$$\Delta W \cdot \Delta t \geq \frac{\hbar}{2}, \tag{2.57}$$

which means that the expectation value of an energy should be evaluated over a sufficiently broad time interval Δt.

A third example results by substituting (2.46) into (2.57) yielding

$$\Delta W \cdot \Delta t = \hbar \left\{ \Delta\omega \left(\left(s + \frac{1}{2} \right) + \omega \Delta s \right\} \Delta t \geq \frac{\hbar}{2} \tag{2.58}$$

which means that, for *very short time intervals Δt, neither frequency ω nor photon number s can be evaluated exactly.*

2.10 Coherent State

In (2.46), the photon number s is introduced as an eigenvalue associated with the eigenstate of the number operator $a^{+}a$ to characterize the optical energy. Meanwhile, the operators corresponding to the electric field $\boldsymbol{E}$ and magnetic field $\boldsymbol{H}$ can be expressed in terms of a^{+} and a as derived from (2.38), (2.39), (2.51), and (2.52):

$$\boldsymbol{E} = j\sqrt{\frac{\hbar\omega}{2\varepsilon}}(a - a^{+})\,\boldsymbol{\Phi}(x, y, z), \tag{2.59}$$

$$\boldsymbol{H} = \frac{1}{\mu}\sqrt{\frac{\hbar}{2\varepsilon\omega}}(a + a^{+})\nabla \times \boldsymbol{\Phi}(x, y, z). \tag{2.60}$$

We need now to find the eigenstates of a^{+} and a to obtain the expectation values of the electric and magnetic fields. As also found in classical mechanics, energy is a time-independent quantity, whereas the electric and magnetic fields are time dependent quantities. Therefore, the representation of both these fields with finite photon number is in principle not simple.

For the eigenstates of a^{+} and a, the following equation for $|\phi\rangle$ has been proposed and is called the *coherent state* or *Glauber state*,

$$|\phi\rangle = \sum_s |s\rangle \left(\frac{N^s e^{-N}}{s!} \right)^{1/2} \exp\left[-j\left(s + \frac{1}{2} \right)(\omega t + \theta). \right] \tag{2.61}$$

Here, N is the average photon number for this state given by

$$\langle\phi|a^{+}a|\phi\rangle = \sum_s \frac{N^s e^{-N}}{s!} = Ne^{-N} \sum_s \frac{N^{s-1}}{(s-1)!} = N. \tag{2.62}$$

The expectation values of the operators a^{+} and a are

$$\langle\phi|a^{+}|\phi\rangle = \sqrt{N}\exp[j(\omega t + \theta)], \tag{2.63}$$

$$\langle\phi|a|\phi\rangle = \sqrt{N}\exp[-j(\omega t + \theta)]. \tag{2.64}$$

The electric and magnetic fields are then represented in this coherent state as

$$\langle\phi|\boldsymbol{E}|\phi\rangle = \left(\frac{2\hbar\omega N}{\varepsilon}\right)^{1/2} \sin(\omega t + \theta)\boldsymbol{\Phi}(x, y, z), \tag{2.65}$$

$$\langle\phi|\boldsymbol{H}|\phi\rangle = \frac{1}{\mu}\left(\frac{2\hbar N}{\varepsilon\omega}\right)^{1/2} \cos(\omega t + \theta)\nabla \times \boldsymbol{\Phi}(x, y, z). \tag{2.66}$$

These two expressions agree with the *classical EM fields.*
The energy of the optical wave calculated from these two expressions is

$$W = \frac{1}{2}\iiint_{cav}\left(\varepsilon\langle\phi|\boldsymbol{E}|\phi\rangle^2 + \mu\langle\phi|\boldsymbol{H}|\phi\rangle^2\right)dxdydz = \hbar\omega N. \tag{2.67}$$

By comparing this result with (2.46), we find that the energy derived from the field expectation values of (2.65) and (2.66) gives the energy component $N\hbar\omega$ with the average photon number N. However, calculations from these field expectation values never produce the zero-point energy $(1/2)\hbar\omega$, which is derived by the direct quantization of the optical wave resulting in (2.46).

The distribution of the photon number in the coherent state follows the Poisson distribution

$$P_s \equiv |\langle s|\phi\rangle|^2 = \frac{N^s e^{-N}}{s!}, \tag{2.68}$$

which are illustrated in Fig. 2.3. This equation is interpreted as meaning that P_s is the probability to measure the photon number of the existing optical wave to be s. After repeating the measurement many times, the average value of the measured photon number approaches N.

2.11 Interaction Between the Optical Wave and a Charged Particle

The Hamiltonian of the optical wave has been derived from Maxwell's classical EM equations presented in (2.41). However, when a charged particle is present along with the optical wave, the dynamics of the particle will be affected by the optical wave. The motion of the charged particle is characterized by another Hamiltonian. The Hamiltonian for the charged particle and wave is of the form

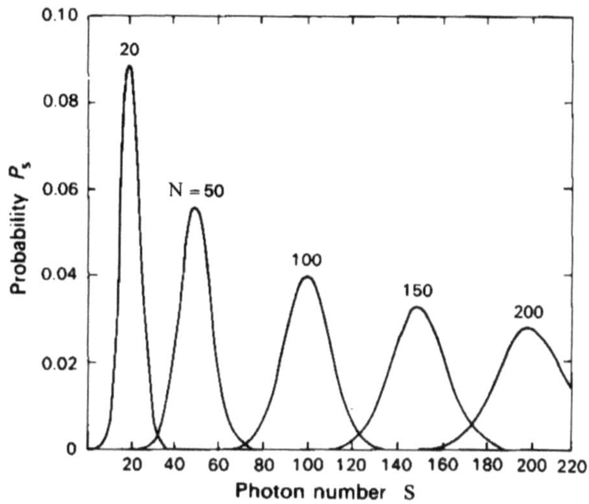

Fig. 2.3 Poisson distribution. S is the measured photon number and N is the averaged photon number after repeating the measurement many times

$$H = \frac{1}{2m}(\boldsymbol{p} - e\boldsymbol{A})^2 + eU(\boldsymbol{r}), \tag{2.69}$$

where m and $\boldsymbol{p}$ are the respective mass and momentum of the charged particle, e its charge, and $\boldsymbol{A}$ and U are the respective vector and scalar potentials describing the optical wave. Equation (2.69) is valid for both classical and quantum mechanics. In quantum mechanics, $\boldsymbol{p}$ and $\boldsymbol{A}$ are treated as operators.

The validity of (2.69) is confirmed by applying Hamilton's principle in classical analytical mechanics to (2.69). We can derive a Newtonian dynamic equation for the charged particle experiencing a *Lorentz force* as

$$m\ddot{\boldsymbol{r}} = e[\boldsymbol{E} + \mu(\dot{\boldsymbol{r}} \times \boldsymbol{H})]. \tag{2.70}$$

The derivation of this equation is given in Appendix 3.

In quantum mechanics, the observable physical quantities such as $\dot{\boldsymbol{r}}$ and $\ddot{\boldsymbol{r}}$ should be obtained as expectation values based on the appropriate quantum mechanical expressions. Using (2.69) and the Schrödinger equation, we can also derive the dynamic equation for the expectation value of the charged particle corresponding to (2.70) as

$$m\frac{d^2\langle \boldsymbol{r}\rangle}{dt^2} = e[\langle \boldsymbol{E}\rangle + \mu(\langle \boldsymbol{v}\rangle \times \langle \boldsymbol{H}\rangle)]. \tag{2.71}$$

The derivation of this equation is given in Appendix 4. Although we have derived a classical Newtonian-like dynamic equation from the Schrödinger equation, we should be careful in its application because it is given after the determination of the expectation values of $\langle \boldsymbol{r}\rangle, \langle \boldsymbol{v}\rangle, \langle \boldsymbol{E}\rangle$ and $\langle \boldsymbol{H}\rangle$, under several assumptions and approximations, as explained in Appendix 4.

2.12 Analysis of Electron Transition Using a Fully Quantized Treatment

As mentioned in Sect. 2.1, the analysis of the interaction between an optical wave and a material (electron) is classified according to three different approaches: fully classical, semi-classical and full quantized. For the fully classical and semi-classical approaches, the dynamics of the electron is represented via polarization, dielectric constant or conductivity, and is introduced into Maxwell's classical EM equations. An analysis of the laser is then performed by applying theories from material science and electronics.

In contrast, for the fully quantized approach, the interaction is analyzed via transition probabilities between energy states of the optical wave and the material. Although an energy conservation rule for the interaction can be derived, the representations of other physical quantities such as the vibrational phase or coherency of the optical wave become complicated. It might be helpful to emphasize that the concept of *coherent light* is founded in *classical mechanics.*

Therefore, the modus operandi of this book is that the development will be based mainly on a semi-classical approach. The required modification resulting from full quantization will be added when necessary.

An analysis of electron transition probability based on a fully quantized treatment is presented in Appendix 5. Readers who are more interested in these quantum mechanical aspects are encouraged to consult Appendix 5 for further details.

Chapter 3
Laser Oscillation

3.1 Representation of Optical Amplification with Laser Polarization

Optical wave amplification is explained in terms of the induced polarization of the laser material, as described in Chap. 1. We shall call this induced polarization *laser polarization*. Optical amplification can be described by *Maxwell's classical EM equations*:

$$\nabla \times \boldsymbol{H} = \frac{\partial \boldsymbol{D}}{\partial t} + \frac{\partial \boldsymbol{P}}{\partial t} + \sigma \boldsymbol{E} = \varepsilon \frac{\partial \boldsymbol{E}}{\partial t} + \frac{\partial \boldsymbol{P}}{\partial t} + \sigma \boldsymbol{E}, \tag{3.1}$$

$$\nabla \times \boldsymbol{E} = -\frac{\partial \boldsymbol{B}}{\partial t} = -\mu_o \frac{\partial \boldsymbol{H}}{\partial t}. \tag{3.2}$$

Here, both the polarization $\boldsymbol{P}$ and the electric displacement $\boldsymbol{D}$ are taken into account simultaneously, but this does not represent a double counting. All materials consist of many electrons and many material properties are the result of linear effects of the collective behavior of these electrons. The laser polarization derives from the electrons in the outermost shell of atoms; the electrons in the inner shell contribute to the electric displacement via the term involving the dielectric constant ε. Conductivity as described by the σ term is also taken into account. In this chapter, ε and σ are real numbers.

Although there are in total six electric and magnetic field components in an exact analysis, we suppose here, for simplicity, that the optical wave in the configuration of Fig. 1.14 can be represented by just the main electric spatial component E_x or E_y.

From (3.1) and (3.2), we obtain a wave equation for E (which here signifies either E_x or E_y) of the electric field:

$$\nabla^2 E - \mu_o \varepsilon(x, y) \frac{\partial^2 E}{\partial t^2} = \mu_o \left(\sigma \frac{\partial E}{\partial t} + \frac{\partial^2 P}{\partial t^2} \right). \tag{3.3}$$

M. Yamada, *Theory of Semiconductor Lasers*,
Springer Series in Optical Sciences 185,
DOI: 10.1007/978-4-431-54889-8_3,

We express the electric field in the form of a traveling wave as

$$E = \sum_p E_p^{(+)}(z)T(x,y)e^{j(\omega_p t - \beta_p z)} + c.c., \tag{3.4}$$

where p is the mode number, ω_p the angular frequency, β_p the propagation constant, $T(x, y)$ a normalized field distribution function in the transverse plane, $E_p^{(+)}(z)$ the amplitude which varies with the propagation along the +z-axis, and $c.c.$ indicates the complex conjugate of the previous term. The values of ω_p and β_p are determined using the eigenfunction $T(x, y)$ in the homogenous equation of (3.3) by putting $\sigma = 0$ and $P = 0$, that is,

$$\left\{\frac{\partial^2}{\partial x^2} + \frac{\partial^2}{\partial y^2} - \beta_p^2 + \mu_o \varepsilon(x,y)\omega_p^2\right\} T(x,y)e^{j(\omega_p t - \beta_p z)} = 0, \tag{3.5}$$

and

$$\int_{-\infty}^{\infty}\int_{-\infty}^{\infty} |T(x,y)|^2 dxdy = 1. \tag{3.6}$$

The laser polarization P is taken to be proportional to the electric field and expressed using the *laser susceptibility* χ as

$$P = \varepsilon_o \sum_p \left\{\chi E_p^{(+)}(z)T(x,y)e^{j(\omega_p t - \beta_p z)} + c.c.\right\}. \tag{3.7}$$

The laser susceptibility is represented as a complex number

$$\chi = \mathrm{Re}\,\chi + j\,\mathrm{Im}\,\chi, \tag{3.8}$$

because the laser polarization develops a phase shift arising from the temporal response to the electric field.

We substitute (3.4) and (3.7) into (3.3) and neglect the second-order term $\partial^2 E_p^{(+)}(z)/\partial z^2$ because the variation of the amplitude $E_p^{(+)}(z)$ is much smoother than the phase variation $\exp[j\beta_p z]$. We also use (3.5) and multiplying by $T^*(x,y)$ take the spatial integration in the transverse plane using (3.6). We then obtain an equation for the variation of the amplitude:

$$\begin{aligned}\frac{\partial E_p^{(+)}(z)}{\partial z} = &-\frac{j\mu_o\varepsilon_o\omega_p^2}{2\beta_p}\int_{-\infty}^{\infty}\int_{-\infty}^{\infty}\chi|T(x,y)|^2 dxdy\, E_p^{(+)}(z) \\ &-\frac{\mu_o\omega_p}{2\beta_p}\int_{-\infty}^{\infty}\int_{-\infty}^{\infty}\sigma|T(x,y)|^2 dxdy\, E_p^{(+)}(z).\end{aligned} \tag{3.9}$$

Because the carrying power of the optical wave is proportional to $\left|E_p^{(+)}(z)\right|^2$, its rate of variation is as follows:

$$\begin{aligned}\frac{\partial\left|E_p^{(+)}(z)\right|^2}{\partial z} &= \frac{\partial E_p^{(+)}(z)}{\partial z}E_p^{(+)*}(z) + E_p^{(+)}(z)\frac{\partial E_p^{(+)*}(z)}{\partial z} \\ &= \left(g_p - \alpha_{loss}\right)\left|E_p^{(+)}(z)\right|^2,\end{aligned} \tag{3.10}$$

where g_p is the *gain coefficient* and α_{loss} is the *loss coefficient*, both defined as

$$g_p = \frac{\sqrt{\mu_o\varepsilon_o}\omega_p}{n_{eq}}\int_{-\infty}^{\infty}\int_{-\infty}^{\infty}\mathrm{Im}\,\chi(x,y)|T(x,y)|^2dxdy, \tag{3.11}$$

$$\alpha_{loss} = \frac{1}{n_{eq}}\sqrt{\frac{\mu_o}{\varepsilon_o}}\int_{-\infty}^{\infty}\int_{-\infty}^{\infty}\sigma(x,y)|T(x,y)|^2dxdy. \tag{3.12}$$

Here, we have defined an *equivalent refractive index* n_{eq} to characterize the propagation constant or velocity of the optical wave along the z direction:

$$\beta_p = \sqrt{\mu_o\varepsilon_o}n_{eq}\omega_p = \frac{n_{eq}}{c}\omega_p. \tag{3.13}$$

If the distribution of the lasing material is restricted solely within the active region, (3.11) is rewritten as

$$g_p = \xi g, \tag{3.14}$$

with

$$g = \frac{\sqrt{\mu_o\varepsilon_o}\omega_p}{n_{eq}}\mathrm{Im}\,\chi \approx \frac{\sqrt{\mu_o\varepsilon_o}\omega_p}{n_r}\mathrm{Im}\,\chi, \tag{3.15}$$

$$\xi = \iint_{act}|T(x,y)|^2dxdy, \tag{3.16}$$

where g is called the *gain coefficient of the plane wave*, ξ the *confinement factor* of the optical wave in the active region, and n_r the refractive index of the active region. $\iint_{act}dxdy$ indicates a two-dimensional integration over a transverse cross-section of the active region. We understand that the amplification is described by the imaginary part of the susceptibility χ.

Let us now determine the contribution of the real part of the susceptibility. We rewrite the amplitude in polar form with absolute value and phase as

$$E_p^{(+)}(z) = \left|E_p^{(+)}(z)\right|e^{j\theta_p(z)}. \tag{3.17}$$

The spatial derivative of the amplitude is then

$$\frac{\partial E_p^{(+)}(z)}{\partial z} = \frac{\partial \left| E_p^{(+)}(z) \right|}{\partial z} e^{j\theta_p(z)} + j\frac{\partial \theta_p(z)}{\partial z} \left| E_p^{(+)}(z) \right| e^{j\theta_p(z)}. \tag{3.18}$$

By substituting this equation into (3.9) and multiplying by $e^{-j\theta_p(z)}$, we can separate terms $\partial \left| E_p^{(+)}(z) \right| / \partial z$ and $\partial \theta_p(z)/\partial z$. The term $\partial \left| E_p^{(+)}(z) \right| / \partial z$ is the same as (3.10); the term $\partial \theta_p(z)/\partial z$ represents a change in the propagation constant. We define the *variation of the propagation constant* $\Delta\beta_p$ to be

$$\Delta\beta_p \equiv -\frac{\partial \theta_p(z)}{\partial z} = \frac{\sqrt{\mu_o \varepsilon_o}\omega_p}{2n_{eq}} \int_{-\infty}^{\infty} \int_{-\infty}^{\infty} \mathrm{Re}\chi(x,y)|T(x,y)|^2 dxdy. \tag{3.19}$$

Using definitions (3.11), (3.12), and (3.19), (3.9) and (3.18) are expressed as

$$\frac{\partial E_p^{(+)}(z)}{\partial z} = \left(\frac{g_p - \alpha_{loss}}{2} - j\Delta\beta_p \right) E_p^{(+)}(z). \tag{3.20}$$

As found by substituting (3.20) into (3.4), $\Delta\beta_p$ gives the variation of the propagation constant from β_p to β_p', i.e.

$$\beta_p' = \beta_p + \Delta\beta_p, \tag{3.21}$$

which can be expressed as a variation in the effective refractive index Δn_{eq},

$$\Delta\beta_p = \sqrt{\mu_o \varepsilon_o}\omega_p \Delta n_{eq}. \tag{3.22}$$

We can then determine the *variation of the refractive index* Δn_r in the active region by the relations

$$\Delta n_{eq} = \xi \Delta n_r, \tag{3.23}$$

$$\Delta n_r \approx \frac{1}{2n_r} \mathrm{Re}\,\chi. \tag{3.24}$$

3.2 Conditions for Laser Oscillation in Traveling Wave Model

A sketch of the *laser cavity* is illustrated in Fig. 3.1. The optical wave propagates along the *active region* being amplified and reflected back by the facet mirrors. Given the power reflectivities of the front and back mirrors, R_f and R_b, respectively, their respective reflectivities for a given electric field are $\sqrt{R_f}$ and $\sqrt{R_b}$.

Denoting the length of the cavity by L, the electric field in the cavity is represented as

$$E = \sum_{p} \Big\{ E_p^{(+)}(0) T(x,y) e^{(g_p - \alpha_{loss}) z/2 - j\beta'_p z + j\omega_p t} + E_p^{(-)}(L) T(x,y) e^{(g_p - \alpha_{loss})(L-z)/2 - j\beta'_p (L-z) + j\omega_p t} \Big\} + c.c. \tag{3.25}$$

where $E_p^{(+)}(0)$ and $E_p^{(-)}(L)$ are amplitudes of the forward and backward propagating waves at z = 0 and $z = L$, respectively. The boundary conditions at the mirrors are

$$E_p^{(+)}(0) e^{(g_p - \alpha_{loss}) L/2 - j\beta'_p L} \sqrt{R_f} = E_p^{(-)}(L), \tag{3.26}$$

$$E_p^{(-)}(L) e^{(g_p - \alpha_{loss}) L/2 - j\beta'_p L} \sqrt{R_b} = E_p^{(+)}(0), \tag{3.27}$$

which when combined yields

$$\Big\{ \sqrt{R_f R_b} e^{(g_p - \alpha_{loss}) L - j2\beta'_p L} - 1 \Big\} E_p^{(+)}(0) = 0. \tag{3.28}$$

This equation has two types of solution. The trivial one, $E_p^{(+)}(0) = 0$, implies that there is no optical wave. The nontrivial one, obtained by equating the bracketed factor to zero, is

$$\sqrt{R_f R_b} e^{(g_p - \alpha_{loss}) L - j2\beta'_p L} = 1. \tag{3.29}$$

Because the right-hand side is a real number, the left-hand side must be a real number and hence the angle from the imaginary part must be

$$2\beta'_p L = 2p\pi, \tag{3.30}$$

where p, called the *longitudinal mode number*, is an integer giving the period number of the standing wave of the cavity. Equation (3.30) is called the *resonance condition* or *phase condition*. The propagation constant, given in terms of the wavelength λ_p, is

$$\beta'_p \approx \beta_p = \frac{2\pi n_{eq}}{\lambda_p}, \tag{3.31}$$

with which the resonance condition is rewritten as

$$\lambda_p = \frac{2L n_{eq}}{p}. \tag{3.32}$$

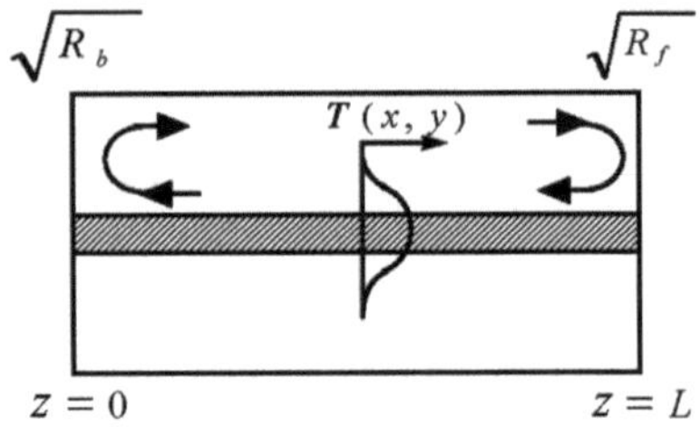

Fig. 3.1 Laser cavity

The absolute value of the left-hand side of (3.29) must be one. Rearranging then gives

$$g_p = \alpha_{loss} + \frac{1}{2L}\ln\left(\frac{1}{R_f R_b}\right), \tag{3.33}$$

which is called the *gain condition* or *amplitude condition*. Lasing operation is achieved when both resonance and gain conditions are satisfied. The criterion for when the laser exhibits single-mode or multimode operation (see Fig. 1.13) will be discussed later after a more detailed examination of the gain coefficient in Chap. 8.

The wavelength separation of adjacent modes is found to be

$$\Delta\lambda = \lambda_{p-1} - \lambda_p = 2Ln_{eq}\left(\frac{1}{p-1} - \frac{1}{p}\right) = \frac{2Ln_{eq}}{p(p-1)} \approx \frac{\lambda^2}{2Ln_{eq}}. \tag{3.34}$$

For the approximation, we have imposed the condition $\lambda \gg \Delta\lambda$.

In some cases, we need to use a more exact equation for the wavelength separation because the effective refractive index n_{eq} in the dispersion relation is not the same for all wavelengths. We expand the refractive index as

$$n_{eq}(\lambda_{p-1}) = n_{eq}(\lambda_p) + \frac{\partial n_{eq}}{\partial \lambda}\Delta\lambda. \tag{3.35}$$

Then the wavelength separation is written as

$$\Delta\lambda = 2L\left\{\frac{n_{eq}(\lambda_{p-1})}{p-1} - \frac{n_{eq}(\lambda_p)}{p}\right\} = \frac{2L}{p(p-1)}\left\{n_{eq}(\lambda_p) + p\frac{\partial n_{eq}}{\partial \lambda}\Delta\lambda\right\}, \tag{3.36}$$

which is then rearranged in the form

$$\Delta\lambda = \frac{\lambda^2}{2Ln_{eq}\left\{1 - \frac{\lambda}{n_{eq}}\frac{\partial n_{eq}}{\partial \lambda}\right\}}. \tag{3.37}$$

3.3 Amplification and Lasing Conditions in Standing Wave Model

The optical wave in the laser cavity has been expressed in the form of traveling waves, and the lasing conditions have been given as solutions for the steady state. However, in many cases, we need to know the temporal dynamics of lasing. Here, we show a way to represent the optical field in the form of a standing wave with a temporally varying amplitude.

The wave equation is the same as (3.3). The electric field is represented as

$$E = \sum_p \widehat{E}_p(t)\Phi_p(x,y,z)e^{j\omega_p t} + c.c.. \tag{3.38}$$

Here, $\widehat{E}_p(t)$ is the *temporally varying amplitude*. $\Phi_p(x,y,z)$ is a normalized three-dimensional field distribution function, which is given as an eigenfunction of the homogenous equation of (3.3) as

$$\left\{\nabla^2 + \mu_o\varepsilon(x,y)\omega_p^2\right\}\Phi_p(x,y,z)e^{j\omega_p t} = 0, \tag{3.39}$$

subject to the orthonormalization condition

$$\int_{cav} \Phi_p(x,y,z)\Phi_q^*(x,y,z)d^3\mathbf{r} = \delta_{p,q}. \tag{3.40}$$

Here, $\int_{cav} d^3\mathbf{r}$ denotes a three-dimensional integration over the whole laser cavity. The resonance condition from (3.30) or (3.32) is assumed already included in the function $\Phi_p(x,y,z)$ with factor $\cos(\beta_p z)$.

The laser polarization is given in terms of the laser susceptibility χ as

$$P = \varepsilon_o\left\{\sum_p \chi\widehat{E}_p(t)\Phi_p(x,y,z)e^{j\omega_p t} + c.c.\right\}. \tag{3.41}$$

When we substitute (3.38) and (3.41) into (3.3), we use a slightly different approximation between the left-hand and right-hand sides of the equation. On the left-hand side, the term with $\partial^2\widehat{E}_p(t)/\partial t^2$ is neglected because it is much smaller than $|\omega_p\partial\widehat{E}_p(t)/\partial t|$. Aided by (3.39) and this approximation, the left-hand side becomes

$$\nabla^2 E - \mu_o\varepsilon(x,y)\frac{\partial^2 E}{\partial t^2} = -2j\mu_o\varepsilon(x,y)\sum_p \omega_p\frac{\partial\widehat{E}_p(t)}{\partial t}\Phi_p(x,y,z)e^{j\omega_p t} + c.c.. \tag{3.42}$$

Because the right-hand side of (3.3) produces only perturbations of the amplitude and phase of the optical wave, both the $\partial^2 \widehat{E}_p(t)/\partial t^2$ and $j\omega_p \partial \widehat{E}_p(t)/\partial t$ terms are neglected leaving just the term with $\omega_p^2 \widehat{E}_p(t)$. The right-hand side is then

$$\mu_o\left(\sigma\frac{\partial E}{\partial t}+\frac{\partial^2 P}{\partial t^2}\right)=\mu_o\sum_p\left(j\omega_p\sigma-\omega_p^2\varepsilon_o\chi\right)\widehat{E}_p(t)\Phi_p(x,y,z)e^{j\omega_p t}+c.c.. \quad (3.43)$$

We multiply (3.42) and (3.43) by $\Phi_p^*(x,y,z)e^{-j\omega_p t}$, take the spatial integrations over the volume of the laser cavity, and perform a temporal average over several periods of $1/\omega_p$. The terms corresponding to the other modes and the complex conjugate are then dropped, giving the temporal variation of the amplitude as

$$\frac{\partial \widehat{E}_p(t)}{\partial t}=\frac{1}{2\varepsilon_{eff}}\left(-j\omega_p\varepsilon_o\chi_{eff}-\sigma_{eff}\right)\widehat{E}_p(t), \quad (3.44)$$

where the following values are defined as spatially averaged effective values:

$$\varepsilon_{eff}\equiv\int_{cav}\varepsilon(x,y)\left|\Phi_p(x,y,z)\right|^2 d^3\mathbf{r}, \quad (3.45)$$

$$\sigma_{eff}\equiv\int_{cav}\sigma(x,y)\left|\Phi_p(x,y,z)\right|^2 d^3\mathbf{r}, \quad (3.46)$$

$$\begin{aligned}\chi_{eff}&\equiv\int_{act}\chi(x,y)\left|\Phi_p(x,y,z)\right|^2 d^3\mathbf{r}\\&=\xi\chi.\end{aligned} \quad (3.47)$$

Here, $\int_{act} d^3\mathbf{r}$ signifies taking the spatial integration over the active region, and ξ is the *field confinement factor* of the optical wave in the active region, defined in (3.16). Because the occupation ratio of the active region along the z-axis is 100 %, the three-dimensional confinement factor is identical to the two-dimensional confinement factor,

$$\xi=\int_{act}\left|\Phi_p(x,y,z)\right|^2 d^3\mathbf{r}. \quad (3.48)$$

With the amplitude expressed in polar form

$$\widehat{E}_p(t)=\left|\widehat{E}_p(t)\right|e^{j\theta_p(t)}. \quad (3.49)$$

and substituted into (3.44), we obtain the relation

$$\begin{aligned}\frac{\partial \widehat{E}_p(t)}{\partial t} &= \frac{\partial \left|\widehat{E}_p(t)\right|}{\partial t} e^{j\theta_p(t)} + j\frac{\partial \theta_p(t)}{\partial t}\left|\widehat{E}_p(t)\right| e^{j\theta_p(t)} \\ &= \frac{1}{2\varepsilon_{eff}}\left(\omega_p\varepsilon_o \mathrm{Im}\chi_{eff} - \sigma_{eff} - j\omega_p\varepsilon_o\mathrm{Re}\chi_{eff}\right)\left|\widehat{E}_p(t)\right| e^{j\theta_p(t)}.\end{aligned} \tag{3.50}$$

Multiplying this equation by $\exp[-j\theta_p(t)]$ and separating the real and imaginary parts, we derive equations for the time variation of the absolute value of the amplitude

$$\frac{\partial \left|\widehat{E}_p(t)\right|}{\partial t} = \frac{1}{2\varepsilon_{eff}}\left(\omega_p\varepsilon_o\mathrm{Im}\chi_{eff} - \sigma_{eff}\right)\left|\widehat{E}_p(t)\right|, \tag{3.51}$$

and the *shift in the angular frequency*

$$\Delta\omega_p \equiv \frac{\partial \theta_p(t)}{\partial t} = -\frac{\omega_p\varepsilon_o\mathrm{Re}\chi_{eff}}{2\varepsilon_{eff}}. \tag{3.52}$$

The rate of change of the square of amplitude $\left|\widehat{E}_p(t)\right|^2$ is obtained from (3.51) as

$$\begin{aligned}\frac{\partial \left|\widehat{E}_p(t)\right|^2}{\partial t} &= 2\left|\widehat{E}_p(t)\right|\frac{\partial \left|\widehat{E}_p(t)\right|}{\partial t} \\ &= \frac{1}{\varepsilon_{eff}}\left(\omega_p\varepsilon_o\mathrm{Im}\chi_{eff} - \sigma_{eff}\right)\left|\widehat{E}_p(t)\right|^2.\end{aligned} \tag{3.53}$$

The stored energy W_{opt} of the optical wave in the cavity is calculated from (3.38),

$$W_{opt} = \frac{1}{2}\int_{cav}\left(\varepsilon E^2 + \mu_o H^2\right)d^3\mathbf{r} = \int_{cav}\varepsilon E^2 d^3\mathbf{r} = \sum_p 2\varepsilon_{eff}\left|\widehat{E}_p(t)\right|^2. \tag{3.54}$$

This classically calculated energy must equal the quantum mechanically determined measurable energy obtained in Chap. 2,

$$W_{opt} = \sum_p S_p\hbar\omega_p. \tag{3.55}$$

Then, the square of the amplitude is re-expressed with the photon number S_p as

$$\left|\widehat{E}_p(t)\right|^2 = \frac{\hbar\omega_p}{2\varepsilon_{eff}}S_p. \tag{3.56}$$

Substituting this equation into (3.52) gives the variation of the *photon number* as

$$\frac{\partial S_p}{\partial t} = \left(G_p - \frac{\sigma_{eff}}{\varepsilon_{eff}}\right) S_p. \tag{3.57}$$

Here, G_p is the *gain coefficient* for the temporal variation which is related to the gain coefficient g_p for the spatial variation,

$$\begin{aligned} G_p &= \frac{\varepsilon_o}{\varepsilon_{eff}} \omega_p \mathrm{Im}\, \chi_{eff} \approx \xi \frac{\omega_p}{n_r^2} \mathrm{Im}\, \chi, \\ &= v g_p \end{aligned} \tag{3.58}$$

where v is the propagating speed of the optical wave in the laser, obtained by dividing the speed of light in vacuum c by the equivalent refractive index,

$$v = \frac{1}{n_{eq}\sqrt{\mu_o \varepsilon_o}} = \frac{c}{n_{eq}}. \tag{3.59}$$

In this standing-wave model, we have not yet accounted for the optical output from the facet mirrors. The optical emission is seen as a type of loss because the optical energy reduces the output. The total variation of the photon number can then be written as

$$\begin{aligned} \frac{dS_p}{dt} &= \frac{\partial S_p}{\partial t} - (\mathit{optical\ output}) \\ &= (G_p - G_{th}) S_p \end{aligned}, \tag{3.60}$$

where G_{th} is called the *threshold gain level*. When the gain coefficient reaches this threshold, laser oscillation is achieved. By analogy with the gain condition of (3.33) in the traveling-wave model, the threshold gain level is given by

$$G_{th} = v\left\{\alpha_{loss} + \frac{1}{2L}\ln\left(\frac{1}{R_b R_f}\right)\right\}. \tag{3.61}$$

Thus G_{th} is interpreted as a decay rate for the optical energy in the laser cavity by absorption into and emission out of the cavity. The inverse value of the threshold gain level is then called the *photon life time* τ_{ph}:

$$\tau_{ph} = 1/G_{th}. \tag{3.62}$$

The amplification gain coefficient comes from the *stimulated emission* by the semiconductor material in the active region. Also occurring in the active region is *spontaneous emission*, which contributes an additional term to (3.60) that is independent of photon number:

$$\frac{dS_p}{dt} = (G_p - G_{th})S_p + C_p, \tag{3.63}$$

where C_p indicates the inclusion of spontaneous emission in the mode p. Although coefficients G_p and C_p are intrinsic to the material, their derivations require prior knowledge and are given later in Chap. 6.

We discuss now the profiles of the optical wave in the traveling-wave model given in Sects. 3.1 and 3.2 and in the standing-wave model given in Sect. 3.3. The transverse field distributions are common to both models, but the longitudinal distributions have slight differences. Even in the traveling-wave model, the optical wave forms a standing wave because of the reflections at the facet mirrors. The longitudinal spatial distribution in the traveling-wave model is

$$E_p^{(+)}(0)e^{(g_p-\alpha_{loss})z/2-j\beta_p z} + E_p^{(-)}(L)e^{(g_p-\alpha_{loss})(L-z)/2-j\beta_p(L-z)}.$$

Specifically, the shape of the envelope of the longitudinal distribution is given by a hyperbolic function. Unfortunately, the mathematical manipulation of a function combining a trigonometric function and a hyperbolic function is not easy. In approximations, the longitudinal distribution in the standing-wave model is given in most cases using only the trigonometric function

$$\Phi_p(x,y,z) = \sqrt{\frac{2}{L}}\cos(\beta_p z + \varphi_p)T(x,y). \tag{3.64}$$

Hence, the optical loss $(v/L)\ln(1/R_bR_f)$ due to optical emission is taken to be uniformly distributed in the longitudinal direction, but not concentrated at the facet mirrors.

3.4 Note on Laser Oscillation

In the traveling-wave model, the laser oscillation conditions are given by the resonance condition of (3.30) and the gain condition of (3.33). The gain condition in the standing-wave model is given by setting $dS_p/dt = 0$ in (3.60) as a steady-state operation and gives

$$G_p = G_{th}. \tag{3.65}$$

Dependence of the gain coefficient and photon number on the injection current is illustrated in Fig. 3.2. Theoretical derivations of these characteristics will be presented in Chap. 6. When the injection current I is zero, the gain coefficient is

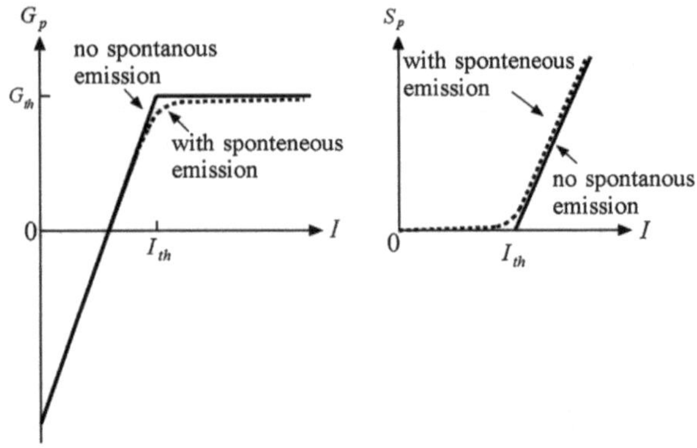

Fig. 3.2 Dependence of the gain coefficient and photon number on injection current

negative. By increasing the injection current, electrons and holes are injected into the active region thereby increasing the gain coefficient. When the gain coefficient reaches threshold as specified in (3.65), laser oscillation is achieved. The injection current at which the gain coefficient becomes equal to the threshold gain level is called the *threshold current* and is denoted I_{th}. By increasing the injection current beyond I_{th}, the photon number increases proportionately with $I–I_{th}$, with the gain coefficient maintained at the threshold level. The above-mentioned characteristics are drawn with solid lines in the figure.

However, we need to pay attention to the fact that the light output comes from the spontaneously emitted light in the active region. The spontaneously emitted light is amplified via stimulated emission. When we take into account this spontaneously emitted light, the variation in photon number S_p is given as in (3.63). By setting $dS_p/dt = 0$ in (3.63), we obtain the relation

$$S_p = \frac{C_p}{G_{th} - G_p}, \tag{3.66}$$

which implies that the gain coefficient G_p never reaches the threshold level G_{th}, and in that case, we can *no longer achieve laser oscillation*. The dependence of the gain coefficient and photon number considering the spontaneous emission are drawn in Fig. 3.2 with dotted lines. We observe a very weak output with small current injection via spontaneous emission. It is important to note that *the laser oscillation and threshold gain level are virtually defined by excluding spontaneous emission*. The experimental measurement of the threshold current I_{th} must be impossible in principle. Therefore, in a real experiment, the threshold current I_{th} is defined as the crossing point between the two extrapolated lines for the gradually increasing optical output from $I = 0$ and the rapidly-increasing optical output, as illustrated in Fig. 3.3.

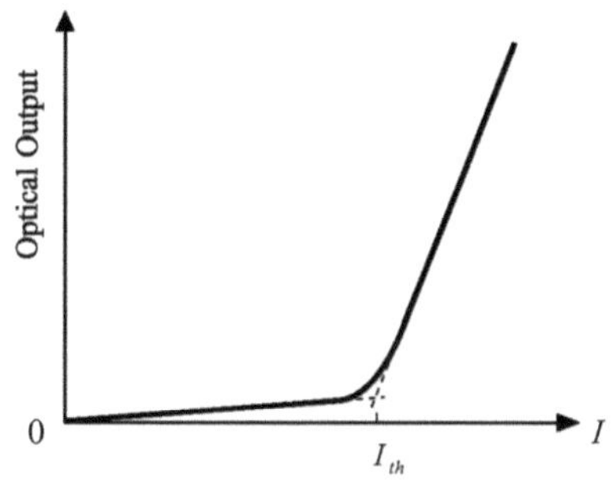

Fig. 3.3 Experimental analysis of the threshold current level

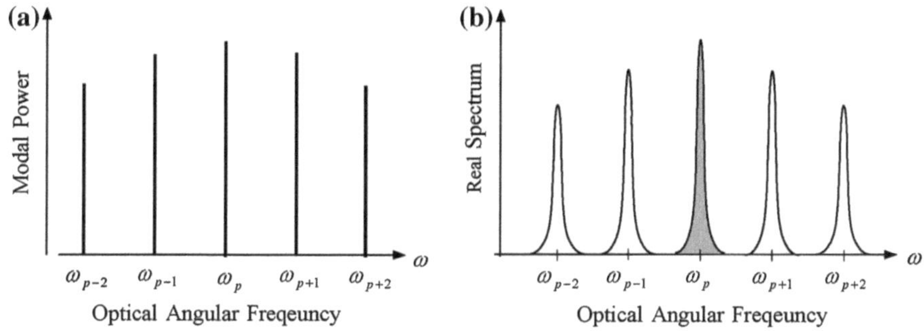

Fig. 3.4 Modal spectrum and real spectrum. **a** Conceptual spectrum of modal power. **b** Experimentally observed real spectrum

3.5 Modal Power and Optical Spectrum

In the operation of the laser, the property of the optical wave is classified according to the mode. Each mode is defined by the spatial distribution and the electric polarization. Orthogonal relations among the mode functions are very helpful in analyzing the theoretically characteristics of laser operation. The angular frequencies of the mode should have *discrete values*, as illustrated in Fig. 3.4a, because the mode values are defined as the eigenvalues of the wave equation given by (3.5), (3.30), and (3.39). However, the real spectrum of the optical wave in the experiment reveals a continuous distribution of optical frequencies or wavelengths, as illustrated in Fig. 3.4b. The difference between the conceptual modal power and real spectrum stems from the fluctuating variation of the instantaneous frequency given by $\varDelta\omega = \partial\theta_p(t)/\partial t$ caused by including the spontaneous emission and by other phenomena. Therefore, the modal power must be evaluated as the *area* under the curve of the experimentally observed power spectrum, and *not the height* of the spectrum.

Chapter 4
Waveguides

4.1 Method to Analyze Two-Dimensional Cross-Section

In the *transverse cross-section* of the laser cavity, the optical wave exhibits a non-trivial *two-dimensional distribution* in regard to refractive index. A schematic of such a waveguide structure is shown in Fig. 4.1a. The refractive index of the *active region* is n_1. The surrounding regions have smaller refractive indices than the active region and function as *cladding regions*. The optical wave is concentrated in the active region but spreads into the cladding regions. This structure has a symmetric configuration in the x direction but an asymmetric configuration in the y direction.

In general, the transverse field distribution in the two-dimensional waveguide can only be analyzed by numerical calculations using, for example, the finite element method. Nevertheless, we shall provide an approximate analysis of these distributions to gain an understanding of the characteristics of the two-dimensional waveguide. We employ a method called the *equivalent refractive index method*.

The wave equation which we shall be using throughout this section is

$$\nabla^2 E - \mu_0 \varepsilon(x, y) \frac{\partial^2 E}{\partial t^2} = 0. \tag{4.1}$$

The dielectric constant is related to the square of the refractive index,

$$\varepsilon_i(x, y) = \varepsilon_o n_i^2, \tag{4.2}$$

where i indexes the regions marked in Fig. 4.1.

The first step in our approximation is to assume that by separation of variables the field distribution function $T(x, y)$, introduced in Chap. 3, can be factored into two, $T_x(x)$ and $T_y(y)$. This assumption produces quantitative errors from the exact

M. Yamada, *Theory of Semiconductor Lasers*,
Springer Series in Optical Sciences 185,
DOI: 10.1007/978-4-431-54889-8_4, © Springer Japan 2014

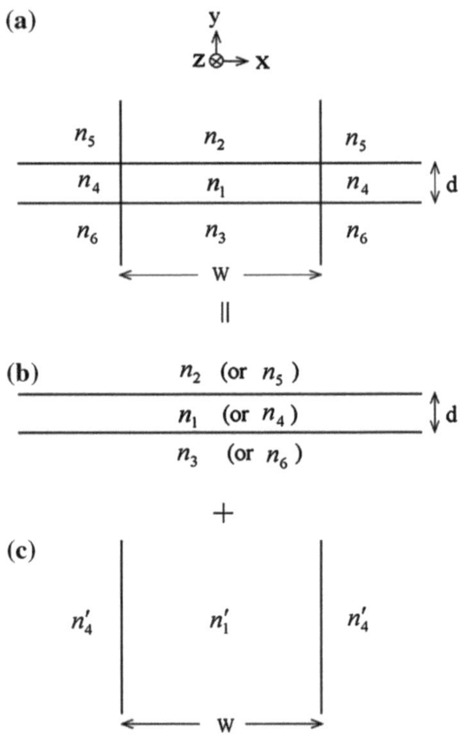

Fig. 4.1 Approximation scheme illustrating the equivalent refractive index method used to analysis the transverse characteristics of the optical wave in the laser cavity. **a** Two-dimensional waveguide. **b** Slab waveguide with y-direction boundaries. **c** Slab waveguide with x-direction boundaries

numerical calculation of $T(x, y)$, especially in regions 5 and 6 where the optical wave distributions are very weak. However, for most cases, this assumption is reliable in evaluating the wave distribution in the active region 1. Hence, the electric field of the forward propagating wave can be put into the form

$$E = E^{(+)}(0)T_x(x)T_y(y)e^{-j\beta z+j\omega t} + c.c.\,. \tag{4.3}$$

Here, the normalization of the distribution functions are defined as

$$\int_{-\infty}^{\infty} |T_x(x)|^2 dx = 1, \tag{4.4}$$

$$\int_{-\infty}^{\infty} |T_y(y)|^2 dy = 1. \tag{4.5}$$

The distribution functions are expressed as *trigonometric functions* in the central region and as *decaying exponential functions* in the outer regions:

$$T_x(x) = \begin{cases} A_i \cos\left(\gamma_{xi} x - \frac{p_x \pi}{2}\right) & \text{for } -\frac{W}{2} \le x \le \frac{W}{2} \\ A_j \cos\left(\gamma_{xi} \frac{W}{2} + \frac{p_x \pi}{2}\right) e^{\gamma_{xj}(x+W/2)} & \text{for } x \le -\frac{W}{2} \\ A_j \cos\left(\gamma_{xi} \frac{W}{2} - \frac{p_x \pi}{2}\right) e^{-\gamma_{xj}(x-W/2)} & \text{for } x \ge \frac{W}{2} \end{cases}, \quad (4.6)$$

where $i = 1, 2, 3, j = 4, 5, 6$, γ_{xi} and γ_{xj} are the propagation constants associated with the x-component of the wave, and $p_x (= 0, 1, 2,\ldots)$ is the transverse mode number;

$$T_y(y) = \begin{cases} B_i \cos(\gamma_{yi} y - \theta_i) & \text{for } -\frac{d}{2} \le y \le \frac{d}{2} \\ B_j \cos\left(\gamma_{yi} \frac{d}{2} + \theta_i\right) e^{\gamma_{yj}(y+d/2)} & \text{for } y \le -\frac{d}{2} \\ B_k \cos\left(\gamma_{yk} \frac{d}{2} - \theta_i\right) e^{-\gamma_{yk}(y-d/2)} & \text{for } y \ge \frac{d}{2} \end{cases}, \quad (4.7)$$

where $i = 1, 4$, $j = 3, 6$, $k = 2, 5$, γ_{yi}, y_{xj}, and γ_{yk} are propagation constants associated with the y-component of the wave, and θ_i is a spatial phase constant characterizing the transverse mode number as well as the asymmetric distribution of the wave along the y-direction.

By substituting (4.3), (4.6), and (4.7) into (4.1), we obtain the following relations among the propagation constants:

$$-\beta^2 - \gamma_{x1}^2 - \gamma_{y1}^2 + \mu_o \varepsilon_o n_1^2 \omega^2 = 0 \quad \text{in region 1,} \quad (4.8)$$

$$-\beta^2 - \gamma_{x1}^2 + \gamma_{y2}^2 + \mu_o \varepsilon_o n_2^2 \omega^2 = 0 \quad \text{in region 2,} \quad (4.9)$$

$$-\beta^2 - \gamma_{x1}^2 + \gamma_{y3}^2 + \mu_o \varepsilon_o n_3^2 \omega^2 = 0 \quad \text{in region 3,} \quad (4.10)$$

$$-\beta^2 + \gamma_{x4}^2 - \gamma_{y4}^2 + \mu_o \varepsilon_o n_4^2 \omega^2 = 0 \quad \text{in region 4,} \quad (4.11)$$

$$-\beta^2 + \gamma_{x5}^2 + \gamma_{y5}^2 + \mu_o \varepsilon_o n_5^2 \omega^2 = 0 \quad \text{in region 5,} \quad (4.12)$$

$$-\beta^2 + \gamma_{x6}^2 + \gamma_{y6}^2 + \mu_o \varepsilon_o n_6^2 \omega^2 = 0 \quad \text{in region 6.} \quad (4.13)$$

The second step is to ignore the variation of the field distribution in the x-direction and treat only the distribution in the y-direction, as illustrated in Fig. 4.1b. In most cases, the active region is designed to be a thin layer so as to produce a lower threshold current and give enough width (i.e., $W \gg d$) to generate sufficient optical power. Hence, the propagation constants in the y-direction become much larger than those in the x-direction, $\gamma_{yi} \gg \gamma_{xi}$, in all regions. Therefore, γ_{xi} in (4.8)–(4.13) can be neglected and *approximated* by

$$-\beta_1^2 - \gamma_{y1}^2 + \mu_o \varepsilon_o n_1^2 \omega^2 = 0 \quad \text{in region 1,} \tag{4.14}$$

$$-\beta_1^2 + \gamma_{y2}^2 + \mu_o \varepsilon_o n_2^2 \omega^2 = 0 \quad \text{in region 2,} \tag{4.15}$$

$$-\beta_1^2 + \gamma_{y3}^2 + \mu_o \varepsilon_o n_3^2 \omega^2 = 0 \quad \text{in region 3,} \tag{4.16}$$

$$-\beta_4^2 - \gamma_{y4}^2 + \mu_o \varepsilon_o n_4^2 \omega^2 = 0 \quad \text{in region 4,} \tag{4.17}$$

$$-\beta_4^2 + \gamma_{y5}^2 + \mu_o \varepsilon_o n_5^2 \omega^2 = 0 \quad \text{in region 5,} \tag{4.18}$$

$$-\beta_4^2 + \gamma_{y6}^2 + \mu_o \varepsilon_o n_6^2 \omega^2 = 0 \quad \text{in region 6.} \tag{4.19}$$

Here, β_1 is the propagation constant along the z-direction of the *slab waveguide* constructed from regions 1, 2, and 3. Similarly, β_4 is another propagation constant along the z-direction of another slab waveguide constructed from regions 4, 5, and 6. The boundaries at $x = \pm W/2$ are ignored as shown in Fig. 4.1b. *The polarization direction* of the optical wave in the original two-dimensional waveguide in Fig. 4.1a is defined in these slab waveguides.

The transverse electric (TE) mode is described by field components E_x, H_y, and H_z. For the TE mode, E_x is the only electric field component from which the other two magnetic components are derived using relations $\partial E_x/\partial z = -j\beta_i E_x = -j\omega\mu_o H_y$ and $\partial E_x/\partial y = j\omega\mu_o H_z$. Values of the propagation constants γ_{yi}, the phase constants θ_i, and the amplitudes B_i, B_j, and B_k are determined from the continuity conditions for E_x and H_z, namely the function $T_y(y)$ in (4.7) and $\partial T_y(y)/\partial y$, at $y = \pm d/2$.

Similarly, the transverse magnetic (TM) mode is restricted to the components H_x, E_y, and E_z. For the TM mode, we evaluate the magnetic component H_x. The other two electric components are related through the relations $\partial H_x/\partial z = -j\beta H_x = j\omega\varepsilon_i E_y$ and $\partial H_x/\partial y = -j\omega\varepsilon_i E_z$. The boundary conditions determine continuity constraints for the magnetic component H_x and the electric component E_z at $y = \pm d/2$.

By reducing the thickness d, the higher-order transverse modes are excluded and only the fundamental TE and TM modes with $p_y = 0$ appear. Exclusion of all higher-order modes is called *waveguide cut-off*. Although two fundamental modes can exist in the laser, the TE mode acquires a larger gain than the TM mode because the field confinement factor ξ into the active region is larger for the TE mode than the TM mode. In addition, the laser polarization is larger for a vibration along the parallel direction in a very thin active region such as in a quantum-well structure, which we shall discuss in Chap. 11. Hence, lasing with only the fundamental TE mode is commonly generated in most lasers.

The third step in the approximation analysis is to take into account the effects from the thickness direction y using *equivalent refractive indices* and applying the model of a slab waveguide for the width direction x, as illustrated in Fig. 4.1c.

The propagation constants β_1 and β_4 are related to the propagating speed of the optical wave in the slab waveguides formed with regions $i = 1, 2$, and 3 and $i = 4$, 5, and 6, respectively. From (4.14) and (4.17), we define the equivalent refractive indices n_1' and n_4' to express these changes in speed, thus

$$\mu_o \varepsilon_o n_1'^2 \omega^2 \equiv \beta_1^2 = \mu_o \varepsilon_o n_1^2 \omega^2 - \gamma_{y1}^2 \quad \text{in region 1,} \tag{4.20}$$

$$\mu_o \varepsilon_o n_4'^2 \omega^2 \equiv \beta_4^2 = \mu_o \varepsilon_o n_4^2 \omega^2 - \gamma_{y4}^2 \quad \text{in region 4.} \tag{4.21}$$

Using these equivalent refractive indices, the original equations (4.8) and (4.11) become

$$-\beta^2 - \gamma_{x1} + \mu_o \varepsilon_o n_1'^2 \omega^2 = 0 \quad \text{in region 1,} \tag{4.22}$$

$$-\beta^2 + \gamma_{x4} + \mu_o \varepsilon_o n_4'^2 \omega^2 = 0 \quad \text{in region 4.} \tag{4.23}$$

These two equations imply a slab waveguide extended along the y direction and have boundaries at $x = \pm W/2$, as in Fig. 4.1c. The distribution function $T_x(x)$ in (4.6) is evaluated using this model.

We need to pay attention here to the relationship between the polarization direction of the optical wave and the boundary condition. As discussed previously with Fig. 4.1b, the TE mode consists of E_x, H_y, and H_z, whereas the TM mode consists of H_x, E_y, and E_z in our treatment. The boundary condition of the TE mode in the current model with Fig. 4.1c is the continuity constraint for the electric displacement $D_x = \varepsilon_o n_i'^2 E_x$ not E_x, and that of the TM mode is the continuity constraint for the electric field E_y not H_x.

The propagation constants γ_{y1} and γ_{y4} in the y-direction become larger by *decreasing* thickness d. If we only reduce the thickness in region 4, the equivalent refractive index of this region becomes *smaller* than that of region 1, even if the *real refractive indices are identical*, i.e., $n_1 = n_4$. This configuration is used in several types of semiconductor lasers, a topic we shall introduce in Sect. 4.3.

4.2 Guiding Modes Taking into Account Gain and Loss

The guiding mode of the optical wave is mainly determined by the refractive index profile developed in the last section. The fractional difference in the refractive index along the thickness direction y is designed to be several percent to produce a small thickness d for the active region. However, the difference along the width x is small so that the width W of the active region can be rather wide. For a *small refractive index difference*, the spatial distribution of the *gain* and *absorption loss* also plays a role in determining the optical wave distribution [1].

We suppose a three-layer symmetric slab waveguide with width W in the x direction, as sketched in Fig. 4.2. The core or active region is labeled 1; cladding regions are labeled 4. These regions have respective equivalent refractive indices n_1' and n_4', introduced in the last section. Aside from the refractive indices, the active region has optical gain g_1 and absorption loss α_1 and other regions have absorption loss α_4. Values of the gain and the loss already include effects from the distribution along the thickness y direction, that is,

$$g_1 = \xi_y g = \xi_y \frac{k_o}{n_1'} \operatorname{Im} \chi, \tag{4.24}$$

where

$$\xi_y = \int_{-d/2}^{d/2} |T_y(y)|^2 dy, \tag{4.25}$$

$$\alpha_i = \int_{-\infty}^{\infty} \alpha(y)|T_y(y)|^2 \, dy = \frac{1}{n_i'}\sqrt{\frac{\mu_o}{\varepsilon_o}} \int_{-\infty}^{\infty} \sigma_i |T_y(y)|^2 dy. \tag{4.26}$$

Here, ξ_y is the field confinement factor in the active region along the y direction and k_o is the propagation constant or wave number of the plane wave in free space,

$$k_o = \sqrt{\mu_o \varepsilon_o}\omega \approx \frac{2\pi}{\lambda}. \tag{4.27}$$

Effects from the gain and loss are taken into account using complex dielectric constants ε_i,

$$\varepsilon_1 = \varepsilon_o n_1'^2 \left\{ 1 + j \frac{g_1 - \alpha_1}{n_1' k_o} \right\}, \tag{4.28}$$

$$\varepsilon_4 = \varepsilon_o n_4'^2 \left\{ 1 - j \frac{\alpha_4}{n_4' k_o} \right\}. \tag{4.29}$$

Equations (4.22) and (4.23), rewritten with these complex dielectric constants, become

$$-\beta^2 - \gamma_{x1}^2 + \mu_o \varepsilon_1 \omega^2 = 0, \tag{4.30}$$

$$-\beta^2 + \gamma_{x4}^2 + \mu_o \varepsilon_4 \omega^2 = 0. \tag{4.31}$$

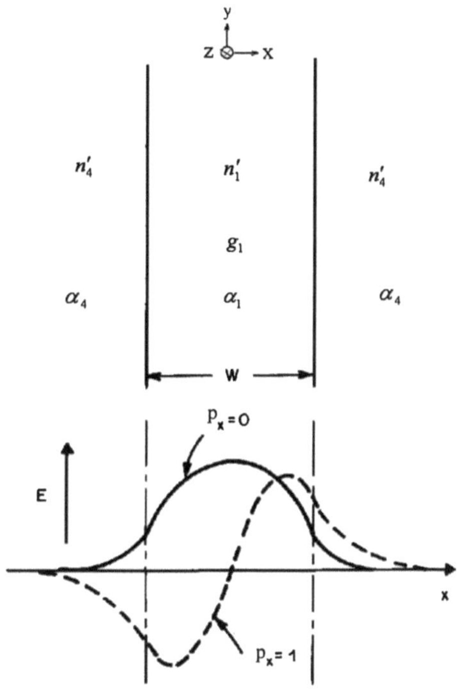

Fig. 4.2 Sketch of the slab waveguide in which gain and loss are taken into account; the electric field distribution of the first two TE modes are also given

Here, the propagation constants β and γ_{xi} are complex numbers; their imaginary parts describe the *amplification* or *attenuation* of the field. Subtracting these two equations gives

$$\gamma_{x1}^2 + \gamma_{x4}^2 = \mu_0(\varepsilon_1 - \varepsilon_4)\omega^2. \tag{4.32}$$

We apply the distribution function $T_x(x)$, as given in (4.6), setting $i = 1$ and $j = 4$, and using the complex form of the amplitudes and the propagation constants. The boundary conditions at $x = \pm W/2$ are determined by the continuity constraints of the electric and magnetic field components. However, the exact boundary condition for the TE mode becomes rather more complicated than that of the TM mode. We then apply approximate continuity constraints for $T_x(x)$ and $\partial T_x(x)/\partial x$ for both TE and TM modes. The boundary condition is given as

$$\tan\left(\gamma_{x1}\frac{W}{2} - \frac{p_x\pi}{2}\right) = \frac{\gamma_{x4}}{\gamma_{x1}}, \tag{4.33}$$

which is rearranged to give

$$1+\tan^2\left(\gamma_{x1}\frac{W}{2}-\frac{p_x\pi}{2}\right)=\frac{1}{\cos^2\left(\gamma_{x1}\frac{W}{2}-\frac{p_x\pi}{2}\right)}$$
$$=1+\frac{\gamma_{x4}^2}{\gamma_{x1}^2}=\frac{\gamma_{x1}^2+\gamma_{x4}^2}{\gamma_{x1}^2}. \tag{4.34}$$

By substituting (4.32) into (4.34), we obtain

$$\cos\left(\gamma_{x1}\frac{W}{2}-\frac{p_x\pi}{2}\right)=\frac{\gamma_{x1}}{\sqrt{\mu_o(\varepsilon_1-\varepsilon_4)}\omega}. \tag{4.35}$$

Although (4.33) has two unknown complex values, γ_{x1} and γ_{x4}, (4.35) has just one unknown complex value, γ_{x1}. Here multiplying the numerator and the denominator of the right-hand side by $W/2$ yields a transcendental equation in the unknown parameter $\gamma_{x1}W/2$.

Because (4.35) is a complex function, the variables, and hence the equation, can be divided into real and imaginary parts. To analyze this equation, we define two new variables T and D such that [1].

$$\left(\frac{W}{2}\right)^2\mu_o(\varepsilon_1-\varepsilon_4)\omega^2\equiv T^2\exp(j\varphi)=T^2\exp\left(j\tan^{-1}D\right), \tag{4.36}$$

where

$$T=\frac{W}{2}\sqrt{\mu_o}\omega\sqrt{|\varepsilon_1-\varepsilon_4|}=\frac{W}{2}n_1'k_o\left[\{2\Delta_n\}^2+\left\{\frac{g_d}{n_1'k_o}\right\}^2\right]^{1/4}, \tag{4.37}$$

$$D=\frac{\mathrm{Im}(\varepsilon_1-\varepsilon_4)}{\mathrm{Re}(\varepsilon_1-\varepsilon_4)}=\frac{g_d}{2n_1'k_o\Delta_n}=\tan\varphi \tag{4.38}$$

with

$$\Delta_n=\frac{n_1'-n_4'}{n_1'}, \tag{4.39}$$

$$g_d=g_1-\alpha_1+\alpha_4. \tag{4.40}$$

Here, Δ_n is the fractional refractive index and g_d is the gain-loss difference. T signifies the strength of the field confinement and is called the *normalized width.* D denotes the contribution ratio of the gain-loss difference compared with the refractive index difference and is called the *relative gain-loss difference*. The relationship between these variables is illustrated in Fig. 4.3 with a representation in the complex plane. As shown in Fig. 4.3a, T^2 is the square of the strength of the confinement. As given by (4.38), D is defined as the ratio of the imaginary and real

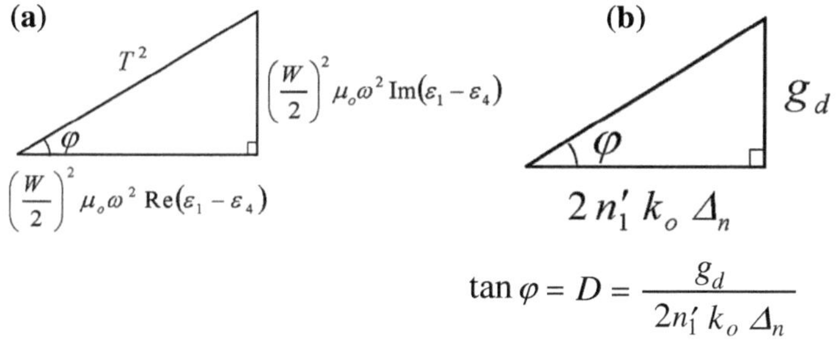

Fig. 4.3 Complex-plane representations for parameters *T* and *D*. **a** Normalized width. **b** Relative gain-loss difference

parts of $(\varepsilon_1 - \varepsilon_4)$. $\varphi = \tan^{-1}D$ determines the angle characterizing both these complex values.

By substituting (4.36) into (4.35), we have

$$\cos\left(\gamma_{x1}\frac{W}{2} - \frac{p_x\pi}{2}\right) = \frac{\gamma_{x1}W/2}{T\exp[j(\tan^{-1}D)/2]}, \tag{4.41}$$

from which we find the complex number $\gamma_{x1}\,W/2$ as a function of T and D. The other propagation constant γ_{x4} is obtained from these values using (4.32) and (4.36),

$$\left(\gamma_{x4}\frac{W}{2}\right)^2 = T^2\exp[j\tan^{-1}D] - \left(\gamma_{x1}\frac{W}{2}\right)^2. \tag{4.42}$$

The gain-loss difference g_d is always positive for semiconductor lasers, but the refractive index difference Δ_n can take both positive and negative values. As Δ_n decreases $+\infty \to 0 \to -\infty$, φ changes smoothly $0 \to \pi/2 \to \pi$, but D changes abruptly, $+0 \to +\infty, -\infty \to -0$. Specifically, D has discontinuity at $\Delta_n \cong 0$ and $\varphi \cong \pi/2$.

Because we have admitted complex-valued solutions, we therefore need to be attentive here that the solutions for the propagation constants always *exist mathematically* even for the *cut-off condition*. The solutions are classified into two types based on the distribution along the x-axis. One gives a damped or converging solution for $x \to \pm\infty$; the other is a diverging solution for $x \to \pm\infty$. The *diverging solution* is called the *cut-off*. The damped or converging solution for $x \to \pm\infty$ is defined as the *guiding condition* and is given mathematically as

$$\mathrm{Re}\,\gamma_{x4} > 0. \tag{4.43}$$

From this inequality, the guiding condition can be express as a function of T and D. Although the derivation is rather complicated, let us examine it in more detail.

We start by assuming that

$$\mathrm{Re}\gamma_{x1} > 0. \tag{4.44}$$

The gain exists only in region 1, where the value has to be larger than the modal gain of the guiding mode. We then find inequality

$$\mathrm{Im}(\mu_o \varepsilon_1 \omega^2) > \mathrm{Im}\beta^2 \tag{4.45}$$

By substituting this inequality into (4.30), we obtain

$$\mathrm{Im}\,\gamma_{x1}^2 > 0. \tag{4.46}$$

The square is given in general as

$$\mathrm{Im}\,\gamma_{x1}^2 = 2\mathrm{Re}\,\gamma_{x1}\mathrm{Im}\gamma_{x1}. \tag{4.47}$$

Using (4.44), (4.46) and (4.47), we have

$$\mathrm{Im}\,\gamma_{x1} > 0. \tag{4.48}$$

Substitution of (4.48) into the boundary condition (4.33) gives the condition

$$\mathrm{Im}\,\gamma_{x4} > 0. \tag{4.49}$$

Equations (4.43) and (4.49) imply the inequality

$$\mathrm{Im}\gamma_{x4}^2 = 2\mathrm{Re}\gamma_{x4}\,\mathrm{Im}\gamma_{x4} \geq 0. \tag{4.50}$$

We substitute (4.50) into (4.42) and separate the imaginary part, giving the guiding condition

$$T^2 \sin\varphi > \mathrm{Im}\left(\gamma_{x1}\frac{W}{2}\right)^2. \tag{4.51}$$

The term $\sin\varphi$ is rewritten in the form

$$\sin\varphi = \sqrt{\frac{1}{1+\cot^2\varphi}} = \frac{|D|}{\sqrt{1+D^2}}. \tag{4.52}$$

Then, the guiding condition becomes

$$T^2\frac{|D|}{\sqrt{1+D^2}} > \mathrm{Im}\left(\gamma_{x1}\frac{W}{2}\right)^2 \text{ for } D \neq +0. \tag{4.53}$$

Equation (4.53) is valid for $D \neq +0$. For $D = +0$, the guiding condition is directly obtained from (4.41) as

$$T > \gamma_{x1} \frac{W}{2} = \frac{p_x \pi}{2} \text{ for } D = +0. \tag{4.54}$$

Equation (4.54) is the same as the guiding condition for conventional slab waveguides constructed with conventional dielectric materials without gain or loss.

The calculated example of the guiding condition shown in Fig. 4.4 is applicable for almost all optical wavelengths because this figure is given with the *normalized width* T and the *relative gain-loss difference* D. The hatched lines indicate boundaries between the *guiding condition* and the *cut-off condition*. The mode number is indicated by p_x. The optical wave can be guided for larger values of T than the boundaries. We find that the optical wave can be guided even if the refractive index difference Δ_n *is zero or negative*, that is $D \leq 0$, if T is large enough. However, the *fundamental mode* $p_x = 0$ shows a *cut-off* for smaller T values when $D < 0$.

It is well known that the refractive index difference should be positive in conventional dielectric waveguides. The optical wave can be guided if helped by being total reflected at the boundary. Total reflection implies that all the energy is reflected back into the core region and no energy escapes through the cladding region. For a waveguide having a gain in the active region, it is not necessary to guarantee total reflection. Even if some energy escapes from the active region, energy is supplied to the region via a contribution associated with a nonzero gain coefficient g_1. Thus, the optical wave can continue propagating.

The guiding mode is classified into three types based on the confinement mechanism. The first is called *index guiding*, where the field is mostly confined by the positive value of the refractive index difference Δ_n in the interval $+0 \leq D < +1$. The second is called *gain guiding*, where the field is constrained by the gain-loss difference g_d and $|D| > 1$. The third is called *anti-index guiding*, where the field is constrained with a negative value of the refractive index difference Δ_n in the interval $-1 < D < -0$.

Although the gain guiding and anti-index guiding modes can propagate along the waveguide, these modes suffer larger losses due to radiation through the cladding regions. In this section, the propagation constant β is a complex number whose imaginary part indicates the gain or loss. Let us examine the imaginary part by setting

$$\beta = \mathrm{Re}\beta + j\,\mathrm{Im}\beta. \tag{4.55}$$

Substitution of (4.55) into (4.30) with (4.28) gives

$$\begin{aligned} \beta^2 &= (\mathrm{Re}\beta)^2 - (\mathrm{Im}\beta)^2 + 2j\,\mathrm{Re}\beta\,\mathrm{Im}\beta \\ &= n_1'^2 k_o^2 - \mathrm{Re}\,\gamma_{x1}^2 + j\left\{n_1' k_o (g_1 - \alpha_1) - \mathrm{Im}\,\gamma_{x1}^2\right\} \end{aligned} \tag{4.56}$$

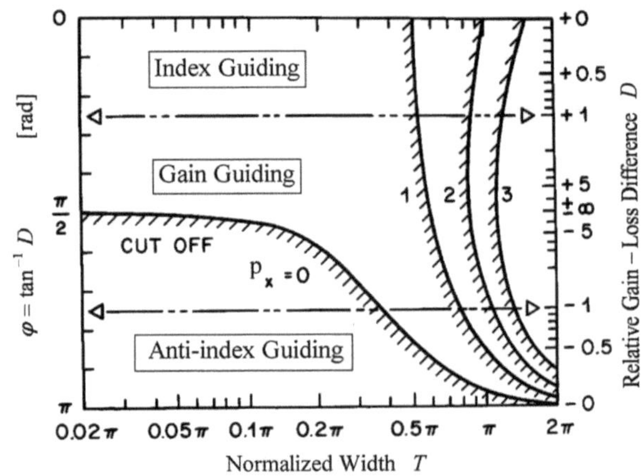

Fig. 4.4 Calculated example of the guiding condition

Because $\text{Re}\beta \gg \text{Im}\beta$, the real part of (4.56) is

$$\text{Re}\beta = \sqrt{n_1'^2 k_o^2 - \text{Re}\gamma_{x1}^2}, \tag{4.57}$$

which is identical to the change in the propagation constant for the conventional dielectric waveguide. The imaginary part of (4.56) is

$$\begin{aligned}\text{Im}\beta &= \frac{1}{2\text{Re}\beta}\left\{n' k_{o1}(g_1 - \alpha_1) - \text{Im}\gamma_{x1}^2\right\} \\ &\approx \frac{1}{2}\left\{g_1 - \alpha_1 - \frac{\text{Im}\gamma_{x1}^2}{n_1' k_o}\right\}, \\ &\equiv \frac{1}{2}(g_1 - \alpha_1 - \alpha_{diff})\end{aligned} \tag{4.58}$$

where

$$\alpha_{diff} = \frac{\text{Im}\gamma_{x1}^2}{n_1' k_o} \tag{4.59}$$

is an additional loss called the *diffraction loss* which corresponds to energy escaping from the active region through the cladding region. Based on a series expansion of (4.41), we give a new approximate expression for the diffraction loss as

$$\alpha_{diff} \approx \frac{\sqrt{2}}{n_1' k_o}\left\{\frac{(p_x+1)\pi}{W}\right\}^2 \sqrt{1 - \frac{D}{|D|\sqrt{1+D^2}} \frac{T^2(T+1/\sqrt{2})}{(T^2+\sqrt{2}T+1)^2}}. \tag{4.60}$$

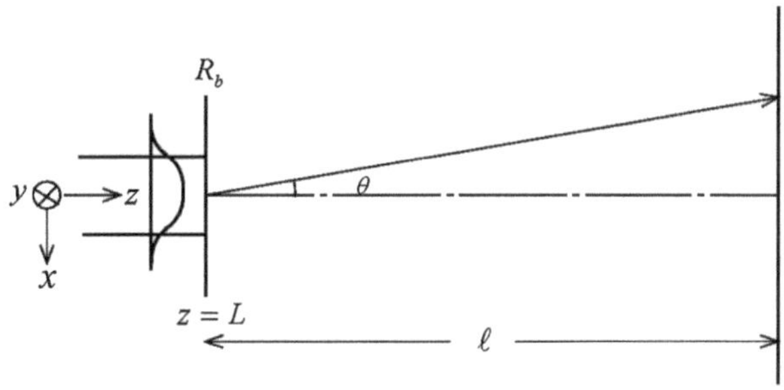

Fig. 4.5 Spatial configuration to measure far-field pattern

This approximation is applicable with an error of several 10 %. We find that diffraction losses become larger for the higher mode numbers p_x and smaller normalized widths T and become largest when approaching $D = -1$.

A large value for the diffraction loss gives a high threshold current for each mode, but produces big threshold differences for the higher-order modes ($p_x \geq 1$). Thus, an increase of a few percent in the diffraction loss is sometimes helpful in suppressing the higher-order modes.

Another property of the field to be determined is the far-field pattern. The field distribution function given in (4.6) or (4.7) is the *near-field pattern* which is formed in the waveguide. The profile of the emitted wave is called the *far-field pattern*. The electric field $E(\theta, L + \ell)$ associated with the far-field pattern is given by an integral transformation of the near-field pattern $E(x, L)$ according to *Huygens' principle*,

$$E(\theta, L + \ell) = \frac{j}{\lambda \ell} e^{-jk_o \ell (1+\theta^2/2)} \int_{-\infty}^{\infty} E(x, L) e^{jk_o \theta x} dx, \tag{4.61}$$

where $E(x, L)$ is the electric field function at the emitting facet ($z = L$), ℓ the distance from the emitting facet to the measuring plane, and θ the angle of deviation from the center axis, as illustrated in Fig. 4.5.

The near-field pattern, equivalent phase plane, and the far-field pattern are illustrated in Fig. 4.6. The *index guiding* mode has very small values for the imaginary part of the propagation constant γ_{x4}. Hence the damping in the cladding region is very smooth. Additionally, the equivalent phase plane is *flat* and the far-field pattern shows a *single peak* for the fundamental mode $p_x = 0$. For *gain guiding* and *anti-index guiding*, the imaginary parts of the propagation constants are not small; $\mathrm{Im}\gamma_{x4}$ indicates the presence of a propagating component along the x-axis, as shown in Fig. 4.6a. The equivalent phase plane in region 4 is $(\mathrm{Re}\beta)z \pm (\mathrm{Im}\,\gamma_{x4})x$ which is *concave*, as indicated in Fig. 4.6b; the optical wave propagates in a direction vertical to the equivalent phase plane. Therefore, the optical output has components whose emitting directions are tilted from the center axis of $\theta = 0$ and produce *interference peaks* beside the main peak, as indicated in Fig. 4.6c. The *side peaks* for gain guiding are not so large, giving in total three peaks including the main peak. Those for anti-index guiding however become larger than the main peak and the optical output exhibits just two peaks in the typical case.

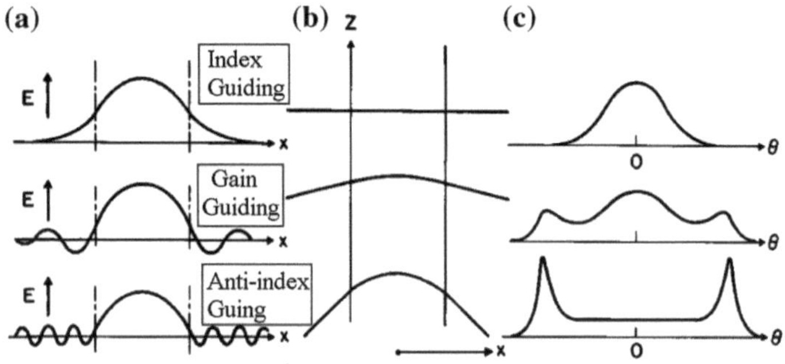

Fig. 4.6 Field pattern of the guided mode. **a** Near-field pattern. **b** Phase plane. **c** Far-field pattern

Table 4.1 Classifications of the guided mode

	Index guiding	Gain guiding	Anti-index guiding
Range of D	$+0 \leq D < +1$	$\lvert D \rvert > 1$	$-1 < D < -0$
Diffraction loss	Small	Relatively large	Large
Threshold level	Low	Relatively high	High
Far-field pattern	Single peak	Single peak or multiple-peaks	Multi-peaks
Fundamental mode operation	$T < \frac{\pi}{2}$	$T < 0.6\pi$ and $D > 0$	Not applicable

Classifications of the guided mode and their features are summarized in Table 4.1. The optimum range for the fundamental mode with a single-peak profile in the far-field pattern is geometrically mapped in Fig. 4.7 with parameters of T and $\varphi = \tan^{-1} D$. The unmarked area represents the suitable range for the operation. The marked region, with smaller T values less than 0.5π and larger φ values greater than 0.3π, is the region in which the far-field pattern exhibits multiple peaks or very high threshold currents. The cut-off boundary determines the limit where higher T values permit the first higher-order mode $p_x = 1$ to propagate. However, higher-order modes might be suppressed owing to larger diffraction losses in the region delineated by the loss difference boundary. Even in the optimum range, higher output with a flat phase plane is beneficial. The operating conditions are designed so that $T \leq \pi/2$ with $D \approx +1$ in most lasers.

Note that the semiconductor laser *tends to oscillate* under gain guiding or anti-index guiding because the *refractive index in the active region is reduced with the injection of electrons and holes.* Changes in refractive index after injection are caused by changes in the laser polarization, Re χ, and *plasma effects* from the

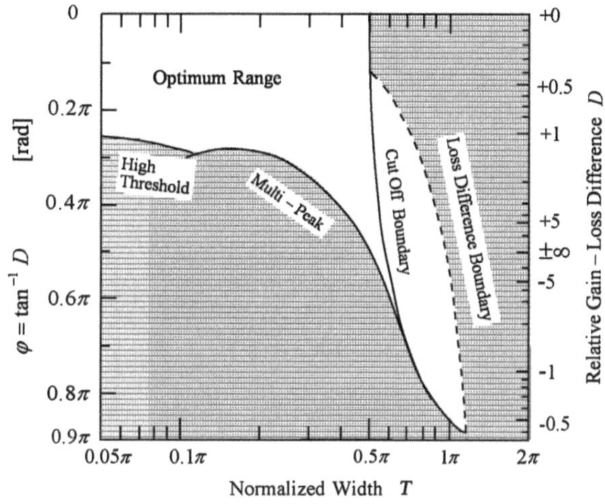

Fig. 4.7 Map of the optimum range for the fundamental mode with a single peak profile in the far-field pattern

electrons and holes. The change brings the relative gain-loss difference into the range $D \leq 0$. Therefore, we need to build a *stripe structure* in the waveguide to *shift the operating point to promote index guiding.*

4.3 Stripe Structures

Stripe structures have been developed to achieve *index guiding* as well as to restrict the area for current injection in the active region. There are various kinds of stripe structures. We show examples of such structures in Fig. 4.8. Hashed areas are the active region and cross-hatched areas are the electrodes.

(a) *Ridge waveguide* structure: The active region is etched out to leave a narrow stripe region. The etched side walls are coated with SiO_2 to protect the region from oxidation. The refractive index difference depends on the semiconductor material in the active region and SiO_2. Because the refractive index difference is large, the stripe width is typically less than several μm.
(b) *Buried hetero*-structure *(BH)*: The etched area is re-grown with a semiconductor material which has a higher band gap with a smaller refractive index than the active region. The refractive index difference depends on the material of the active region and that of the re-grown region.
(c) *Channeled substrate planar (CSP)*: The substrate is etched prior to growing the active region. The layers of the active region and others are then grown. In this structure, the material of the active region is uniform in the horizontal direction x. The area just above the prepared channel is used as the active region following electron injection. The specific feature of this structure is the difference in distance to the substrate from the active region along the thickness direction y. The field distribution of the optical wave along the thickness

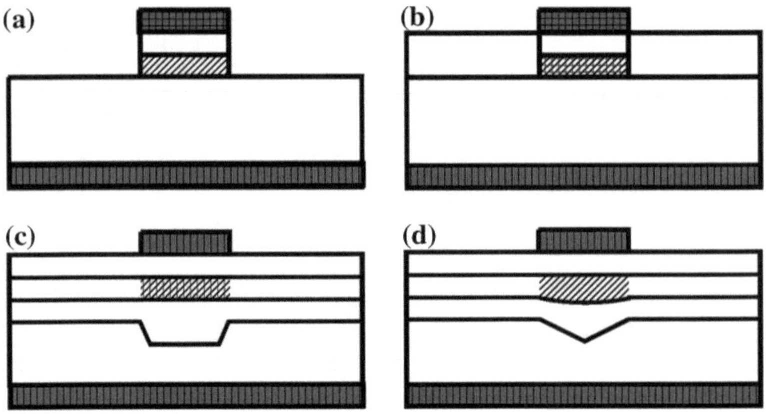

Fig. 4.8 Examples of stripe structures. **a** Ridge wave-guide. **b** Buried heterostructure (BH). **c** Channeled substrate planar (CSP). **d** V-channel substrate inner-stripe (VSIS)

direction y becomes sharper in the surrounding region than that in the active region because reflections at the boundary to the substrate are stronger when the boundary is closer. Therefore, the condition $\gamma_{y1} < \gamma_{y4}$ applies in Fig. 4.1, and the *equivalent refractive indices* satisfy $n_1^{'} > n_4^{'}$ in (4.20) and (4.21).

(d) *V-channel substrate inner-stripe* (VSIS): The fabrication process and operating mechanism are similar to those of the CSP laser explained in (c). An additional property of the VSIS laser is that the active region is slightly thicker than the surrounding regions because the crystal growing rate is faster just above the etched channel. The field distributions along thickness y direction satisfy $\gamma_{y1} < \gamma_{y4}$ producing a more enhanced result with a higher difference in the equivalent refractive indices $n_1^{'} - n_4^{'} > 0$.

Bibliography

1. M. Yamada, Transverse and longitudinal mode control in semiconductor injection lasers. IEEE J. Quantum Electron **QE-19**(9), 1365–1380 (1983)

Chapter 5
Density Matrix of a Semiconductor Material

The quantum mechanical properties of a single electron are described using the Schrödinger equation. The wave function obtained from this equation has a statistical character and determines the expectation value of a physical quantity based upon the spatial distribution of the single electron. However, most physical quantities in electronics, such as the electric current and optical gain in lasers, are determined by the *group dynamics* of the electrons not by single electrons. The dynamics of the group and that of a single electron have slight differences, which appear as statistical fluctuations in the collective behavior of the many-electron system. Distinct electrons located at same spatial position can not take identical eigenstate simultaneously known as the *Pauli exclusion principle*, whose property is not included in the Schrödinger equation. Analysis that takes in account both quantum mechanical properties and the statistical behavior is called *quantum statistics*.

The *density matrix* approach is one means of treating quantum statistics problems as developed in several textbooks, some of which are listed in the bibliography in Chap. 1. The *rate equation* (or dynamic equation) used in the analysis of the laser is derived from the density matrix.

5.1 Quantum Statistics

We label each electron in an *ensemble* (group) of electrons with index ν. The quantum mechanical state of the νth electron is represented as

$$\left|\Psi^{(\nu)}\right\rangle = \sum_n C_n^{(\nu)}|n\rangle. \tag{5.1}$$

Here $|n\rangle$ is the nth *energy eigenstate* for a single electron and $C_n^{(\nu)}$ is a time-dependent coefficient that describes the electron dynamics. Because the *completeness* of any quantum mechanical state based on a *set* of energy eigenstates is

M. Yamada, *Theory of Semiconductor Lasers*,
Springer Series in Optical Sciences 185,
DOI: 10.1007/978-4-431-54889-8_5,

guaranteed, the energy eigenstates are common for all electrons and $|n\rangle$ need not be labeled with v. The different behaviors of each electron is included in $C_n^{(v)}$. If we suppose a quantum mechanical operator A, the expectation value of this operator on the vth electron is

$$\left\langle \Psi^{(v)} |A| \Psi^{(v)} \right\rangle = \sum_{n,m} C_m^{(v)*} C_n^{(v)} \langle m|A|n\rangle, \tag{5.2}$$

where $\sum_{n,m}$ denotes a double summation over both n and m.

In the next step, we examine the average of the expectation values over the ensemble. This averaged value is called an *ensemble average*. The contributions from all electrons are not identical in general. We then introduce a *sampling probability* or *contributing probability* of the vth electron $P^{(v)}$, which must be normalized to unity over the ensemble,

$$\sum_v P^{(v)} = 1. \tag{5.3}$$

The restriction rule such as the *Pauli exclusion principle* can be kept by this normalization.

The expectation value $\langle A\rangle$ of the operator A for the ensemble is

$$\begin{aligned} \langle A\rangle &= \sum_v P^{(v)} \left\langle \Psi^{(v)} |A| \Psi^{(v)} \right\rangle \\ &= \sum_{v,n,m} P^{(v)} C_m^{(v)*} C_n^{(v)} \langle m|A|n\rangle. \end{aligned} \tag{5.4}$$

Here, we introduce a simplifying notation:

$$A_{mn} = \langle m|A|n\rangle, \tag{5.5}$$

$$\rho_{nm} = \langle n|\rho|m\rangle = \sum_v P^{(v)} C_m^{(v)*} C_n^{(v)}. \tag{5.6}$$

We imagine here a matrix ρ, called the *density matrix*, which consists of *matrix elements* ρ_{nm} as depicted in Fig. 5.1. Using this notation, (5.4) is rewritten

$$\langle A\rangle = \sum_{n,m} \rho_{nm} A_{mn} = \sum_{n,m} \langle n|\rho|m\rangle \langle m|A|n\rangle. \tag{5.7}$$

Because $\sum_m |m\rangle\langle m|$ is the representation of the *identity operator* in the energy basis, as explained in Appendix 6, the summation falls out and (5.7) is reduced to the form

Fig. 5.1 Density matrix

$$\rho = \begin{bmatrix} \rho_{11}\,\rho_{12} & \cdots & \cdots & \cdots & \cdots \\ \rho_{21}\,\rho_{22} & \cdots & \cdots & \cdots & \cdots \\ \cdots \quad \cdots & \cdots & \cdots & \cdots & \cdots \\ \cdots \quad \cdots & \cdots & \cdots & \rho_{nm} & \cdots \\ \cdots \quad \cdots & \cdots & \cdots & \cdots & \cdots \end{bmatrix}$$

$$\langle A \rangle = \sum_n \langle n | \rho A | n \rangle. \tag{5.8}$$

Note the summation is over the diagonal elements of the matrix ρA. We call such a summation the *trace* of the matrix and write Tr( ). Then, (5.8) is rewritten as

$$\langle A \rangle = \mathrm{Tr}(\rho A). \tag{5.9}$$

Clearly the trace of the density matrix itself is unity,

$$\mathrm{Tr}(\rho) = 1. \tag{5.10}$$

Readers are recommended to prove (5.10) for themselves. This equation means that the total number of electrons is constant.

5.2 Dynamic Equation of the Density Matrix

As shown in (5.8) or (5.9), we can determine the *statistical expectation value* if we know the properties of the density matrix, without needing to delve into details of $P^{(v)}$ or $\left|\Psi^{(v)}\right\rangle$. Let us derive the dynamic equation of the density matrix.

The defining equation of the matrix element is rearranged as follows:

$$\begin{aligned} \rho_{nm} &= \sum_v P^{(v)} C_m^{(v)*} C_n^{(v)} \\ &= \sum_v P^{(v)} \left\langle \Psi^{(v)} | m \right\rangle \left\langle n | \Psi^{(v)} \right\rangle \\ &= \sum_v \left\langle n | \Psi^{(v)} \right\rangle P^{(v)} \left\langle \Psi^{(v)} | m \right\rangle \\ &= \langle n | \left\{ \sum_v \left| \Psi^{(v)} \right\rangle P^{(v)} \left\langle \psi^{(v)} \right| \right\} | m \rangle. \end{aligned} \tag{5.11}$$

The last expression gives a bracketed operator sandwiched between bra $\langle n|$ and ket $|m\rangle$. We can now define the *density matrix* in the *operator form*

$$\rho = \sum_{\nu} \left|\Psi^{(\nu)}\right\rangle P^{(\nu)} \left\langle \psi^{(\nu)}\right|. \tag{5.12}$$

The time derivative of this equation is

$$\frac{d\rho}{dt} = \sum_{\nu} \left\{ \frac{d\left|\Psi^{(\nu)}\right\rangle}{dt} P^{(\nu)} \left\langle \psi^{(\nu)}\right| + \left|\Psi^{(\nu)}\right\rangle P^{(\nu)} \frac{d\left\langle \psi^{(\nu)}\right|}{dt} \right\}. \tag{5.13}$$

Here, we are assuming that the sampling probability $P^{(\nu)}$ has no time dependency. The time derivative of the state vector is given by the Schrödinger equation

$$j\hbar \frac{d\left|\Psi^{(\nu)}\right\rangle}{dt} = H\left|\Psi^{(\nu)}\right\rangle. \tag{5.14}$$

so that (5.13) is rewritten as

$$\begin{aligned} \frac{d\rho}{dt} &= \sum_{\nu} \left\{ \frac{H}{j\hbar} \left|\Psi^{(\nu)}\right\rangle P^{(\nu)} \left\langle \psi^{(\nu)}\right| - \left|\Psi^{(\nu)}\right\rangle P^{(\nu)} \left\langle \psi^{(\nu)}\right| \frac{H}{j\hbar} \right\} \\ &= \frac{1}{j\hbar}(H\rho - \rho H) \\ &= \frac{1}{j\hbar}[H, \rho]. \end{aligned} \tag{5.15}$$

This is the *dynamic equation of the density matrix*. The equation describing stimulated emission is obtained from this dynamic equation using the Hamiltonian associated with both electrons and the optical wave in a laser cavity.

5.3 Energy State in Semiconductor Material

A conduction electron in a semiconductor material spreads over a large number of atoms, which form the *energy band*. The *wave function* of the electron is represented by *Bloch functions*

$$|n\rangle = |\varphi_n(r)\rangle, \tag{5.16a}$$

$$\varphi_n(\boldsymbol{r}) = \frac{1}{\sqrt{V_e}} e^{jk_n r} u_n(\boldsymbol{r}). \tag{5.16b}$$

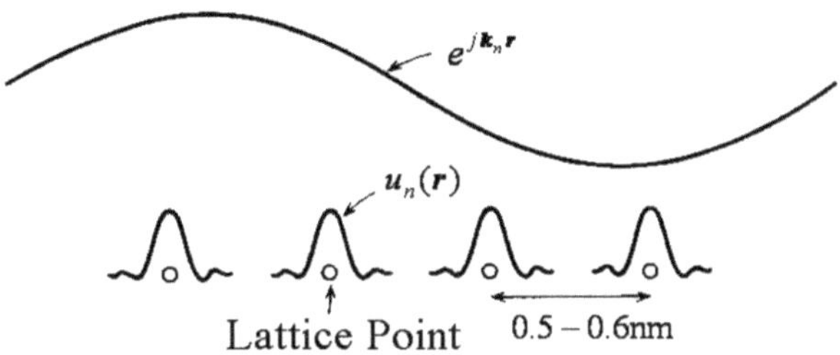

Fig. 5.2 Electron wave function in a periodic potential

Here, V_e is the volume occupied by an electron, $\boldsymbol{k}_n$ the *wave number* and $u_n(\boldsymbol{r})$ a *periodic function* repeating at the lattice points, as illustrated in Fig. 5.2.

Originally, the Bloch function is obtained based on the assumption that the electron is spread over an infinitely wide space. Assuming a finite volume, V_e is written as

$$V_e = \Delta x \Delta y \Delta z. \tag{5.17}$$

An uncertainty relation exists between the coordinate position x and wave number k_x of the electron

$$\Delta x \Delta k_x \geq \frac{1}{2}. \tag{5.18}$$

The uncertainty in the electron's energy then becomes $\Delta W = \hbar^2 (\Delta k_x)^2 / 2m_c \geq \hbar^2 / \left[8 (\Delta x)^2 m_c\right]$. That is to say, to ascribe a simultaneous representation of position and energy of an electron is difficult in principle. However, we can regard the electron as *localized in finite space* V_e because of mutual collisions among electrons. For the semiconductor GaAs, the length of spread is estimated to be $\Delta x = 10-30\,\text{nm}$, corresponding to an energy uncertainty of $\Delta W = 0.3-1.0\,\text{meV}$. This value is *much smaller than* the *thermal energy* $k_B T \approx 28$ meV of an electron at room temperature.

Furthermore, the spread of Δx in a semiconductor material is *much shorter than* typical *optical wavelengths* of $\lambda / n_{eq} = 200-450\,\text{nm}$. Therefore, an electron in a semiconductor material can be regarded as *a spatially localized particle whose wave function has Bloch function characteristics associated with the energy band structure.*

5.4 Interaction Hamiltonian

Whether interpreted classically or quantum mechanically, the *Hamiltonian* of a charged particle interacting with an electromagnetic wave is given as

$$H = \frac{1}{2m}(\boldsymbol{p} - e\boldsymbol{A})^2 + U(\boldsymbol{r}). \tag{5.19}$$

Here, m and $\boldsymbol{p}$ are the mass and momentum of the charged particle, $U(\boldsymbol{r})$ is the potential energy for the charged particle, and $\boldsymbol{A}$ the vector potential of the electromagnetic wave. A discussion of this Hamiltonian can be found in Appendices 3 and 4. The Hamiltonian is separated into two parts: the *principle Hamiltonian* H_o, which corresponds to the energy of the charged particle, and the *interaction Hamiltonian* H_{int}, which consists of cross terms involving the charged particle and the electromagnetic wave. That is,

$$H_o = \frac{1}{2m}\boldsymbol{p}^2 + U(\boldsymbol{r}), \tag{5.20}$$

$$H_{\text{int}} = -\frac{e}{2m}(\boldsymbol{p} \cdot \boldsymbol{A} + \boldsymbol{A} \cdot \boldsymbol{p}) + \frac{e^2}{2m}\boldsymbol{A}^2. \tag{5.21}$$

The principle Hamiltonian H_o determines the energy levels for the bands of semiconductor material associated with eigenvalue W_n of eigenstate $|n\rangle$. That is, the eigensystem is

$$H_o|n\rangle = W_n|n\rangle. \tag{5.22}$$

The last term A^2 in H_{int} represents the simultaneous interaction of two photons, but this probability is much smaller than the other term, $(\boldsymbol{p} \cdot \boldsymbol{A} + \boldsymbol{A} \cdot \boldsymbol{p})$, and is thus neglected.

Because of the applied electromagnetic wave, we suppose that the motion of the charged particle is *periodical* about a fixed point. Its position can be written as

$$\boldsymbol{r} = \boldsymbol{r}_o e^{j\omega t} + \boldsymbol{r}_o^* e^{-j\omega t}. \tag{5.23}$$

From classical Newtonian mechanics, the momentum of this charged particle is then

$$\boldsymbol{p} = m\dot{\boldsymbol{r}} = m\left(j\omega \boldsymbol{r}_o e^{j\omega t} - j\omega \boldsymbol{r}_o^* e^{-j\omega t}\right). \tag{5.24}$$

We set the vector potential to be

$$\boldsymbol{A} = \boldsymbol{A}_o e^{j\omega t} + \boldsymbol{A}_o^* e^{-j\omega t}. \tag{5.25}$$

Then the electric field is written as

$$\boldsymbol{E} = -\frac{\partial \boldsymbol{A}}{\partial t} = -j\omega \boldsymbol{A}_o e^{j\omega t} + j\omega \boldsymbol{A}_o^* e^{-j\omega t}. \tag{5.26}$$

Substitution of (5.24) and (5.25) into (5.21) gives

$$\begin{aligned} H_{\text{int}} &= -\frac{e}{m} m\{j\omega \boldsymbol{r}_o \cdot \boldsymbol{A}_o^* - j\omega \boldsymbol{r}_o^* \cdot \boldsymbol{A}_o\} \\ &= -e\boldsymbol{r} \cdot \boldsymbol{E} \\ &= -\boldsymbol{R} \cdot \boldsymbol{E}. \end{aligned} \tag{5.27}$$

Here,

$$\boldsymbol{R} = e\boldsymbol{r} \tag{5.28}$$

is the *dipole moment.*

Although (5.27) is derived classically, the position vector $\boldsymbol{r}$ and the dipole moment $\boldsymbol{R}$ are treated as *quantum mechanical operators.* $\boldsymbol{E}$ though is retained as a classical electric field. With that assumption, (5.27) is called the *dipole approximation* with H_{int} denoting the energy increase resulting from the dipole's interaction with the electric field. The original interaction Hamiltonian in (5.21), given with vector potential $\boldsymbol{A}$, includes a magnetic force term, as is proved in Appendices 3 and 4. Because we have supposed a periodic motion around a fixed point in (5.22), the magnetic force is neglected.

Using the dynamic equation of the density matrix, (5.15) and (5.27), *stimulated emission* is then described by

$$\left.\frac{\partial \rho}{\partial t}\right|_{stim} = \frac{1}{j\hbar}[(H_o - \boldsymbol{R} \cdot \boldsymbol{E}), \rho]. \tag{5.29}$$

5.5 Dipole Moment

We now determine the properties of the dipole by which the electron transition is characterized. As a first step, we verify that the diagonal elements of the dipole moment in an isotropic material are zero. Specifically, restricting to just a single coordinate position x, the one-dimensional integration of the dipole moment is given by

$$\langle n|R_x|n\rangle = e\int x|\varphi_n(x)|^2 dx = 0, \tag{5.30}$$

where we note that the integral vanishes because $|\varphi_n(x)|^2$ is an even function but x is an odd function under space inversion. The analysis holds analogously for the other coordinates y and z. Thus, the *diagonal matrix elements for the dipole moment are zero,*

$$\langle n|\boldsymbol{R}|n\rangle = 0. \tag{5.31}$$

The next step involves evaluating the off-diagonal elements of the dipole moment. We denote here an energy level state in the conduction band by $|b\rangle$ and that in the valence band by $|a\rangle$. With the electron wave function given in (5.16), the off-diagonal matrix elements of the dipole moment are given by

$$\langle b|\boldsymbol{R}|a\rangle = \frac{e}{V_e}\int_{V_e} e^{-j\boldsymbol{k}_b\boldsymbol{r}} u_b^*(\boldsymbol{r})\boldsymbol{r}\, u_a(\boldsymbol{r}) e^{j\boldsymbol{k}_a\boldsymbol{r}} d^3\boldsymbol{r}. \tag{5.32}$$

Here, $\int_{V_e} d^3\boldsymbol{r}$ indicates a three-dimensional volume integration for a single electron. The functions $u_b(\boldsymbol{r})$ and $u_a(\boldsymbol{r})$ are periodic functions with respect to the points of the lattice and satisfy

$$u_b(\boldsymbol{r}) = u_b(\boldsymbol{r}_m + \boldsymbol{r}), \tag{5.33}$$

where $\boldsymbol{r}_m$ is a lattice vector corresponding to the mth lattice point. Note that $e^{j\boldsymbol{k}_b\boldsymbol{r}}$ and $e^{j\boldsymbol{k}_a\boldsymbol{r}}$ varying more moderately than $u_b(\boldsymbol{r})$ and $u_a(\boldsymbol{r})$, as illustrated in Fig. 5.2. The functions $u_b(\boldsymbol{r})$ of the conduction band and $u_a(\boldsymbol{r})$ of the valence band are taken to be orthogonal. The spatial integration in (5.32) is then divided into an integration about a single lattice point and a summation over all lattice points as

$$\langle b|\boldsymbol{R}|a\rangle = \frac{e}{V_e}\sum_m e^{j(\boldsymbol{k}_a-\boldsymbol{k}_b)\cdot\boldsymbol{r}_m}\int_{prim} u_b^*(\boldsymbol{r})\,\boldsymbol{r}\, u_a(\boldsymbol{r}) d^3\boldsymbol{r}, \tag{5.34}$$

where $\int_{prim} d^3\boldsymbol{r}$ means the spatial integration over the *primitive cell* belonging to each lattice point.

The summation over all lattice points retains only those terms for which the wavenumbers coincide, $\boldsymbol{k}_b = \boldsymbol{k}_a$, called the *wave number selection rule* or the *momentum conservation rule*. This rule is illustrated in Fig. 5.3 and mathematically gives the term in (5.34)

$$\frac{1}{V_e}\sum_m e^{j(\boldsymbol{k}_a-\boldsymbol{k}_b)\cdot\boldsymbol{r}_m} = \delta_{\boldsymbol{k}_b,\boldsymbol{k}_a}. \tag{5.35}$$

The wavenumber selection rule is very important in evaluating not only the linear lasing gain but also complicated nonlinear phenomena. The matrix element of the dipole moment, therefore, reduces to an evaluation of the spatial integration over the primitive cell,

$$\langle b|\boldsymbol{R}|a\rangle = e\int_{prim} u_b^*(\boldsymbol{r})\,\boldsymbol{r}\, u_a(\boldsymbol{r}) d^3\boldsymbol{r}\delta_{\boldsymbol{k}_b,\boldsymbol{k}_a}. \tag{5.36}$$

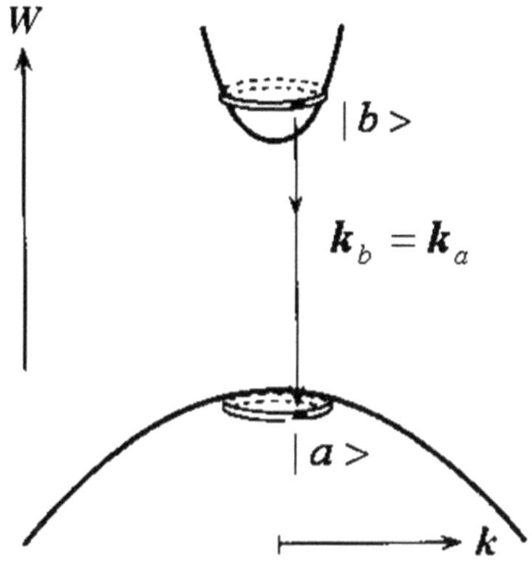

Fig. 5.3 Wave number selection rule

Expressions for the periodic functions $u_b(\boldsymbol{r})$ and $u_a(\boldsymbol{r})$ are developed using the so-called *k-p perturbation method* [1]. Properties of the conduction and valence bands, such as the band-gap energy and the effective masses, are characterized with these periodic functions. Therefore, the *expectation value of the dipole moment* can be expressed in terms of these values and are found to be

$$|R_{ba}|^2 = \frac{\hbar^2 e^2}{6 m_o W_{ba}^2} \left(\frac{m_o}{m_c} - 1 \right) \frac{W_g (W_g + \Delta)}{W_g + (2/3)\Delta}. \tag{5.37}$$

Here, m_o is the mass of an electron in free space, m_c the *effective mass* in the conduction band, W_g the *band gap*, and Δ the *spin-orbit splitting* in the valence band, as illustrated in Fig. 5.4.

Calculated values of the square of the expectation value of the dipole moment for several III–V compound materials are given in a log-normal plot in Fig. 5.5. The energy levels around the band edges were taken into account. We find that these squared values $|R_{ba}|^2$ are nearly inversely proportional to the square of the energy difference W_{ba} and hence almost independent of the type of III–V compound material.

5.6 Additional Phenomena

The diagonal elements ρ_{nn} of the density matrix signify the occupation probability of the electron in the energy state $|n\rangle$. In contrast, the off-diagonal elements ρ_{mn}, $m \neq n$, determine the amount of *polarization* as the electron spreads over energy states $|m\rangle$ and $|n\rangle$. Although the stimulated emission is obtained from (5.29), we

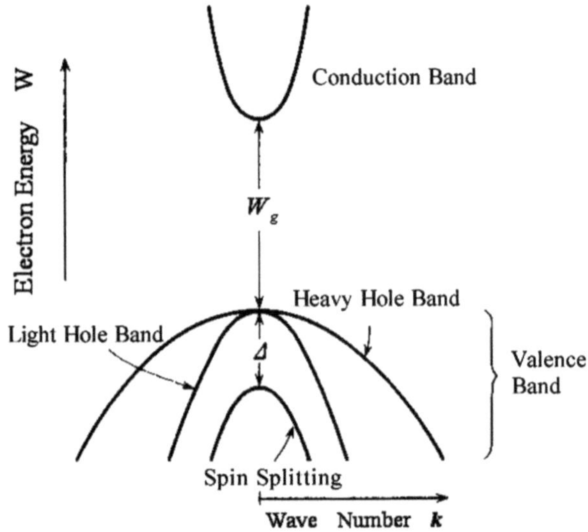

Fig. 5.4 Details of the energy levels in the band structure

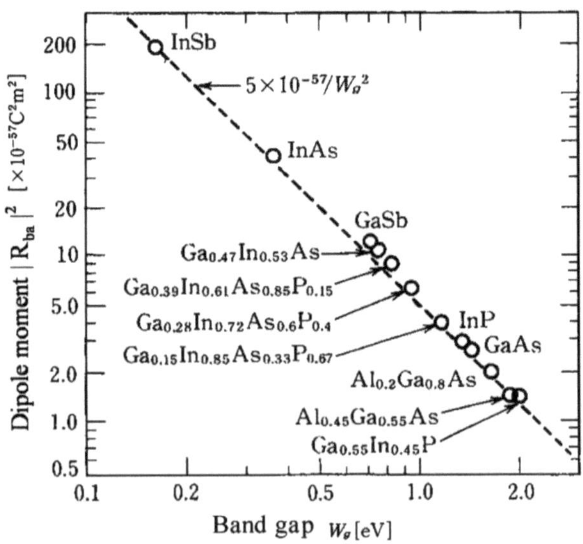

Fig. 5.5 Calculated values of the dipole moment [2] (Copyright, 1985, IEEE)

need to add more terms to the dynamic equation of the density matrix, to account for electron injection, spontaneous emission, and electric relaxation effects.

Pumping of the semiconductor laser refers to the injection of current into the active region. This effect can be simply added to the dynamic equation via

$$\left.\frac{\partial \rho}{\partial t}\right|_{pump} = \Lambda. \tag{5.38}$$

$$\langle m|\Lambda|n\rangle = \Lambda_n \delta_{m,n}. \tag{5.39}$$

Here, operator Λ describes the *electron injection* and is supposed to have only diagonal elements.

Two other effects are damping or decay of the electrons. In the derivation of (5.15), we have assumed that the sampling probability $P^{(v)}$ has no time dependency and the Hamiltonian H is common to all electrons. However, are these assumptions correct? The probability $P^{(v)}$ must have a temporal fluctuation and the Hamiltonian must differ for each electron. This type of non-uniformity must give rise to some *relaxation effect* for the electron wave or a change in energy. Below, we shall study how to introduce the relaxation effect.

We assume here that the Hamiltonian consists of a Hermitian part $H^{(e)}$ and a non-Hermitian part,

$$H = H^{(e)} - j\hbar\Gamma/2, \tag{5.40}$$

with eigenvalues

$$\left\langle m|H^{(e)}|n\right\rangle = \hbar\omega_n \delta_{m,n}, \tag{5.41}$$

$$\langle m|\Gamma|n\rangle = \frac{1}{\tau_n}\delta_{m,n}. \tag{5.42}$$

By substituting (5.40) into the Schrödinger equation in (5.14), the state vector is represented as

$$\left|\Psi^{(v)}(t)\right\rangle = \left|\Psi^{(v)}(0)\right\rangle e^{(-j\omega_n - 1/2\tau_n)t}. \tag{5.43}$$

Thus, relaxation or decay is taken into account by introducing a *non-Hermitian part* to the Hamiltonian. The corresponding dynamic equation of the density matrix in (5.15) becomes

$$\begin{aligned}\frac{d\rho}{dt} &= \frac{1}{j\hbar}(H\rho - \rho H^{+}) \\ &= \frac{1}{j\hbar}\left(H^{(e)}\rho - \rho H^{(e)}\right) - \frac{1}{2}(\Gamma\rho + \rho\Gamma).\end{aligned} \tag{5.44}$$

We find from (5.44) that stimulated emission is given by the imaginary terms whereas electron relaxation can be introduced as real numbers.

Because spontaneous emission can be thought of as a reduction in the number of electrons in the conduction band and a concomitant increase in the valence band, we express the dynamic equations for spontaneous emission as

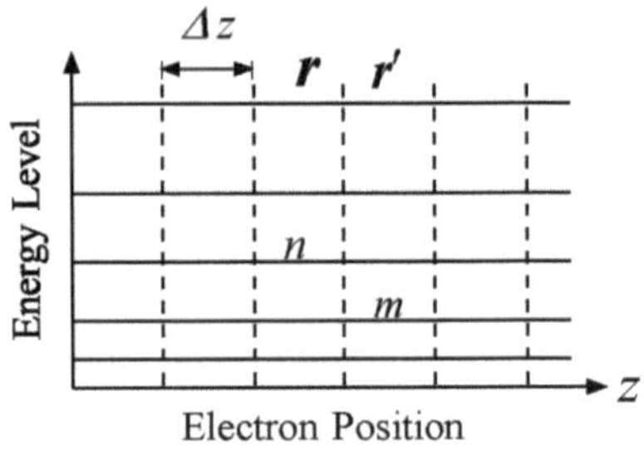

Fig. 5.6 Spatial position and energy levels for electrons

$$\left.\frac{\partial \rho}{\partial t}\right|_{spon} = -\frac{1}{2}\{(\rho - \tilde{\rho})\Gamma_{sp} + \Gamma_{sp}(\rho - \tilde{\rho})\}, \tag{5.45}$$

$$\langle m|\Gamma_{sp}|n\rangle = \frac{1}{\tau_s}\delta_{m,n}. \tag{5.46}$$

Here, Γ_{sp} is an operator corresponding to *spontaneous emission*. It has only diagonal elements with expectation value $1/\tau_s$, where τ_s is the *electron life time* whose value is of the order $10^{-9}\,\text{s} = 1\,\text{ns}$. $\tilde{\rho}$ is the electron distribution function at *thermal equilibrium* and carries a single suffix $\tilde{\rho}_n$ to denote that this is not a matrix element but a scalar function. Although $\tilde{\rho}_n$ is smaller than ρ_{nn} in the conduction band, $\tilde{\rho}_n$ is larger than ρ_{nn} in the valence band. Thus, we need to pay attention here that the density matrix of a semiconductor laser is defined originally for electrons and not for holes, even those in the valence band.

5.7 Intraband Relaxation

Another relaxation effect is called *intraband relaxation*. This relaxation is caused by scattering or collisions among electrons within time intervals smaller than $10^{-13}\,\text{s} = 0.1\,\text{ps}$. Although the energy level of each electron changes through scattering, the total number of electrons in the energy band does not change.

In our model, an electron is assumed to be in an energy state $|n\rangle$ located within a box of volume $V_e = \Delta x\,\Delta y\,\Delta z$, as illustrated in Fig. 5.6. We assume that an electron at position $\boldsymbol{r}$ in state $|n\rangle$ with probability $\rho_{nn}(\boldsymbol{r})$ transits to position $\boldsymbol{r}'$ in state $|m\rangle$ with probability $\rho_{mm}(\boldsymbol{r}')$ by a collision whose scattering probability is $S_{nm}(\boldsymbol{r}, \boldsymbol{r}')$. The transition probability is then

$$\begin{aligned} &S_{nm}(\mathbf{r}, \mathbf{r}')[\rho_{nn}(\mathbf{r})\{1 - \rho_{mm}(\mathbf{r}')\} - \rho_{mm}(\mathbf{r}')\{1 - \rho_{nn}(\mathbf{r})\}] \\ &\quad = S_{nm}(\mathbf{r}, \mathbf{r}')[\rho_{nn}(\mathbf{r}) - \rho_{mm}(\mathbf{r}')]. \end{aligned} \tag{5.47}$$

By summing over the positions $\boldsymbol{r}'$ and states $|m\rangle$ of all final transition states in the semiconductor material, the change in $\rho_{nn}(\mathbf{r})$ is given by

$$\left.\frac{\partial \rho_{nn}(\boldsymbol{r})}{\partial t}\right|_{\text{int}} = -\int \sum_{m} S_{nm}(\boldsymbol{r},\boldsymbol{r}')[\rho_{nn}(\boldsymbol{r}) - \rho_{mm}(\boldsymbol{r}')]d^3\boldsymbol{r}'. \tag{5.48}$$

The first term on the right-hand side yields a summation and an integration over just the transition coefficient. It depends on the *intraband relaxation time* τ_n as

$$\int \sum_{m} S_{nm}(\boldsymbol{r},\boldsymbol{r}')d^3\boldsymbol{r}' \equiv \frac{1}{\tau_n}. \tag{5.49}$$

The spatial integration and summation of the second term on the right-hand side of (5.48) performed over the whole space and on all energy levels except at the singular point $\boldsymbol{r}$ with the initial state $|n\rangle$. Hence, this term can be characterized by the initial state and defines a new distribution function $\tilde{\tilde{\rho}}_n(\boldsymbol{r})$ called *quasi-equilibrium distribution function*:

$$\int \sum_{m} S_{nm}(\boldsymbol{r},\boldsymbol{r}')\, \rho_{mm}(\boldsymbol{r}')d^3\boldsymbol{r}' \equiv \frac{\tilde{\tilde{\rho}}_n(\boldsymbol{r})}{\tau_n}. \tag{5.50}$$

Note that, although indicated with a single suffix n, $\tilde{\tilde{\rho}}_n(\boldsymbol{r})$ is not a matrix element but a scalar function. Equation (5.48) is then rewritten as

$$\left.\frac{\partial \rho_{nn}(\boldsymbol{r})}{\partial t}\right|_{\text{int}} = -\frac{\rho_{nn}(\boldsymbol{r}) - \tilde{\tilde{\rho}}_n(\boldsymbol{r})}{\tau_n}. \tag{5.51}$$

We examine the electron dynamics associated with the off-diagonal elements in similar manner as above. We assume that an electron undergoes a transition from an initial state associated with off-diagonal element $\rho_{nm}(\boldsymbol{r})$ to a final state associated with $\rho_{lk}(\boldsymbol{r}')$ and the transition has scattering probability $S_{nm,lk}(\boldsymbol{r},\boldsymbol{r}')$. Then, the dynamic equation becomes

$$\left.\frac{\partial \rho_{nm}(\boldsymbol{r})}{\partial t}\right|_{\text{int}} = -\int \sum_{lk} S_{nm,lk}(\boldsymbol{r},\boldsymbol{r}')[\rho_{nm}(\boldsymbol{r}) - \rho_{lk}(\boldsymbol{r}')]d^3\boldsymbol{r}'. \tag{5.52}$$

The relaxation time for the off-diagonal element is defined as

$$\int \sum_{lk} S_{nm,lk}(\boldsymbol{r},\boldsymbol{r}')d^3\boldsymbol{r}' \equiv \frac{1}{\tau_{nm}}. \tag{5.53}$$

Now, we note that the off-diagonal element $\rho_{lk}(\boldsymbol{r}')$ has a rotating phase of $\exp[j\omega_{kl}t]$ as we shall show later. Then, after summing over all possible energy levels, the second term on the right-hand side of (5.52) becomes zero,

$$\int \sum_{lk} S_{nm,lk}(\boldsymbol{r},\boldsymbol{r}')\rho_{lk}(\boldsymbol{r}')d^3\boldsymbol{r}' = 0. \tag{5.54}$$

Hence, (5.52) is reduced to

$$\left.\frac{\partial \rho_{nm}(\boldsymbol{r})}{\partial t}\right|_{\text{int}} = -\frac{\rho_{nm}(r)}{\tau_{nm}}. \tag{5.55}$$

Moreover, the relaxation of the *off-diagonal element* is given as an *arithmetic mean* of the relaxation of the *initial* and *final* levels, that is

$$\frac{1}{\tau_{nm}} = \frac{1}{2}\left(\frac{1}{\tau_n} + \frac{1}{\tau_m}\right). \tag{5.56}$$

Equations (5.51) (5.55), and (5.56) are unified as a single operator equation

$$\left.\frac{\partial \rho}{\partial t}\right|_{\text{int}} = -\frac{1}{2}\{(\rho - \tilde{\tilde{\rho}})\Gamma_{in} + \Gamma_{in}(\rho - \tilde{\tilde{\rho}})\}, \tag{5.57}$$

with

$$\langle m|\Gamma_{in}|n\rangle = \frac{1}{\tau_n}\delta_{m,n}. \tag{5.58}$$

Here Γ_{in} is the operator determining the *intraband relaxation effect.*

The complete *density matrix equation* thus formulated for semiconductor laser dynamics is

$$\begin{aligned}\frac{d\rho}{dt} &= \left.\frac{\partial \rho}{\partial t}\right|_{stim} + \left.\frac{\partial \rho}{\partial t}\right|_{\text{int}} + \left.\frac{\partial \rho}{\partial t}\right|_{spon} + \left.\frac{\partial \rho}{\partial t}\right|_{pump} \\ &= \frac{1}{j\hbar}[(H_o - \boldsymbol{R}\cdot\boldsymbol{E}),\rho] - \frac{1}{2}\{(\rho - \tilde{\tilde{\rho}})\Gamma_{in} + \Gamma_{in}(\rho - \tilde{\tilde{\rho}})\}. \\ &\quad -\frac{1}{2}\{(\rho - \tilde{\rho})\Gamma_{sp} + \Gamma_{sp}(\rho - \tilde{\rho})\} + \Lambda\end{aligned} \tag{5.59}$$

Many features of the semiconductor laser, such as lasing gain and nonlinear phenomena, are obtained from this equation.

The introduction of the intraband relaxation effect via Γ_{int} with the quasi-equilibrium distribution function $\tilde{\tilde{\rho}}$ allows for a special scheme in semiconductor laser operations. Several types of electron distributions in the conduction band are illustrated in Fig. 5.7. There, $\tilde{\rho}_n$ represents the electron distribution under *thermal equilibrium*, that is, the distribution without electron injection. Following electron injection into the conduction band, the electron distribution ρ_{nn} increases. During this increase, several electrons transition to the valence band through *spontaneous*

Fig. 5.7 Electron distributions in the conduction band

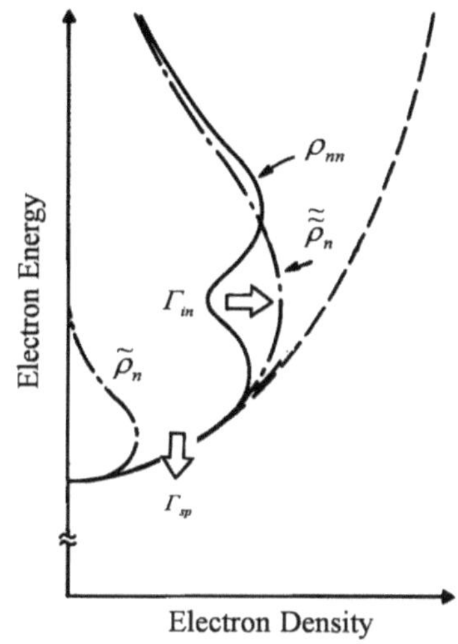

emission as described by the Γ_{sp} dynamics. When laser oscillations begin, many electrons make transitions to the valence band through *stimulated emission* under restricted optical frequencies. The electron distribution decreases as a consequence of the limited number of energy levels. We call this effect *spectral hole burning.* Electron scattering or collisions tend to smooth out this burning through electron transitions among the energy levels in the conduction band. This effect is called the *intra-band relaxation* effect. If we look at the electron distribution ρ_{nn}, this distribution seems to asymptote towards $\tilde{\tilde{\rho}}_n$. Hence we call $\tilde{\tilde{\rho}}$ the *quasi-equilibrium distribution.*

Bibliography

1. E.O. Kane, Bnad structure of indium antimonide. J. Phys. Chem. Solid **1**, 249–261 (1957)
2. M. Asada, Y. Suematsu, Density-matrix theory of semiconductor lasers with relaxation broadening model—gain and gain- suppression in semiconductor lasers. IEEE J. Quantum Electron. **QE-21**(5), 432–442 (1985)

Chapter 6
Gain Coefficient and Rate Equation

6.1 Laser Polarization

Express the electric field in its simplest form

$$E = E_\omega e^{j\omega t} + E_\omega^* e^{-j\omega t}, \tag{6.1}$$

the laser polarization P induced by this electric field is then written in classical form as

$$P = \varepsilon_o\{\chi(\omega)E_\omega e^{j\omega t} + \chi^*(\omega)E_\omega^* e^{-j\omega t}\} \tag{6.2}$$

or in quantum form given by the density matrix ρ

$$P = N_t\langle R\rangle = N_t\mathrm{Tr}(\rho R) = N_t\sum_{m,n}\rho_{mn}R_{nm}, \tag{6.3}$$

where N_t is the total electron density including both the conduction and valence bands. Enforcing the wave number selection rule, the double summation is reduced to a single summation giving

$$P = N_t\sum_{b \text{ or } a}(\rho_{ba}R_{ab} + \rho_{ab}R_{ba}). \tag{6.4}$$

Here, $\sum_{b \text{ or } a}$ indicates that the summation is over either levels b in the conduction band or levels a in the valence band.

The dynamic equation of the density matrix for semiconductor materials has been given in (5.59) as

$$\begin{aligned}\frac{d\rho}{dt} =& \frac{1}{j\hbar}(H_o\rho - \rho H_o) - \frac{1}{j\hbar}(RE\rho - \rho RE)\\ &-\frac{1}{2}\{(\rho - \tilde{\tilde{\rho}})\Gamma_{in} + \Gamma_{in}(\rho - \tilde{\tilde{\rho}})\} - \frac{1}{2}\{(\rho - \tilde{\rho})\Gamma_{sp} + \Gamma_{sp}(\rho - \tilde{\rho})\} + \Lambda.\end{aligned} \tag{6.5}$$

M. Yamada, *Theory of Semiconductor Lasers*,
Springer Series in Optical Sciences 185,
DOI: 10.1007/978-4-431-54889-8_6, © Springer Japan 2014

Here, the dipole moment R picks up a spatial component in the same direction as the electric field E. Because the energy state is time independent, the dynamic equation for the matrix element is given as

$$\frac{d\rho_{mn}}{dt} = \frac{d\langle m|\rho|n\rangle}{dt} = \left\langle m\left|\frac{d\rho}{dt}\right|n\right\rangle, \tag{6.6}$$

which we separate into four types depending on which band the states belong:

$$\frac{d\rho_{bb}}{dt} = \frac{j}{\hbar}(R_{ba}\rho_{ab} - \rho_{ba}R_{ab})E - \left(\frac{1}{\tau_b} + \frac{1}{\tau_s}\right)\rho_{bb} + \frac{\tilde{\tilde{\rho}}_b}{\tau_b} + \frac{\tilde{\rho}_b}{\tau_s} + \Lambda_b, \tag{6.7}$$

$$\frac{d\rho_{aa}}{dt} = \frac{j}{\hbar}(R_{ab}\rho_{ba} - \rho_{ab}R_{ba})E - \left(\frac{1}{\tau_a} + \frac{1}{\tau_s}\right)\rho_{aa} + \frac{\tilde{\tilde{\rho}}_a}{\tau_a} + \frac{\tilde{\rho}_a}{\tau_s} + \Lambda_a, \tag{6.8}$$

$$\frac{d\rho_{ab}}{dt} = \left(j\omega_{ba} - \frac{1}{\tau_{in}}\right)\rho_{ab} + j\frac{R_{ab}}{\hbar}(\rho_{bb} - \rho_{aa})E, \tag{6.9}$$

$$\frac{d\rho_{ba}}{dt} = \left(-j\omega_{ba} - \frac{1}{\tau_{in}}\right)\rho_{ba} + j\frac{R_{ba}}{\hbar}(\rho_{aa} - \rho_{bb})E, \tag{6.10}$$

where ω_{ba} is the angular frequency corresponding to the energy difference between energy level b in the conduction band and that of a in the valence band

$$\omega_{ba} = \frac{W_b - W_a}{\hbar} = \frac{W_{ba}}{\hbar} > 0, \tag{6.11}$$

and τ_{in} is the effective intraband relaxation time

$$\frac{1}{\tau_{in}} = \frac{1}{2}\left(\frac{1}{\tau_b} + \frac{1}{\tau_a}\right) + \frac{1}{\tau_s}. \tag{6.12}$$

In (6.5), the electric field E is seemed as if an operator, but is factored from the calculation for the matrix element in (6.7)–(6.10), because, as already discussed in Sect. 5.3, E can be regarded classically as being spatially uniform with a sufficiently long optical wavelength compared with the range of spread for an electron wave.

If we assume $E = 0$ in (6.7)–(6.10), the diagonal elements of both ρ_{bb} and ρ_{aa} are constant under temporal variation, whereas the off-diagonal elements ρ_{ab} and ρ_{ba} have variations $\exp\{(j\omega_{ba} - 1/\tau_{in})t\}$ and $\exp\{(-j\omega_{ba} - 1/\tau_{in})t\}$, respectively.

We next find solutions of the off-diagonal elements, subject to the presence of the electric field, by supposing the diagonal elements are constant. The *off-diagonal element* ρ_{ab} is taken to be of the form

$$\rho_{ab}(t) = w(t)e^{(j\omega_{ba} - 1/\tau_{in})t}, \tag{6.13}$$

with $w(t)$ an unknown amplitude. By substituting (6.13) into (6.9) with (6.1), we obtain the variation of $w(t)$ as

$$\frac{\partial w(t)}{\partial t} = j\frac{R_{ab}}{\hbar}(\rho_{bb} - \rho_{aa})\left[E_\omega e^{\{j(\omega-\omega_{ba})+1/\tau_{in}\}t} + E_\omega^* e^{\{-j(\omega+\omega_{ba})+1/\tau_{in}\}t}\right]. \quad (6.14)$$

Integrating over time t produces

$$w(t) = j\frac{R_{ab}}{\hbar}(\rho_{bb} - \rho_{aa})\left[E_\omega \frac{e^{\{j(\omega-\omega_{ba})+1/\tau_{in}\}t} - 1}{(\omega - \omega_{ba}) + 1/\tau_{in}} + E_\omega^* \frac{e^{\{-j(\omega+\omega_{ba})+1/\tau_{in}\}t} - 1}{-j(\omega + \omega_{ba}) + 1/\tau_{in}}\right]. \quad (6.15)$$

Substitution of this equation into (6.13) and taking $t \gg \tau_{in}$ gives the steady state solution

$$\rho_{ab}(t) = j\frac{R_{ab}}{\hbar}(\rho_{bb} - \rho_{aa})\left[\frac{E_\omega e^{j\omega t}}{j(\omega - \omega_{ba}) + 1/\tau_{in}} + \frac{E_\omega^* e^{-j\omega t}}{-j(\omega + \omega_{ba}) + 1/\tau_{in}}\right]. \quad (6.16)$$

The denominator $j(\omega - \omega_{ba}) + 1/\tau_{in}$becomes small when $\omega = \omega_{ba}$, whereas $-j(\omega + \omega_{ba}) + 1/\tau_{in}$ is always very large because $\omega_{ba} > 0$. Hence, (6.16) can be simplified to

$$\rho_{ab}(t) = \frac{jR_{ab}(\rho_{bb} - \rho_{aa})}{\hbar\{j(\omega - \omega_{ba}) + 1/\tau_{in}\}} E_\omega e^{j\omega t} \quad (6.17)$$

$\rho_{ba}(t)$ can also be simplified and is the complex conjugate of $\rho_{ab}(t)$,

$$\rho_{ba}(t) = \rho_{ab}^*(t) = -\frac{jR_{ba}(\rho_{bb} - \rho_{aa})}{\hbar\{j(\omega_{ba} - \omega) + 1/\tau_{in}\}} E_\omega^* e^{-j\omega t}. \quad (6.18)$$

To obtain the *laser susceptibility*, we substitute (6.17) and (6.18) into (6.4) and compare with (6.2). We have

$$\chi(\omega) = \frac{jN_t}{\varepsilon_o} \sum_{b\,\mathrm{or}\,a} \frac{|R_{ba}|^2(\rho_{bb} - \rho_{aa})}{j(\hbar\omega - W_{ba}) + \hbar/\tau_{in}}. \quad (6.19)$$

The real and imaginary parts of the susceptibility are

$$\mathrm{Re}\,\chi(\omega) = \frac{N_t}{\varepsilon_o} \sum_{b\,\mathrm{or}\,a} \frac{|R_{ba}|^2(\rho_{bb} - \rho_{aa})(\hbar\omega - W_{ba})}{(\hbar\omega - W_{ba})^2 + (\hbar/\tau_{in})^2}, \quad (6.20)$$

$$\mathrm{Im}\,\chi(\omega) = \frac{N_t}{\varepsilon_o} \sum_{b\,\mathrm{or}\,a} \frac{|R_{ba}|^2(\rho_{bb} - \rho_{aa})(\hbar/\tau_{in})}{(\hbar\omega - W_{ba})^2 + (\hbar/\tau_{in})^2}. \quad (6.21)$$

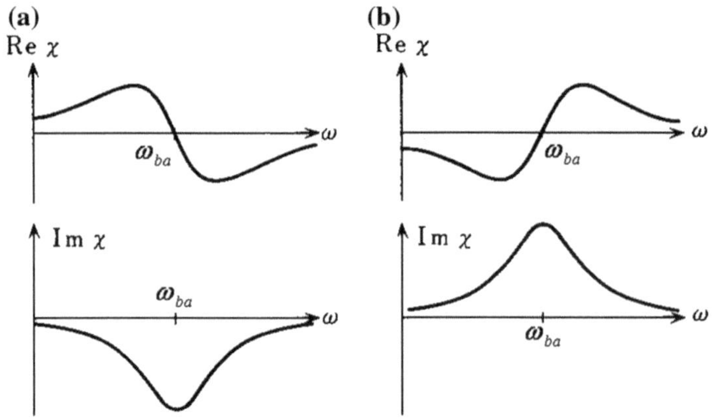

Fig. 6.1 Dispersion relations of the laser susceptibility for a two energy-level model. **a** $\rho_{bb} < \rho_{aa}$. **b** $\rho_{bb} > \rho_{aa}$

Hence, the *gain coefficient of the plane wave*, defined in (3.15), is

$$
\begin{aligned}
g(\omega) &= \frac{\sqrt{\mu_o \varepsilon_o}\omega}{n_r} \operatorname{Im} \chi(\omega) \\
&= \sqrt{\frac{\mu_o}{\varepsilon_o}} \frac{\omega N_t}{n_r} \sum_{b \text{ or } a} \frac{|R_{ba}|^2 (\rho_{bb} - \rho_{aa})(\hbar/\tau_{in})}{(\hbar\omega - W_{ba})^2 + (\hbar/\tau_{in})^2}.
\end{aligned}
\tag{6.22}
$$

6.2 Amplification Mechanism

When there is no electron injection, the number of electrons in the conduction band is much smaller than that in the valence band; that is, $\rho_{bb} < \rho_{aa}$ corresponding to conditions for *thermal equilibrium*. The gain coefficient $g(\omega)$ in (6.22) is then negative, indicating that optical absorption is occurring. By increasing the number of electrons in the conduction band and holes in the valence band, we then have $\rho_{bb} > \rho_{aa}$ corresponding to *population inversion*. The gain coefficient $g(\omega)$ becomes positive and under population inversion, the optical wave can be amplified.

Let us examine the mechanism of optical amplification by analyzing the vibration phases for the electric field and polarization. We suppose for simplicity that the laser material consists of only two energy levels b and a. The dispersion relations of the laser susceptibility $\chi(\omega)$ are depicted in Fig. 6.1, with *resonance angular frequency* of $\omega_{ba} = (W_b - W_a)/\hbar$. When an optical wave propagates within the laser material, a laser polarization P is induced that interacts with the incident electric field E_i, resulting in an attenuation or amplification of the optical output E_{out}, as illustrated in the top diagram of Fig. 6.2. The bottom two diagrams

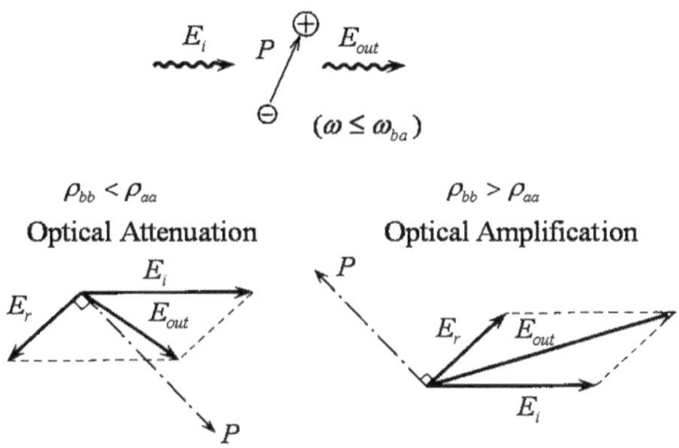

Fig. 6.2 Schematic explaining the optical attenuation or amplification by analyzing vibration phases of the polarization and electric field

of Fig. 6.2 are *complex plane* representations of the phase relation between the electric field and the laser polarization. We suppose here the *angular frequency* ω of the *incident electric field* is slightly smaller than the resonance angular frequency, $\omega \leq \omega_{ba}$.

When $\rho_{bb} < \rho_{aa}$, the induced polarization $P = \varepsilon_o(\mathrm{Re}\chi + j\mathrm{Im}\chi)E_i$ is phase delayed with respect to E_i because $\mathrm{Re}\chi > 0$ and $\mathrm{Im}\chi < 0$, as indicated with P situated in the 4th quadrant in the lower-left diagram. *Optical radiation* E_r from this laser polarization P has a $\pi/2$ phase delay. Here E_r is determined from (3.7) and (3.9) as

$$E_r = -\frac{j}{2n_r}\sqrt{\frac{\mu_o}{\varepsilon_o}}\,\omega P \Delta z,$$

where Δz is the length of interaction, and E_r extends into the 3rd quadrant. The *output* E_{out} after interaction is given by complex number addition $E_{out} = E_i + E_r$ in the complex plane. Then, length E_{out} becomes shorter than E_i, $|E_{out}| < |E_i|$, as observed in the diagram, implying an attenuation or absorption of the optical wave.

For population inversion, $\rho_{bb} > \rho_{aa}$, the laser polarization P extends into the 2nd quadrant because $\mathrm{Re}\chi < 0$ and $\mathrm{Im}\chi > 0$, as indicated in the lower-right diagram. The optical radiation E_r from this laser polarization P has also $\pi/2$ phase delay, as given by (3.7) and (3.9), and is located in the 1st quadrant. Therefore, the length of $E_{out} = E_i + E_r$ becomes longer than E_i, $|E_{out}| > |E_i|$, implying an amplification or emission of the optical wave.

If we suppose the angular frequency ω of the incident electric field E_i is slightly larger than the resonance angular frequency, $\omega \geq \omega_{ba}$, the laser polarization P and the optical radiation E_r extend into the 3rd and 2nd quadrants, respectively, if $\rho_{bb} < \rho_{aa}$. Then, the output $E_{out} = E_i + E_r$ is smaller than E_i and thus attenuates. If $\rho_{bb} > \rho_{aa}$, the laser polarization P and the optical radiation E_r are located in the 1st and the 4th quadrants, respectively, resulting in amplification with $E_{out} = E_i + E_r$.

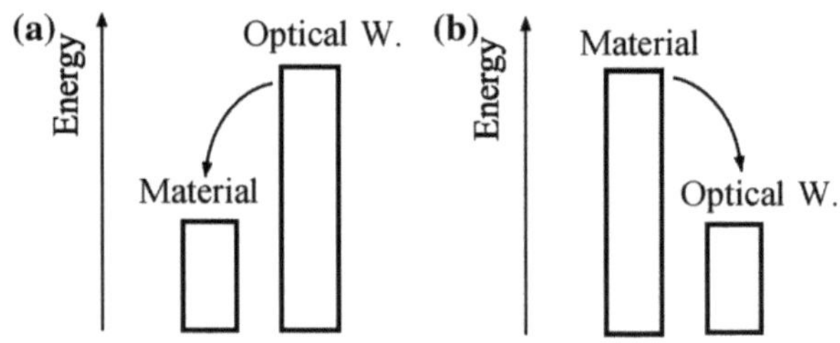

Fig. 6.3 Illustration of energy transfer between optical wave and material. **a** Optical absorption. **b** Optical emission

As seen from the above analysis, whether attenuation or amplification occurs depends on whether the phase of the induced laser polarization P is delayed if $\mathrm{Im}\,\chi < 0$ or advanced if $\mathrm{Im}\,\chi > 0$ with respect to the incident electric field E_i. An analysis of the dynamics for induced polarization from classical Newtonian mechanics is given in Appendix 7. The susceptibility $\chi(\omega)$ obtained is given in (A.7.17), which corresponds to conditions $\rho_{aa} = 1$ and $\rho_{bb} = 0$ in (6.19). In particular, the vibration phase of the induced polarization is always delayed in Newtonian mechanics, with the material absorbing all incident optical energy. This strange circumstance comes from the fact that Newtonian mechanics is founded on a causal relation as discussed in Sect. 2.2.

Although the analytical process based on quantum mechanics appear difficult, the results obtained may be easily understood. Both optical absorption and emission are illustrated in Fig. 6.3. Stored energy in the material is represented by the term $\rho_{bb} - \rho_{aa}$. When the energy of the material is low, with $\rho_{bb} < \rho_{aa}$, the energy of the optical wave is regarded to be higher than that of the material. Energy is transferred from the optical wave to the material, resulting in optical absorption. When the energy stored in the material is high, with $\rho_{bb} > \rho_{aa}$, the energy of the optical wave may be regarded as lower than that stored in the material. Energy is transferred from the material to the optical wave, resulting in optical emission. The direction of energy transfer is towards achieving the condition $\rho_{bb} = \rho_{aa}$. When the population in the material reaches a balance, $\rho_{bb} = \rho_{aa}$, there is neither optical absorption nor optical emission; that is, the material is transparent. The vibration phase of the polarization P is suitably tuned to achieve this balanced condition.

Because we can obtain a classical Newtonian dynamic equation from the Schrödinger equation, as demonstrated in Appendix 4, some readers might have questions about the difference between quantum mechanics and classical mechanics. It should be noted here that, because observable physical quantities must be "real numbers", we have dropped terms involving the imaginary unit in rewriting (A.4.21) as (A.4.23). However, these imaginary terms do have possible physical effects under different situations. Stimulated emission can be induced by these dropped terms. It could be said that classical mechanics looks only at a part of the physics and not the whole.

6.3 Representation by Continuous Energy Levels in Semiconductor Material

The gain coefficient g(ω) in (6.22) is given by summing over all energy levels either in the conduction band or in the valence band. The energy levels are originally given as discrete levels, because the occupation volume V_e of a single electron is assumed finite. However, a representation of the gain with continuous energy levels is more convenient to treat as it is similar to that in conventional semiconductor physics. In this section, we replace the summation with an energy integration. Although this procedure is rather tedious, it is essential in understanding certain features of the semiconductor laser.

We suppose the *extent of the electron localization* is ℓ. Then the occupation volume V_e is written as

$$V_e = \ell^3. \tag{6.23}$$

The wave number k_x in the x direction is determined from the *periodic boundary condition*

$$k_x \ell = 2m_x \pi, \tag{6.24}$$

where m_x is an integer. Then, the separation Δk_x between adjacent wave numbers is

$$\Delta k_x = \frac{2\pi}{\ell}. \tag{6.25}$$

A similar result follows for the other two components k_y and k_z. Therefore, the energy levels are indicated at grid points whose separation is $2\pi/\ell$ in the three-dimensional *wave number space*, as illustrated in Fig. 6.4. States having the same value of $k = \sqrt{k_x^2 + k_y^2 + k_z^2}$ have the same energy value and hence belong to the same energy level. The volume of the sphere with radius k in the wave number space is

$$\int d^3\mathbf{k} = 4\pi \int_o^k k^2 dk. \tag{6.26}$$

Dividing this volume by $(2\pi/\ell)^3$ gives the number of energy levels within radius k. Then the *density of states* $g_c(W_b)$ is derived as

$$2 \times \left(\frac{\ell}{2\pi}\right)^3 \int d^3\mathbf{k} = \frac{\ell^3}{\pi^2} \int_0^{k_b} k^2 dk \equiv V_e \int_{W_c}^{W_b} g_c(W_b) dW_b, \tag{6.27}$$

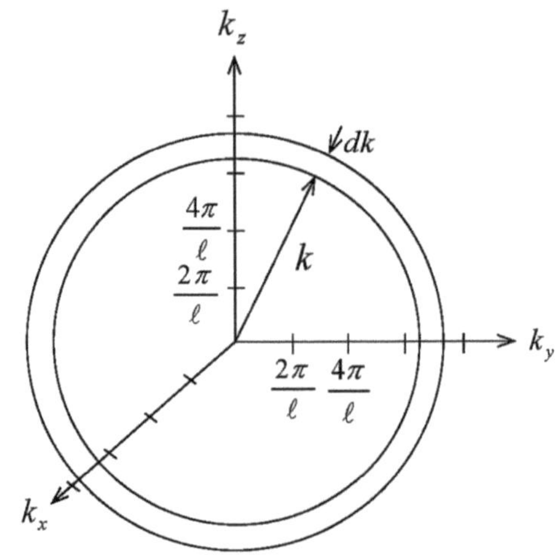

Fig. 6.4 Three-dimensional wave number space

where W_c is the *energy of the level at the edge of the conduction band*, and the factor 2 occurs to account for the electron spin polarization associated with each energy state. Thus,

$$g_c(W_b)dW_b = \frac{1}{\pi^2}k_b^2 dk_b. \tag{6.28}$$

Meanwhile, the electron energy W_b for a conduction band state is given by

$$W_b = W_c + \frac{\hbar^2 k_b^2}{2m_c}. \tag{6.29}$$

where m_c is the *effective mass* in the conduction band. Rearranging and differentiating, we obtain

$$k_b = \frac{\sqrt{2m_c(W_b - W_c)}}{\hbar}, \tag{6.30}$$

$$dk_b = \frac{1}{2\hbar}\sqrt{\frac{2m_c}{W_b - W_c}}dW_b. \tag{6.31}$$

We then have

$$g_c(W_b)dW_b = \frac{(2m_c)^{3/2}}{2\pi^2\hbar^3}\sqrt{W_b - W_c}\,dW_b. \tag{6.32}$$

By a similar argument, the density of states for the valence band is

$$g_v(W_a)dW_a = \frac{(2m_v)^{3/2}}{2\pi^2\hbar^3}\sqrt{W_v - W_a}\,dW_a, \tag{6.33}$$

where W_v and m_v are the respective *energy of the edge states* and *effective mass* for the valence band. Although heavy-hole and light-hole bands exist, as is illustrated in Fig. 5.4, we have simplified the treatment to take into account only the heavy holes here.

Summations over the electron densities existing in the conduction and valence bands are written as

$$\begin{aligned} N_t \sum_{b\,\mathrm{or}\,a} \rho_{bb} &= \frac{2}{(2\pi)^3}\iint f_c(k_b)\delta(\boldsymbol{k}_b - \boldsymbol{k}_a)d^3\boldsymbol{k}_b d^3\boldsymbol{k}_a \\ &= \int_{W_c}^{\infty} g_c(W_b) f_c(W_b)\,dW_b \\ &\equiv N_c \end{aligned}, \tag{6.34}$$

$$\begin{aligned} N_t \sum_{b\,\mathrm{or}\,a} \rho_{aa} &= \frac{2}{(2\pi)^3}\iint f_v(k_a)\delta(\boldsymbol{k}_b - \boldsymbol{k}_a)d^3\boldsymbol{k}_b d^3\boldsymbol{k}_a \\ &= \int_{-\infty}^{W_v} g_v(W_a) f_v(W_a)\,dW_a \\ &\equiv N_v \end{aligned}. \tag{6.35}$$

Here, N_c and N_v are *electron densities* in the conduction and the valence band, respectively, whereas $f_c(W_b)$ and $f_v(W_a)$ are the *Fermi-Dirac distribution functions* characterized by the *quasi-Fermi levels* μ_c and μ_v in the conduction and valence bands, respectively. Specifically, we have

$$f_c(W_b) = \frac{1}{1+\exp\left(\frac{W_b-\mu_c}{k_B T}\right)}, \tag{6.36}$$

$$f_v(W_a) = \frac{1}{1+\exp\left(\frac{W_a-\mu_c}{k_B T}\right)}. \tag{6.37}$$

For the next step, we change the variables of integration W_b and W_a to $W_{ba} = W_b - W_a$. The energy level W_b has been given in (6.29). The energy level W_a in the valence band is

$$W_a = W_v - \frac{\hbar^2 k_a^2}{2m_v}. \tag{6.38}$$

Using this equation with the wave number selection rule of $\boldsymbol{k}_b = \boldsymbol{k}_a$, (6.29) is written as

$$W_b = W_c + \frac{m_v}{m_c}(W_v - W_a), \tag{6.39}$$

and rearranged to give

$$\begin{aligned} W_b &= W_c + \frac{m_v}{m_c}(W_c - W_c + W_v - W_b + W_b - W_a) \\ &= W_c + \frac{m_v}{m_c}(W_c - W_g - W_b + W_{ba}), \end{aligned} \tag{6.40}$$

where W_g is the *band gap* given by

$$W_g = W_c - W_v. \tag{6.41}$$

Equation (6.40) is rearranged once more

$$W_b - W_c = \frac{m_v}{m_c + m_v}(W_{ba} - W_g). \tag{6.42}$$

Substitution of this equation into (6.39) gives

$$W_v - W_a = \frac{m_c}{m_c + m_v}(W_{ba} - W_g). \tag{6.43}$$

We now define the reduced state density $g_{cv}(W_{ba})$ for a pair of electrons at energy levels b and a in the following relation of

$$\begin{aligned} \frac{2}{(2\pi)^3}\delta(\boldsymbol{k}_b - \boldsymbol{k}_a)d^3\boldsymbol{k}_b d^3\boldsymbol{k}_a &= g_c(W_b)\,dW_b \\ &= g_v(W_a)\,dW_a \\ &= g_{cv}(W_{ba})\,dW_{ba} \end{aligned} \tag{6.44}$$

Substituting (6.42) into (6.32) gives the *reduced state density* as

$$g_{cv}(W_{ba})dW_{ba} = \left(\frac{2m_c m_v}{m_c + m_v}\right)^{3/2}\frac{1}{2\pi^2\hbar^3}\sqrt{W_{ba} - W_g}\,dW_{ba}. \tag{6.45}$$

Interestingly, by comparing (6.45), (6.32), and (6.33), the reduced state density implies a vibrational motion of two pairs of particles with effective masses m_c and m_v, as represented by *reduced mass* m_R given by

$$\frac{1}{m_R} = \frac{1}{m_c} + \frac{1}{m_v} = \frac{m_c + m_v}{m_c m_v}. \tag{6.46}$$

Our picture of the optical emission via an electric dipole, explained in Sect. 1.1, well fits the quantum mechanical treatment introduced using the electron wave in the semiconductor material.

The *Fermi–Dirac distribution functions* are rewritten by substitutions of (6.42) and (6.43) into (6.36) and (6.37)

$$f_c(W_{ba}) = \frac{1}{1+\exp\left(\frac{\frac{m_v}{m_c+m_v}(W_{ba}-W_g)-\mu_c+W_c}{k_B T}\right)}, \tag{6.47}$$

$$f_v(W_{ba}) = \frac{1}{1+\exp\left(\frac{-\frac{m_c}{m_c+m_v}(W_{ba}-W_g)-\mu_v+W_v}{k_B T}\right)}. \tag{6.48}$$

Using these functions, the gain coefficient $g(\omega)$ in (6.22) is written as [1]

$$g(\omega) = \sqrt{\frac{\mu_o}{\varepsilon_o}}\frac{\omega}{n_r}\int_{W_g}^{\infty}\frac{|R_{ba}|^2 g_{cv}(W_{ba})\{f_c(W_{ba})-f_v(W_{ba})\}(\hbar/\tau_{in})}{(\hbar\omega-W_{ba})^2+(\hbar/\tau_{in})^2}dW_{ba}. \tag{6.49}$$

The variables essential to describe laser operations are the quasi-Fermi levels μ_c and μ_v in f_c (W_{ba}) and f_v (W_{ba}). The difference between these two quasi-Fermi levels arises from the *applied voltage* V_{appl} to the active region from the p and n sides,

$$e\,V_{appl} = \mu_c - \mu_v. \tag{6.50}$$

Values of μ_c and μ_v are determined to satisfy the *neutral charge condition*

$$\begin{aligned} N_c - (N_t - N_v) &= \int_{W_g}^{\infty} g_{cv}(W_{ba})\{f_c(W_{ba},\mu_c)+f_v(W_{ba},\mu_v)-1\}dW_{ba} \\ &= 0. \end{aligned} \tag{6.51}$$

Here, N_t is the total electron density including the conduction and valence bands. The *electron densities* in the conduction and valence bands are obtained from

$$N_c = N_t\sum_{b\,\mathrm{or}\,a}\rho_{bb} = \int_{W_g}^{\infty} g_{cv}(W_{ba})f_c(W_{ba},\mu_c)dW_{ba}, \tag{6.52}$$

$$N_v = N_t\sum_{b\,\mathrm{or}\,a}\rho_{aa} = \int_{-\infty}^{W_v} g_{cv}(W_{ba})f_v(W_{ba},\mu_v)dW_{ba}. \tag{6.53}$$

Fig. 6.5 Gain coefficient for GaAs. **a** photon energy dispersion. **b** Variation of peak gain with electron density

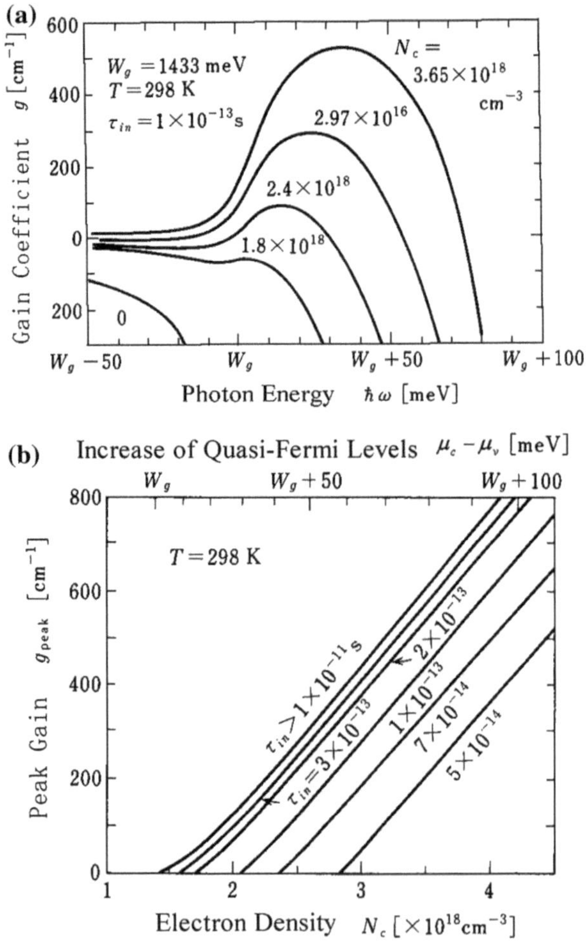

The *hole density* in the valence band is

$$N_t - N_v = \int_{-\infty}^{W_v} g_{cv}(W_{ba})\{1 - f_v(W_{ba}, \mu_v)\}dW_{ba}. \tag{6.54}$$

Calculated examples of the gain coefficient g in GaAs are shown in Fig. 6.5, where Fig. 6.5a shows the dispersion on the photon energy $\hbar\omega$, and Fig. 6.5b shows the variation of the peak gain with the injected electron density N_c. The parameter values used for the plots are $m_c = 0.067\,m_o$, $m_v = 0.45\,m_o$, $W_g = 1424\,\text{meV}$, $\Delta = 340\,\text{meV}$, and $|R_{ba}|^2 = 3.41 \times 10^{-57}\,\text{C}^2\text{m}^2$. If the electron injection density N_c is zero, the gain coefficient is negative, indicating absorption. Although there is no energy state in the band gap, the gain $g(\omega)$ or absorption profile shows a tail in the band gap $\hbar\omega < W_g$, as shown in Fig. 6.5a, owing to

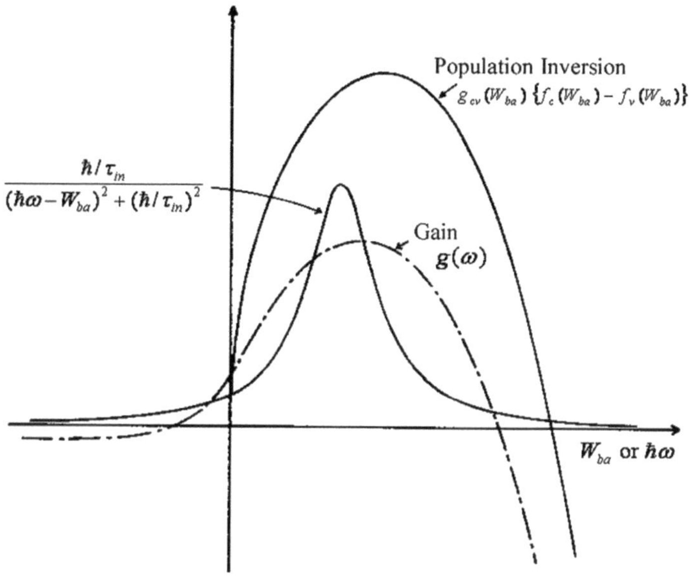

Fig. 6.6 Comparison of dispersion functions

broadening from intraband relaxation $\hbar/\tau_{in}$. The gain $g(\omega)$ is given as a convolution between the population inversion and the Lorentz-type broadening function. These different distribution functions are illustrated in Fig. 6.6.

The peak value of the gain coefficient g_{peak} increases almost linearly with electron injection density N_c for GaAs, as shown in Fig. 6.5b. The gain coefficient becomes smaller if the intraband relaxation time τ_{in} becomes smaller. If the relaxation time is longer than 10^{-11} s, the dispersion from relaxation in (6.49) is approximated by a delta function

$$\frac{\hbar/\tau_{in}}{(\hbar\omega - W_{ba})^2+(\hbar/\tau_{in})^2} \approx \pi\delta(\hbar\omega - W_{ba}), \tag{6.55}$$

and the gain coefficient can be approximated as

$$g(\omega) \approx \sqrt{\frac{\mu_o}{\varepsilon_o}}\frac{\pi\omega}{n_{eff}}|R_{ba}|^2 g_{cv}(\hbar\omega)\{f_c(\hbar\omega) - f_v(\hbar\omega)\}. \tag{6.56}$$

However, this approximation is not suitable because the intraband relaxation time is estimated to be $\tau_{in} = (0.7 - 1.0) \times 10^{-13}$ s.

Another calculated example of the gain coefficient in $In_{0.53}Ga_{0.47}As$ is shown in Fig. 6.7. The parameter values used are $m_c = 0.041\,m_o$, $m_c = 0.42\,m_o$,

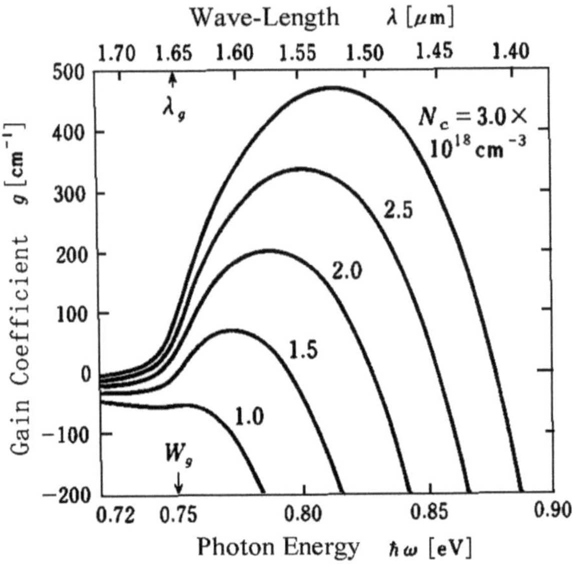

Fig. 6.7 Gain coefficient in $In_{0.53}Ga_{0.47}As$, ($T = 300\,\mathrm{K}, \tau_{in} = 1 \times 10^{-13}\,\mathrm{s}$, $W_g = 0.75\,\mathrm{eV}$, $\lambda_g = 1.65\,\mu\mathrm{m}$) [2] (Copyright, 1985, IEEE)

$W_g = 750\,\mathrm{meV}$, $\Delta = 420\,\mathrm{meV}$, and $|R_{ba}|^2 = 1.15 \times 10^{-56} \times (W_g/W_{ba})^2\,\mathrm{C^2m^2}$. The wavelength corresponding to the band gap $W_g = 75\,\mathrm{meV}$ is $\lambda_g = 1.65\,\mu\mathrm{m}$.

The variation of the peak gain with electron injection density in various materials are shown in Fig. 6.8.

6.4 Deviation of Refractive Index

The real part of the laser susceptibility gives the *dispersion characteristic* of the *dielectric constant* or the *refractive index* around the band gap, and is calculated from an energy integration [1],

$$\mathrm{Re}\chi(\omega) = \frac{1}{\varepsilon_0}\int_{W_g}^{\infty} \frac{|R_{ba}|^2 g_{cv}(W_{ba})\{f_c(W_{ba}) - f_v(W_{ba})\}(\hbar\omega - W_{ba})}{(\hbar\omega - W_{ba})^2 + (\hbar/\tau_{in})^2}\, dW_{ba}. \quad (6.57)$$

A calculated example of the real part of the laser susceptibility in GaAs is shown in Fig. 6.9.

For a simple two-level model, $\mathrm{Re}\chi(\omega)$ must become zero at the resonance angular frequency $\omega_{ba} = W_{ba}/\hbar$, as sketched in Fig. 6.1. However, for semiconductor materials, $\mathrm{Re}\,\chi(\omega)$ exhibits a complicated profile, as shown in Fig. 6.9, because the reduced state density $g_{cv}(W_{ba})$ gives an asymmetric distribution for the energy W_{ba}. The variation of $\mathrm{Re}\,\chi(\omega)$ with electron injection density gives the

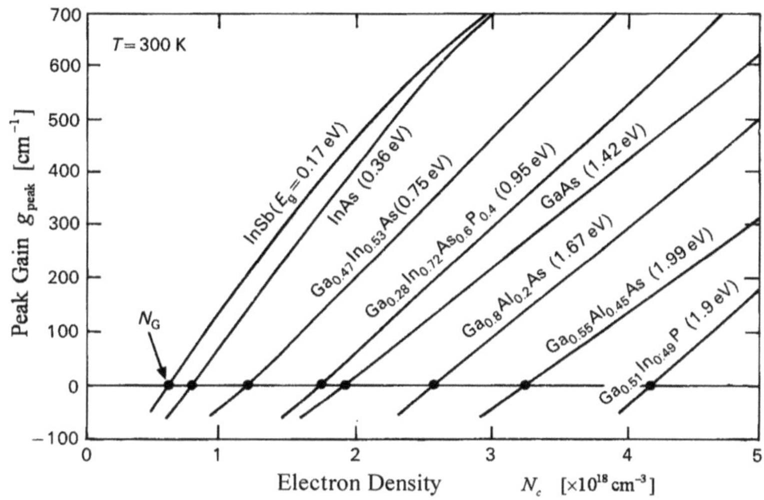

Fig. 6.8 Variation of peak gain with electron density for various materials [2] (Copyright, 1985, IEEE)

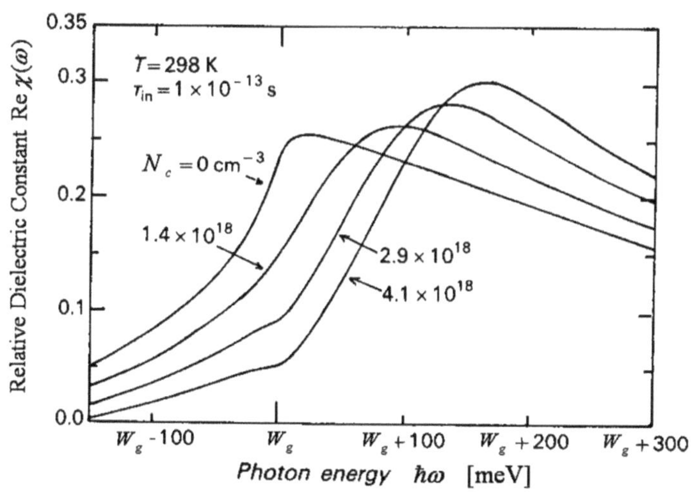

Fig. 6.9 Dispersion characteristics of the relative dielectric constant around the band gap

variation of the relative dielectric constant Δn_r as already discussed with reference to (3.24). As found by comparing Figs. 6.5a and 6.9, the variation of Re $\chi(\omega)$ around the gain peak is not negligible.

As a general principle, Re $\chi(\omega)$ and Im $\chi(\omega)$ are related by an integral transformation that yields the so-called *Kramers–Krönig relation*. The derivation of this relation based on the density matrix treatment is given in Appendix 8.

6.5 Plasma Effect

Lasing polarization P is characterized by electron transitions from the conduction band to the valence band that follow from electrons interacting with the optical wave. Apart from these dynamics, there is an effect, called the *plasma effect*, by which the injected electrons and holes are excited by the optical wave and make transitions not to another band but remain in the original band. Unfortunately, a quantum mechanical analysis of this effect in semiconductor materials has not been established yet, because this effect is not caused by electron transitions between eigenenergy levels. As discussed in Sect. 6.2, classical mechanics is applicable to topics related to the excitations of charged particles. We have analyzed the plasma effect based on classical mechanics; this is presented in Appendix 9.

This plasma effect produces a *reduction in the refractive index* and *optical absorption*, called *free carrier absorption*. From the calculations in Appendix 9 for GaAs, the extent of refractive-index reduction is almost 1/3 of that by laser polarization and the extent of free carrier absorption is 5–7 cm^{-1}.

6.6 Rate Equation for the Electron Density

The time variation of the *electron density* is obtained from the dynamic equations of the diagonal elements of the density matrix as in (6.7) and (6.8). We take the summation over all energy levels $|b\rangle$ and multiply (6.7) by N_t,

$$\begin{aligned} N_t \sum_b \frac{d\rho_{bb}}{dt} = & N_t \sum_b \left\{ \frac{j}{\hbar}(R_{ba}\rho_{ab} - \rho_{ba}R_{ab})E \right\} - N_t \sum_b \frac{\rho_{bb} - \tilde{\tilde{\rho}}_b}{\tau_b} \\ & - N_t \sum_b \frac{\rho_{bb} - \tilde{\rho}_b}{\tau_s} + N_t \sum_b \Lambda_b \end{aligned} . \tag{6.58}$$

The first term on the right-hand side of this equation is rearranged using (6.17) and (6.18),

$$\begin{aligned} N_t \sum_b \frac{j}{\hbar}(R_{ba}\rho_{ab} - \rho_{ba}R_{ab})E = & - N_t \sum_b \frac{|R_{ba}|^2(\rho_{bb} - \rho_{aa})}{\hbar^2} \left\{ \frac{E_\omega e^{j\omega t}}{j(\omega - \omega_{ba}) + 1/\tau_{in}} + \frac{E_\omega^* e^{-j\omega t}}{j(\omega_{ba} - \omega) + 1/\tau_{in}} \right\} \\ & \times \left\{ E_\omega e^{j\omega t} + E_\omega^* e^{-j\omega t} \right\} \\ = & - N_t \sum_b \frac{|R_{ba}|^2(\rho_{bb} - \rho_{aa})(\hbar/\tau_{in})}{\hbar\left\{ (\hbar\omega - W_{ba})^2 + (\hbar/\tau_{in})^2 \right\}} 2|E_\omega|^2 \\ = & - \frac{\varepsilon_o}{\hbar} \mathrm{Im}\chi(\omega) 2|E_\omega|^2 . \end{aligned} \tag{6.59}$$

The second term is zero because the total number of electrons in the conduction band does not change under the intraband relaxation effect,

$$\sum_b \frac{\rho_{bb} - \tilde{\tilde{\rho}}_b}{\tau_b} = 0. \tag{6.60}$$

The third term describes the reduction in electron number from spontaneous emission. We define the electron densities $\tilde{N}_c$ and $\tilde{N}_v$ in the conduction and valence bands at *thermal equilibrium* as

$$\tilde{N}_c \equiv N_t \sum_b \tilde{\rho}_b, \tag{6.61}$$

$$\tilde{N}_v \equiv N_t \sum_v \tilde{\rho}_v. \tag{6.62}$$

The last term in (6.58) describes *current injection* with current I defined by

$$I = eV_a N_t \sum_b \Lambda_b = -eV_a N_t \sum_a \Lambda_a, \tag{6.63}$$

where V_a is the volume of the active region. Using these results, (6.58) becomes

$$\frac{dN_c}{dt} = -\frac{\varepsilon_o}{\hbar}\mathrm{Im}\chi(\omega)2|E_\omega|^2 - \frac{N_c - \tilde{N}_c}{\tau_s} + \frac{I}{eV_a}. \tag{6.64}$$

As introduced in (3.38), the electric field in the cavity is expressed as a standing wave

$$E = \widehat{E}_p(t)\Phi_p(x, y, z)e^{j\omega_p t} + c.c., \tag{6.65}$$

for which the amplitude is related to the photon number in the cavity, as in (3.56),

$$|\widehat{E}_p(t)|^2 = \frac{\hbar\omega_p}{2\varepsilon_{eff}} S_p. \tag{6.66}$$

Then, (6.64) is expressed in terms of the photon number as

$$\frac{dN_c}{dt} = -\frac{\omega_p}{n_{eff}^2}\mathrm{Im}\chi(\omega)S_p|\Phi_p(x, y, z)|^2 - \frac{N_c - \tilde{N}_c}{\tau_s} + \frac{I}{eV_a}. \tag{6.67}$$

By taking a spatial integration over the active region and dividing by V_a again, we obtain the *rate equation* for the *electron density*,

$$\frac{dN_c}{dt} = -G_p\frac{S_p}{V_a} - \frac{N_c - \tilde{N}_c}{\tau_s} + \frac{I}{eV_a}. \tag{6.68}$$

Here, G_p is the *gain coefficient* of the mode p introduced in (3.58) as

$$G_p = \xi G = \frac{\omega_p}{n_{eff}^2} \operatorname{Im} \chi_{eff}(\omega), \tag{6.69}$$

with a *gain coefficient* G for a plane wave of

$$G = \frac{\omega_p}{n_r^2} \operatorname{Im} \chi(\omega). \tag{6.70}$$

The gain coefficient can be approximately given based on the numerically calculated values as

$$G = a\left\{N - N_g - b(\lambda_p - \lambda_o)^2\right\}, \tag{6.71}$$

where a is called the *tangential coefficient*, N_g the *transparent electron density*, b a coefficient giving the frequency dispersion of the gain, and λ_o the wavelength at the gain peak.

The variation of electron density in the *valence band* is similarly derived and is given as

$$\frac{dN_v}{dt} = G_p \frac{S_p}{V_a} + \frac{\tilde{N}_v - N_v}{\tau_s} - \frac{I}{eV_a}. \tag{6.72}$$

6.7 Spontaneous Emission and Electron Life Time

As previously discussed in Sect. 1.1, with (1.3)–(1.5) and results of Appendix 5, the electron transition probability from the upper level $|b\rangle$ to the lower level $|a\rangle$ is proportional to $f_b(1 - f_a)(S + 1)$, and that from the lower level to the upper level is proportional to $f_a(1 - f_b)S$. The stimulated emission is then proportional to $(f_b - f_a)S$ and the spontaneous emission is proportional to $f_b\,(1 - f_a)$. The term G_pS_p in (6.68) represents the stimulated emission. Therefore, the spontaneous emission probability $T_{sp}(\omega)$ is given by replacing $\{f_c(W_{ba}) - f_v(W_{ba})\}$ in the gain coefficient with $f_c(W_{ba})\{1 - f_v(W_{ba})\}$ as [3]

$$T_{sp}(\omega) = \frac{\omega}{\varepsilon_o n_r^2} \int_{W_g}^{\infty} \frac{|R_{ba}|^2 g_{cv}(W_{ba}) f_c(W_{ba})\{1 - f_v(W_{ba})\}(\hbar/\tau_{ba})}{(\hbar\omega - W_{ba})^2 + (\hbar/\tau_{ba})^2} dW_{ba}, \tag{6.73}$$

where

$$\frac{1}{\tau_{ba}} = \frac{1}{2}\left(\frac{1}{\tau_a} + \frac{1}{\tau_b}\right). \tag{6.74}$$

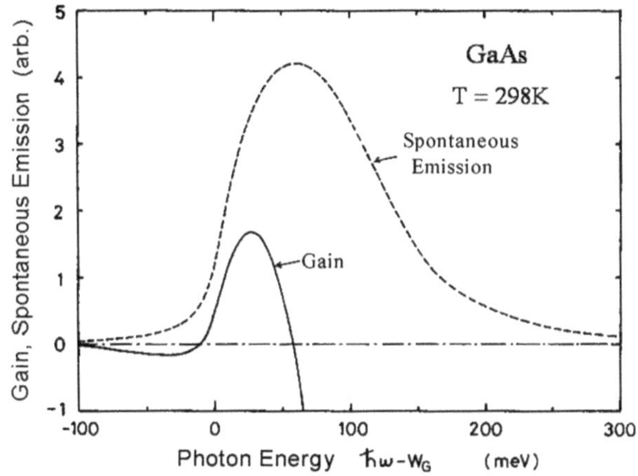

Fig. 6.10 Example of numerically calculated lasing gain and spontaneous emission for GaAs

For the gain coefficient, we use τ_{in} as the relaxation time. However, it might be better to use relaxation time τ_{ba} for the spontaneous emission instead of τ_{in} in (6.73) because τ_{in} already includes the effect of spontaneous emission, as defined in (6.12).

An example of a numerically calculated gain coefficient G and spontaneous emission T_{sp} for GaAs is shown in Fig. 6.10. Although the gain coefficient becomes negative for high photon energies, spontaneous emission is positive over all photon energies.

Spontaneous emission works as a seed for laser light which is amplified via lasing gain. When the injection current I is lower than the threshold current I_{th}, we can observe amplified spontaneous emission emitted from the facet of the semiconductor laser. The theoretically calculated result and experimentally measured data are shown in Fig. 6.11. We find from this figure that the intensity tail of the optical emission is well explained using intraband relaxation time with relaxation time $\tau_{in} = 7 \times 10^{-14}$s for GaAs [3].

Several results of spectral measurements for amplified spontaneous emission are given in Appendix 10. Experimental data shown in Fig. 6.11 were obtained based on the method presented in Appendix 10. Another measuring method for the gain profile, called the *Hakki-Paoli method* [4], is also included in that appendix.

Equation (6.73) gives the probability that light of a single frequency ω is emitted. There are many other zero-point optical fields with different ω surrounding an electron-hole pair. These optical fields consist of not only guided modes which propagate along the active region, but also *radiation modes* that propagate outwardly by crossing the cladding regions. A single spontaneous emission from an electron-hole pair is performed by sharing the contribution to all these optical fields. If we define the *mode density* to be $D(\omega)\, d\omega$, the *electron lifetime* τ_s by spontaneous emission is defined as

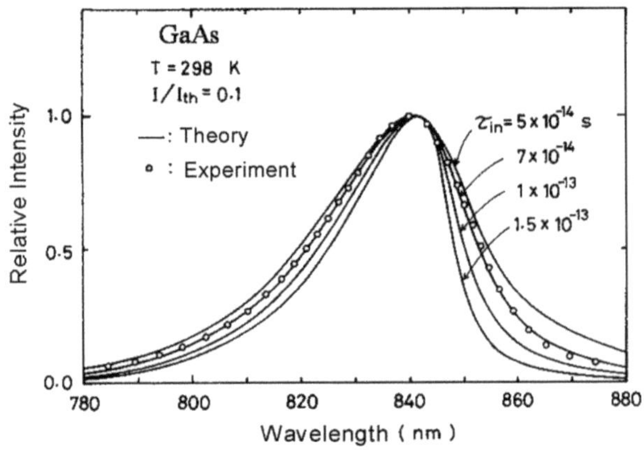

Fig. 6.11 Optical emission profile from the end facet of a GaAs laser under operating conditions of lower injection current than threshold

$$\int_0^\infty T_{sp}(\omega)D(\omega)d\omega = \frac{N_c - \tilde{N}_c}{\tau_s}. \tag{6.75}$$

Then, we need to examine the mode density *D(ω)* including the radiation modes. We consider a cubic box in a semiconductor laser surrounding a portion of the active region, as illustrated in Fig. 6.12. The length of its sides L_r is assumed much longer than the wavelength λ/n_r of the optical wave, thicker than the thickness *d*, and wider than the width *W* of the active region. All existing optical modes including both guided and radiation modes are defined within this box. The boundaries between the active region and the cladding regions are ignored in this model because reflections there are not sufficient to confine the radiation modes.

The spatial distribution of an optical mode in the hypothetical box of side length L_r is represented as

$$\Phi(x, y, z) = \sqrt{\frac{8}{L_r^3}} \sin(k_x x) \sin(k_y y)\ \sin(k_z z), \tag{6.76}$$

where k_x, k_y, and k_z are optical-mode wave numbers. The boundary condition imposed requires the field to be zero at the walls of the box, thereby giving 100 % reflection. Hence,

$$k_x L_r = p_x \pi, \tag{6.77}$$

where p_x is a positive integer.

The number of optical modes is counted throughout the volume in wave-number space, as shown in Fig. 6.13. Our current analysis is very similar to that for electron density treated in Sect. 6.3. Equation (6.77) corresponds to (6.24), except for an additional factor 2. The reason for this difference is that the discussion in Sect. 6.3 has been based on the forward traveling wave given by (5.16), but the current analysis is based on a standing wave defined by (6.76).

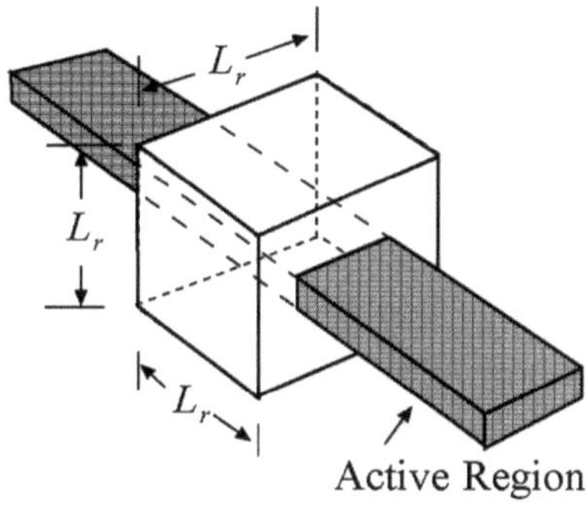

Fig. 6.12 Hypothetical box to count all optical field modes contributing to spontaneous emission

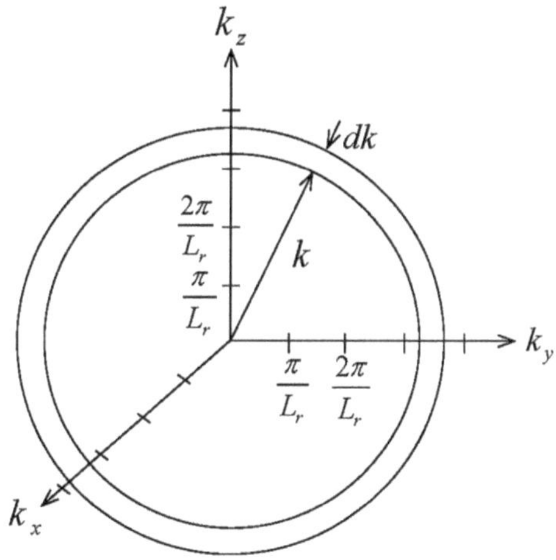

Fig. 6.13 Wave number space

The modes exist at grid points which are separated by $\Delta k = \pi/L_r$ and in the region with positive values of the wave number k. The frequency ω of each mode is simply proportional to the wave number k as

$$\omega = \sqrt{\frac{k_x^2 + k_y^2 + k_z^2}{\varepsilon\,\mu_o}} = \frac{k}{\sqrt{\varepsilon\,\mu_o}}. \tag{6.78}$$

Then, the number of modes between ω and $\omega + d\omega$ is the same as that between k and $k + dk$. The electric field can take two orthogonal polarizations for each ω. The region corresponding to positive values of the wave number components represents 1/8 of the sphere of radius k. Therefore, the number of modes totals

$$4\pi k^2 dk \times \left(\frac{L_r}{\pi}\right)^3 \times 2 \times \frac{1}{8} = \frac{L_r^3}{\pi^2}(\varepsilon\mu_o)^{3/2}\omega^2 d\omega \equiv L_r^3 D(\omega)d\omega. \tag{6.79}$$

Hence, the number of modes is proportional to the originally supposed volume L_r^3. We then find the *mode density* per unit volume to be

$$D(\omega)d\omega = \frac{(\varepsilon\mu_o)^{3/2}}{\pi^2}\omega^2 d\omega. \tag{6.80}$$

The electron life-time τ_s in (6.75) can be calculated from (6.52), (6.73), (6.75), and (6.80). As we need to make a double integration over W_{ba} and ω for $T_{sp}(\omega)$, we apply an approximation using (6.55) to make it simpler because the distribution of $f_c(W_{ba})\{1 - f_v(W_{ba})\}$ is much smoother than the Lorentz distribution $(\hbar/\tau_{ba})/\left\{(\hbar\omega - W_{ba})^2 + (\hbar/\tau_{ba})^2\right\}$ and we only need the total value of the emission,

$$T_{sp}(\omega) \approx \frac{\pi\omega}{\varepsilon}|R_{ba}|^2 g_{cv}(W_{ba}) f_c(W_{ba})\{1 - f_v(W_{ba})\}. \tag{6.81}$$

Thus, the *electron life-time* τ_s is expressed as [1]

$$\tau_s = \frac{\pi\hbar^4 \int_{W_g}^{\infty} g_{cv}(W_{ba})\{f_c(W_{ba}) - \tilde{f}_c(W_{ba})\}dW_{ba}}{\mu_o^{3/2}\varepsilon_o^{1/2} n_{eff} \int_{W_g}^{\infty} |R_{ba}|^2 W_{ba}^3\, g_{cv}(W_{ba}) f_c(W_{ba})\{1 - f_v(W_{ba})\}dW_{ba}}. \tag{6.82}$$

Numerically calculated data for the electron life-time in GaAs crystal, shown in Fig. 6.14, yield an approximate expression

$$\tau_s = 2.85 \times \frac{10^7 T}{N_c}, \tag{6.83}$$

where T is the absolute temperature in Kelvins [K] and N_c is the electron density in units of $[\mathrm{cm}^{-3}]$.

The electron transition from the conduction band to the valence band by spontaneous emission is alternatively called *radiative recombination*. However, an electron can make a transition without any optical wave emission; this process is called *non-radiative recombination* for which there are many mechanisms. One such mechanism produces the *Auger effect* and is appreciable in the InGaAsP crystal. In such instances, the effective electron life-time τ_s' is written more generally as

$$\frac{1}{\tau_s'} = \frac{1}{\tau_s} + \frac{1}{\tau_{nr}}, \tag{6.84}$$

where τ_{nr} is the non-radiative recombination time.

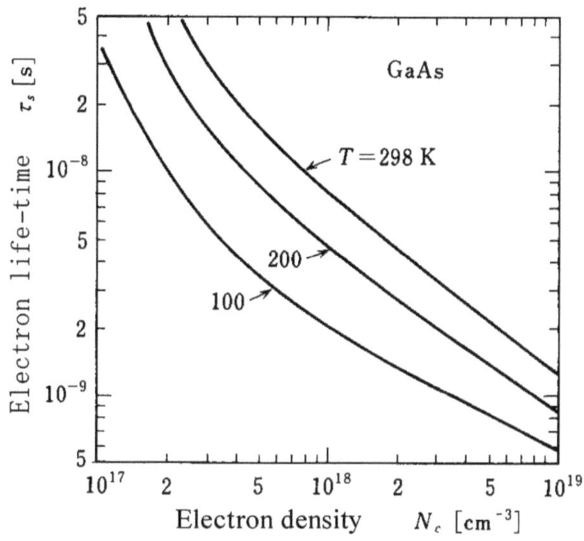

Fig. 6.14 Electron life-time in GaAs

6.8 Rate Equation for the Photon Number Including Spontaneous Emission

As seen in the last section, spontaneous emission is generated by all possible optical fields surrounding an electron-hole pair. Only a part of the spontaneous emission yields a lasing mode p. Denoting this fraction by coefficient C, called *spontaneous emission factor*, the variation in photon number is then written

$$\frac{dS_p}{dt} = (G_p - G_{th})S_p + C\frac{(N_c - \tilde{N}_c)V_a}{\tau_s}. \tag{6.85}$$

Here, we discuss the various contributions. The stimulated emission is given by $G_p\, S_p$ in (6.68) and an approximate expression of the gain is given by (6.71) with (6.69). Therefore, terms ξaNS_p and ξaN_gS_p must correspond to optical emission and absorption, respectively, in the expression of $(f_b - f_a)S_p$ for the stimulated emission. Then, the fraction of spontaneous emission contributing to the mode p must be approximately expressed by $\xi\, a\, N$. Therefore, the rate equation for the photon number S_p including the spontaneous emission is written

$$\frac{dS_p}{dt} = (G_p - G_{th})S_p + \xi aN. \tag{6.86}$$

By comparing (6.85) and (6.86), we find that the spontaneous emission factor is given by

$$C = \frac{\xi a N \tau_s}{V_a(N_c - \tilde{N}_c)} \approx \frac{\xi a \tau_s}{V_a}. \tag{6.87}$$

In conventional semiconductor lasers, typical numerical values for this factor are $10^{-7}$to 10^{-5}.

Readers might feel a little confused about spontaneous emission: one electron-hole pair generates one photon as the spontaneous emission indicated by $(N_c - \tilde{N}_c)V_a/\tau_s$. This emission is prompted by the *cooperative behavior of a large number of optical modes*. Only a fraction $C \approx 10^{-6}$ enter the lasing mode. Therefore, should the optical energy in the lasing mode be counted as the photon? Note, $C = 10^{-6}$ is not an integer!

The number of photons from spontaneous emission in the lasing mode should in principle be either "one" or "zero". A value of $C = 10^{-6}$ signifies the statistical probability that one photon is detected. The difference between "one" and "10^{-6}" is measured as a fluctuation and is the intrinsic cause of "noise" in the laser.

Bibliography

1. M. Yamada, H. Ishiguro, Gain calculation of undoped GaAs injection laser taking account of electronic intra-band relaxation. Japan. J. Appl. Phys. **20**(7), 1279–1288 (1981)
2. M. Asada, Y. Suematsu, Density-matrix theory of semiconductor lasers with relaxation broadening model—gain and gain- suppression in semiconductor lasers. IEEE J. Quantum Electron. **QE-21**(5), 432–442 (1985)
3. M. Yamada, H. Ishiguro, H. Nagato, Estimation of the intra-band relaxation time in undoped AlGaAs injection laser. Japan. J. Appl. Phys. **19**(1), 135–142 (1980)
4. B.W. Hakki, T.L. Paoli, Gain spectra in GaAs double-heterostructure injection lasers. J. Appl. Phys. **46**(3), 1299–1306 (1974)

Chapter 7
Typical Operating Characteristics

7.1 Threshold Current

The basic equations characterizing lasing operations are *rate equations* for the *photon number* S_p and the *injected electron density* in the conduction band $N = N_c - \tilde{N}_c$; these have been given in (6.86) and (6.68), respectively,

$$\frac{dS_p}{dt} = (G_p - G_{th})S_p + \xi a N, \tag{7.1}$$

$$\frac{dN}{dt} = -G_p \frac{S_p}{V_a} - \frac{N}{\tau_s} + \frac{I}{eV_a}. \tag{7.2}$$

The gain coefficient G_p is approximately written as

$$G_p = \xi a(N - N_g), \tag{7.3}$$

where ξ is the field confinement factor within the active region and has been defined using (3.48), N_g the electron injection density at which the gain coefficient becomes positive, and a a coefficient of proportionality.

By increasing the injection current I from zero, the lasing gain G_p increases as observed in Fig. 3.2; the photon number S_p also rises from the *threshold current* I_{th}. However, the threshold current is defined by *excluding the spontaneous emission*,

$$\frac{dS_p}{dt} = (G_p - G_{th})S_p. \tag{7.4}$$

The solution to the steady state $dS_p/dt = 0$ of this equation is either $S_p = 0$ corresponding to a non-oscillating state or $G_p = G_{th}$ with $S_p \neq 0$ corresponding to *laser oscillation*. The threshold electron density N_{th} is given by relation

$$G_{th} = \xi a(N_{th} - N_g) = v\left\{\alpha_{loss} + \frac{1}{2L}\ln\left(\frac{1}{R_b R_f}\right)\right\}. \tag{7.5}$$

M. Yamada, *Theory of Semiconductor Lasers*,
Springer Series in Optical Sciences 185,
DOI: 10.1007/978-4-431-54889-8_7, © Springer Japan 2014

The threshold current I_{th} is obtained from (7.2) by putting $dN/dt = 0$, $S_p = 0$, and $N = N_{th}$ giving

$$I_{th} = \frac{eN_{th}V_a}{\tau_s}. \tag{7.6}$$

By substituting (7.5) into this equation, the threshold current I_{th} depends on guiding loss α_{loss} and mirror reflectivities R_b and R_f,

$$I_{th} = \frac{e\,V_a}{\tau_s}\left[\frac{v}{\xi\,a}\left\{\alpha_{loss} + \frac{1}{2L}\ln\left(\frac{1}{R_b\,R_f}\right)\right\} + N_g\right]. \tag{7.7}$$

As found from this equation, the threshold current I_{th} is reduced by increasing the value of the tangential coefficient a. Also, it seems that the volume V_a of the active region should be small. However, we need to pay attention to the fact that the field confinement factor ξ also changes with V_a because ξ is mostly determined by the structure along the thickness direction y, as discussed in Sect. 4.1. Furthermore, the injection of electrons and holes is performed in the y direction. We can then determine the property of the threshold current density J_{th} given by

$$J_{th} = \frac{I_{th}}{WL} = \frac{ed}{\tau_s}\left[\frac{v}{\xi a}\left\{\alpha_{loss} + \frac{1}{2L}\ln\left(\frac{1}{R_bR_f}\right)\right\} + N_g\right], \tag{7.8}$$

where WL is the area of the electrode or the active region in the x–z plane. Calculated examples of the threshold current density J_{th} for various active region thicknesses d are given in Fig. 7.1 for the InGaAsP and AlGaAs lasers having the double hetero(DH)-junction structure. The dependence of ξ on thickness d has been taken into account.

If d is sufficiently large, the optical field is confined in the active region giving $\xi \rightarrow 1$, and J_{th} increases almost linearly with thickness d. In contrast, if d is sufficiently small, the field confinement ξ becomes too small; hence J_{th} increases again. Therefore, the thickness d of the active region has an optimum value. Conventional semiconductor lasers are designed to minimize the threshold current density J_{th}.

7.2 Quantum Efficiency

In the ideal case, one electron in the conduction band emits a photon by either spontaneous emission or stimulated emission and makes a transition to the valence band. However, not all electrons emit a photon because of possible undesirable mechanisms such as non-radiative recombination. We define several types of *quantum efficiency* to determine the quality of a laser.

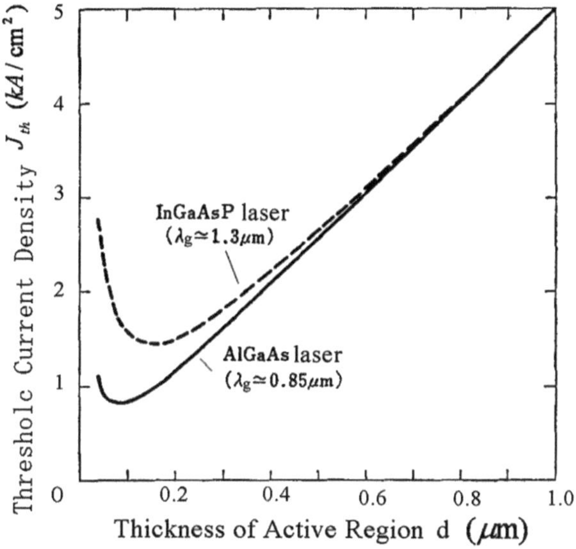

Fig. 7.1 Calculated example of the threshold current density [1]

Because we cannot measure directly photon numbers in a laser cavity, we examine the relationship between the output optical power and photon number. We consider a laser cavity, as depicted in Fig. 1.8, with cavity length L and electric field reflectivities $\sqrt{R_f}$ and $\sqrt{R_b}$ for the front and back facets, respectively. The electric field in the laser is represented in the form of a traveling wave, as given by (3.25),

$$\begin{aligned} E = \sum_p \Big\{ & E_p^{(+)}(0)T(x,y)e^{(g_p-\alpha_{loss})z/2-j\beta_p z+j\omega_p t} \\ & + E_p^{(-)}(L)T(x,y)e^{(g_p-\alpha_{loss})(L-z)/2-j\beta_p(L-z)z+j\omega_p t}\Big\} + c.c. \end{aligned} \tag{7.9}$$

The optically stored energy W_{opt} in the laser cavity is

$$\begin{aligned} W_{opt} &= \int_{cav} \varepsilon E^2 d^3\boldsymbol{r} \\ &= \sum_p \frac{2\varepsilon_{eff}}{g_p - \alpha_p}\left\{\left|E_p^{(+)}(0)\right|^2 + \left|E_p^{(-)}(L)\right|^2\right\}\left\{e^{(g_p-\alpha_{loss})L} - 1\right\} \end{aligned} \tag{7.10}$$

The initial amplitudes are related to the amplitude at the front facet as

$$\left|E_p^{(+)}(L)\right|^2 = \left|E_p^{(+)}(0)\right|^2 e^{(g_p-\alpha_{loss})L}, \tag{7.11}$$

$$\left|E_p^{(+)}(L)\right|^2 R_f = \left|E_p^{(-)}(L)\right|^2. \tag{7.12}$$

As given in (3.33), the gain and loss coefficients as expressed in terms of the reflectivities based on the oscillation condition are

$$g_p - \alpha_{loss} = \frac{1}{L}\ln\left(\frac{1}{\sqrt{R_f R_b}}\right) = \frac{1}{2L}\ln\left(\frac{1}{R_f R_b}\right). \tag{7.13}$$

Then, the optical energy is written with $\left|E_p^{(+)}(L)\right|^2$ as

$$W_{opt} = \sum_p \frac{4\varepsilon_{eff} L}{\sqrt{R_b}\ln\left(\frac{1}{R_f R_b}\right)}\{\sqrt{R_f} + \sqrt{R_b}\}\{1 - \sqrt{R_f R_b}\}\left|E_p^{(+)}(L)\right|^2. \tag{7.14}$$

This value must be the optical energy as found from quantum mechanics:

$$W_{opt} = \sum_p S_p \hbar\omega_p. \tag{7.15}$$

The optical output power P_f from the front facet is

$$\begin{aligned} P_f &= (1 - R_f)\sum_p \int_{-\infty}^{\infty}\int_{-\infty}^{\infty}\left\{E_p^{(+)}(L)H_p^{(+)*}(L) + c.c.\right\}|T(x,y)|^2 dxdy \\ &= 2(1 - R_f)\sum_p \sqrt{\frac{\varepsilon_{eff}}{\mu_o}}\left|E_p^{(+)}\right|^2 \\ &= \frac{v(1 - R_f)\sqrt{R_b}}{2L(\sqrt{R_f} + \sqrt{R_b})(1 - \sqrt{R_f R_b})}\ln\left(\frac{1}{R_f R_b}\right)\sum_p S_p \hbar\omega_p \end{aligned} \tag{7.16}$$

Similarly, the optical output power P_b from the back facet is

$$P_b = \frac{v(1 - R_b)\sqrt{R_f}}{2L(\sqrt{R_f} + \sqrt{R_b})(1 - \sqrt{R_f R_b})}\ln\left(\frac{1}{R_f R_b}\right)\sum_p S_p \hbar\omega_p. \tag{7.17}$$

Therefore, the total optical output power P_t from both facets is the sum

$$P_t = P_f + P_b = \frac{v}{2L}\ln\left(\frac{1}{R_f R_b}\right)\sum_p S_p \hbar\omega_p. \tag{7.18}$$

Now, we return to determining the quantum efficiency. The most intrinsic efficiency is the *internal quantum efficiency* η_i defined as

$$\eta_i = \frac{\text{Generated photon number}}{\text{Injected electron number}}. \tag{7.19}$$

The photon number S_p at the steady state is derived from (7.2) by putting $G_p = G_{th}$ and $N/\tau_s = I_{th}/eV_a$. When we count many modes, the total photon number is given by the injection current I. As not all the electrons emit photons, the total photon number depends on the internal quantum efficiency η_i,

$$\sum_p S_p = \eta_i \frac{I - I_{th}}{eG_{th}}. \tag{7.20}$$

Despite the inability to measure photon number S_p in the laser cavity, we can measure the variation of the optical output power P_t with injection current I, as given in Fig. 7.2. The output power increases almost linearly with I–I_{th} based on stimulated emission. The output power is a product of the photon energy $\hbar\omega$ and the number of photons emitted per unit time. That is, $P_t/\hbar\omega$ is the emitted photon number per unit time and I/e is the number of electrons injected into the active region per unit time. We can then define another quantum efficiency η_d, called the *differential quantum efficiency* , which gives the efficiency of converting injected electrons into contributing to the output power,

$$\eta_d \equiv \frac{d(P_t/\hbar\omega)}{d(I/e)} = \frac{e}{\hbar\omega}\frac{dP_t}{dI}. \tag{7.21}$$

From (7.5), (7.18) and (7.20), the differential quantum efficiency η_d is related to the internal quantum efficiency η_i in a simple relation

$$\eta_d = \frac{ev}{2L}\ln\left(\frac{1}{R_f R_b}\right)\sum_p \frac{dS_p}{dI} = \frac{\eta_i v}{2LG_{th}}\ln\left(\frac{1}{R_f R_b}\right) = \frac{\eta_i}{1 + \frac{2L\alpha_{loss}}{\ln\left(\frac{1}{R_f R_b}\right)}}. \tag{7.22}$$

Hence, the internal quantum efficiency η_i can be estimated from the experimentally measurable differential quantum efficiency η_d from

$$\eta_i = \eta_d \left\{ 1 + \frac{2L\alpha_{loss}}{\ln\left(\frac{1}{R_f R_b}\right)} \right\}. \tag{7.23}$$

For the differential quantum efficiency, the power consumption up to threshold current I_{th} is not taken into account. The quantum efficiency including operations below the threshold current is defined as the *external quantum efficiency* η_{ex};

$$\eta_{ex} = \frac{\text{Output photon number}}{\text{Injected electron number}} = \frac{e}{\hbar\omega}\frac{P_t}{I} = \eta_d\left(1 - \frac{I_{th}}{I}\right). \tag{7.24}$$

The analysis so far supposes that the applied voltage to the active region is the same as the photon energy $\hbar\omega$. However, we need higher voltages than the photon

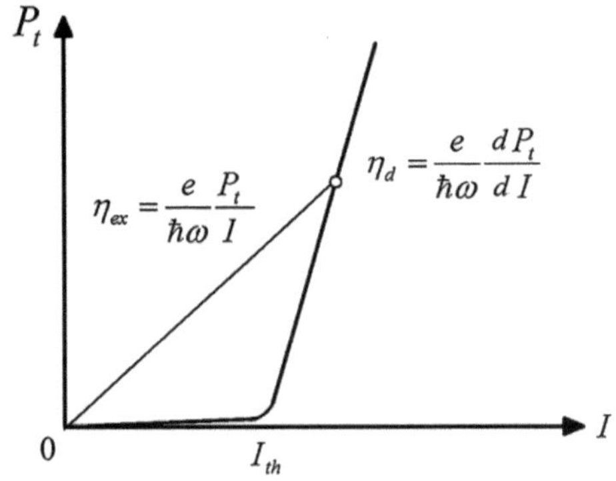

Fig. 7.2 Output power versus injection current

energy to inject electrons into the active region owing to an increase in for example the quasi-Fermi level and ohmic loss. Then, the *device efficiency* η_{dev} is defined based on the actual energy consumed

$$\eta_{dev} = \frac{P_t}{IV_{dev}} \approx \eta_{ex}\frac{\hbar\omega}{eV_{dev}}, \tag{7.25}$$

where V_{dev} is the *applied voltage* between the two electrodes.

7.3 Direct Modulation

One example of the application of the rate equation is the analysis of direct modulation in the semiconductor laser. The simplest method, called *direct modulation*, to introduce an electric signal on the laser light is to modulate the injection current I. We add a sinusoidal modulation which gives

$$I = \bar{I} + I_M\cos(\Omega t) = \bar{I} + \frac{I_M}{2}e^{j\Omega t} + \frac{I_M}{2}e^{-j\Omega t}, \tag{7.26}$$

where $\bar{I}$ is the DC component of the injection current, I_M the amplitude of the modulation current, and Ω the angular frequency of the modulation. We suppose that the DC component is higher than the threshold current $\bar{I} > I_{th}$ and the modulated current never falls below the threshold level, $I_M < \bar{I} - I_{th}$. The electron injection density and photon number are modulated and written as

$$N = \bar{N} + \frac{N_M}{2}e^{j\Omega t} + \frac{N_M^*}{2}e^{-j\Omega t}, \tag{7.27}$$

$$S_p = \bar{S} + \frac{S_M}{2} e^{j\Omega t} + \frac{S_M^*}{2} e^{-j\Omega t}. \tag{7.28}$$

Here the "bar" denotes the DC component and subscript "M" identifies quantities related to the modulation. Modulated components are supposed to be much smaller than the corresponding DC components; that is, $|N_M| \ll \bar{N}$ and $|S_M| \ll \bar{S}$. We substitute (7.27) and (7.28) into (7.3) and (7.4), and determine separate relations for the DC and modulated terms:

$$\xi a(\bar{N} - N_g) = G_{th} = \frac{1}{\tau_{ph}}, \tag{7.29}$$

$$j\Omega S_M = \xi a \bar{S} N_M. \tag{7.30}$$

We also substitute (7.26)–(7.28) into (7.2) with (7.3) to get relations:

$$\frac{\xi a(\bar{N} - N_g)\bar{S}}{V_a} = \frac{\bar{I}}{eV_a} - \frac{\bar{N}}{\tau_s}, \tag{7.31}$$

$$\left(j\Omega + \frac{\xi a \bar{S}}{V_a} + \frac{1}{\tau_s}\right) N_M + \frac{\xi a (\bar{N} - N_g)}{V_a} S_M = \frac{I_M}{e V_a}. \tag{7.32}$$

From (7.29) and (7.31), the DC component of the photon number is given as

$$\bar{S} = \frac{\bar{I} - I_{th}}{eG_{th}}, \tag{7.33}$$

where I_{th} is the threshold current defined in (7.6)

$$I_{th} = \frac{eN_{th}V_a}{\tau_s}, \tag{7.34}$$

with threshold electron density of

$$N_{th} = \bar{N}. \tag{7.35}$$

By substituting (7.29) and (7.31) into (7.30) and (7.32), the modulation characteristics of the photon number S_M and the electron density N_M by the injection current I_M become:

$$\frac{S_M}{I_M} = \frac{\tau_{ph}}{e\left\{1 - \left(\frac{\Omega}{\Omega_R}\right)^2 + j\left(\frac{\Omega}{\Omega_R}\right)\left(\Omega_R \tau_{ph} + \frac{1}{\Omega_R \tau_s}\right)\right\}}, \tag{7.36}$$

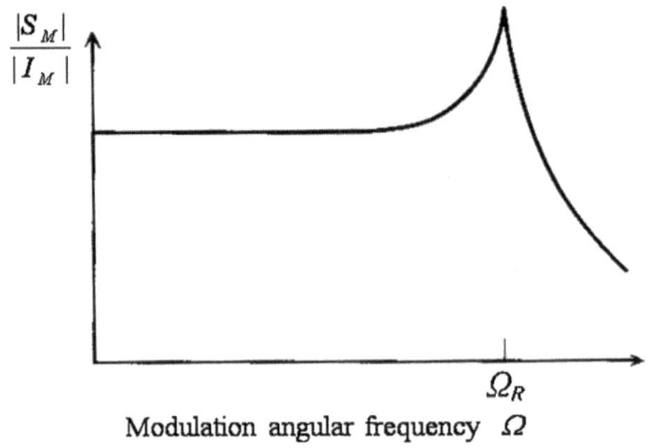

Fig. 7.3 Characteristic of direct modulation

$$\frac{N_M}{I_M} = \frac{1}{e\Omega_R^2 V_a \left\{1 - \left(\frac{\Omega}{\Omega_R}\right)^2 + j\left(\frac{\Omega}{\Omega_R}\right)\left(\Omega_R \tau_{ph} + \frac{1}{\Omega_R \tau_s}\right)\right\}}. \tag{7.37}$$

Here, Ω_R is called the *resonance angular frequency* and is given by

$$\Omega_R = \sqrt{\frac{\xi a G_{th} \bar{S}}{V_a}}. \tag{7.38}$$

This equation can be rewritten using (7.29) and (7.33) as

$$\Omega_R = 2\pi f_R = \sqrt{\frac{1}{\tau_s \tau_{ph}} \frac{\bar{I} - I_{th}}{I_{th} - I_g}}, \tag{7.39}$$

where f_R is called the *relaxation oscillation frequency* and I_g a current level corresponding to the transparent electron density N_g,

$$I_g = \frac{e N_g V_a}{\tau_s}. \tag{7.40}$$

The modulation characteristic, illustrated in Fig. 7.3, shows resonance at $\Omega = \Omega_R$ and a rapid decay at higher frequencies. We view f_R as the band width for direct modulation. f_R is a function of the operating current $\bar{I}$, but is applicable up to several GHz under conventional operations.

Bibliography

1. H. Yonezu, *Optical Communication Device Engineering* (Kogaku Tosyo Co. Ltd., Tokyo, 1984), p. 182, ISBN: 4-3054-20194-2213 (in Japanese)

Part II
Advanced Theory for Mode Competition and Noise

Chapter 8
Nonlinear Gain

8.1 Redefinition of the Gain Coefficient

The lasing condition is achieved when gain reaches a threshold level. However, a pressing issue is whether the gain spectrum is fixed when the laser is in oscillation. Its resolution is given by an analysis of the nonlinear property of lasing gain. Nonlinear effects treated in this book stem from electron dynamics and not from the non-isotropic properties of the material.

We now redefine the gain coefficient including contributions from spontaneously emitted light. We introduce *laser polarization* P_{st} and *spontaneous emission* P_{sp} into Maxwell's classical wave equation:

$$P = P_{st} + P_{sp}, \tag{8.1}$$

$$\nabla^2 E - \mu_o \varepsilon \frac{\partial^2 E}{\partial t^2} = \mu_o \sigma \frac{\partial \mathrm{E}}{\partial t} + \mu_o \frac{\partial^2 P}{\partial t^2}. \tag{8.2}$$

The electric field is represented as a sum of lasing modes p,

$$E = \sum_p \widehat{E}_p(t) \Phi_p(\boldsymbol{r}) e^{j\omega_p t} + c.c., \tag{8.3}$$

where $\widehat{E}_p(t)$ is assumed to be a slowly varying amplitude and $\Phi_p(\boldsymbol{r})$ is a three-dimensional normalized distribution function obtained as the normalized eigenfunction of the equation without loss and gain,

$$\left\{\nabla^2 + \mu_o \varepsilon \omega_p^2\right\} \Phi_p(\boldsymbol{r}) e^{j\omega_p t} = 0, \tag{8.4}$$

$$\int_{cav} \Phi_p(\boldsymbol{r}) \Phi_q^*(\boldsymbol{r}) d^3\boldsymbol{r} = \delta_{p,q}. \tag{8.5}$$

M. Yamada, *Theory of Semiconductor Lasers*,
Springer Series in Optical Sciences 185,
DOI: 10.1007/978-4-431-54889-8_8, © Springer Japan 2014

We substitute (8.3) into (8.2), multiply by $\Phi_p^*(\boldsymbol{r})^{-j\omega_p t}$, perform a spatial integration $\int_{cav} d^3r$ over the laser cavity, take a time average for Δt which is several periods of $1/\omega_p$, and use (8.4) assuming $|\partial^2 \widehat{E}_p(t)/\partial t^2| \ll |\omega_p \partial \widehat{E}_p(t)/\partial t|$. We then obtain an equation for the variation of the amplitude:

$$\frac{\partial \widehat{E}_p(t)}{\partial t} = -\frac{\sigma_{eff}}{2\varepsilon_{eff}} \widehat{E}_p(t) - \frac{j\omega_p}{2\varepsilon_{eff}\Delta t} \int_{t-\Delta t}^{t} \int_{act} P\Phi_p^*(\boldsymbol{r}) e^{-j\omega_p t} d^3\boldsymbol{r}\, dt, \tag{8.6}$$

where

$$\varepsilon_{eff} \equiv \int_{cav} \varepsilon(\boldsymbol{r}) \left|\Phi_p(\boldsymbol{r})\right|^2 d^3\boldsymbol{r}, \tag{8.7}$$

$$\sigma_{eff} \equiv \int_{cav} \sigma(\boldsymbol{r}) \left|\Phi_p(\boldsymbol{r})\right|^2 d^3\boldsymbol{r}. \tag{8.8}$$

The temporal variation for the power is

$$\begin{aligned} \frac{\partial \left|\widehat{E}_p(t)\right|^2}{\partial t} &= \frac{\partial \widehat{E}_p}{\partial t} \widehat{E}_p^* + \widehat{E}_p \frac{\partial \widehat{E}_p^*}{\partial t} \\ &= -\frac{\sigma_{eff}}{\varepsilon_{eff}} \left|\widehat{E}_p(t)\right|^2 \\ &\quad + \mathrm{Im}\left\{ \frac{\omega_p}{\varepsilon_{eff} \widehat{E}_p \Delta t} \int_{t-\Delta t}^{t} \int_{act} P_{st} \Phi_p^*(\boldsymbol{r}) e^{-j\omega_p t} d^3\boldsymbol{r}\, dt \right\} \left|\widehat{E}_p\right|^2 \\ &\quad + \mathrm{Im}\left\{ \frac{\omega_p \widehat{E}_p^*}{\varepsilon_{eff}\Delta t} \int_{t-\Delta t}^{t} \int_{act} P_{sp} \Phi_p^*(\boldsymbol{r}) e^{-j\omega_p t} d^3\boldsymbol{r}\, dt \right\}. \end{aligned} \tag{8.9}$$

The optical power is expressed in terms of *photon number* S_p associated with the laser cavity as already derived in (3.56),

$$S_p = \frac{2\,\varepsilon_{eff}}{\hbar\omega_p} \left|\widehat{E}_p\right|^2. \tag{8.10}$$

By including the additional loss from the optical output through the facets, (8.9) is changed to a rate equation for the photon number,

$$\frac{dS_p}{dt} = \left(G_p - G_{th}\right) S_p + C_p. \tag{8.11}$$

Here, G_p is the gain coefficient, G_{th} the threshold gain level, and C_p the term that accounts for spontaneous emission. Specifically, we have

$$G_p = \mathrm{Im}\left\{ \frac{\omega_p}{\varepsilon_{eff} \widehat{E}_p \Delta t} \int_{t-\Delta t}^{t} \int_{act} P_{st} \Phi_p^*(\mathbf{r}) e^{-j\omega_p t} d^3\mathbf{r}\, dt \right\}, \tag{8.12}$$

$$G_{th} = \frac{\sigma_{eff}}{\varepsilon_{eff}} + \frac{c}{2Ln_{eq}} \ln\left(\frac{1}{R_b R_f}\right) = v\left\{ \alpha_{loss} + \frac{1}{2L} \ln\left(\frac{1}{R_b R_f}\right) \right\}, \tag{8.13}$$

$$C_p = \frac{2}{\hbar} \mathrm{Im}\left\{ \frac{\widehat{E}_p^*}{\Delta t} \int_{t-\Delta t}^{t} \int_{act} P_{sp} \Phi_p^*(\mathbf{r}) e^{-j\omega_p t} d^3\mathbf{r}\, dt \right\}. \tag{8.14}$$

8.2 Perturbation Expansion of Density Matrix

In the dynamic equation of the *density matrix* for the semiconductor laser, given in (5.59) or (6.5), we find that the density matrix ρ is a function of RE and ρ itself, specifically, ρ is a *nonlinear function* of RE. We can then expand the density matrix according to powers of RE or E,

$$\rho = \rho^{(0)} + \rho^{(1)}(E) + \rho^{(2)}(E^2) + \rho^{(3)}(E^3) + \cdots + \rho^{(i)}(E^i) + \cdots. \tag{8.15}$$

Substituting this equation into (5.59) or (6.5) and equating terms of the same power in RE, we obtain the following set of equations [1]:

$$\begin{aligned} \frac{d\rho^{(0)}}{dt} &= \frac{1}{j\hbar}(H_o \rho^{(0)} - \rho^{(0)} H_o) - \frac{1}{2}\left\{ (\rho^{(0)} - \tilde{\tilde{\rho}})\Gamma_{in} + \Gamma_{in}(\rho^{(0)} - \tilde{\tilde{\rho}}) \right\} \\ &\quad - \frac{1}{2}\left\{ (\rho^{(0)} - \tilde{\rho})\Gamma_s + \Gamma_s(\rho^{(0)} - \tilde{\rho}) \right\} + \Lambda, \end{aligned} \tag{8.16}$$

$$\begin{aligned} \frac{d\rho^{(i)}}{dt} &= \frac{1}{j\hbar}\left(H_o \rho^{(i)} - \rho^{(i)} H_o\right) - \frac{1}{j\hbar}\left(RE\rho^{(i-1)} - \rho^{(i-1)} RE\right) \\ &\quad - \frac{1}{2}\left\{ \rho^{(i)}(\Gamma_{in} + \Gamma_s) + (\Gamma_{in} + \Gamma_s)\rho^{(i)} \right\} \quad \text{for } i \geq 1. \end{aligned} \tag{8.17}$$

Equation (8.17) provides a recursion equation for the i-th order term given in terms of the $(i-1)$-th order term with RE in a single time step. (8.16) gives the *zeroth-order* terms which are independent of RE. Some readers might doubt that $\tilde{\tilde{\rho}}$ must be a nonlinear function of E because $\tilde{\tilde{\rho}}$ is derived from ρ. However, $\tilde{\tilde{\rho}}$ can be treated as a zeroth-order term in our treatment because our *perturbation expansion* is based on *coherent interactions* with RE, which implies maintaining optical-frequency vibrations. $\tilde{\tilde{\rho}}$ arises from the *scattering* among electrons resulting in a random vibration phase.

The dynamic equation for the *diagonal elements* of $\rho^{(0)}$ are

$$\frac{d\rho_{bb}^{(0)}}{dt} = -\frac{\rho_{bb}^{(0)} - \tilde{\tilde{\rho}}_b}{\tau_b} - \frac{\rho_{bb}^{(0)} - \tilde{\rho}_b}{\tau_s} + \Lambda_b \quad \text{in the conduction band,} \tag{8.18}$$

$$\frac{d\rho_{aa}^{(0)}}{dt} = -\frac{\rho_{aa}^{(0)} - \tilde{\tilde{\rho}}_a}{\tau_a} - \frac{\rho_{aa}^{(0)} - \tilde{\rho}_a}{\tau_s} + \Lambda_a \quad \text{in the valence band.} \tag{8.19}$$

For the off-diagonal elements, we have

$$\frac{d\rho_{ab}^{(0)}}{dt} = \left(j\omega_{ba} - \frac{1}{\tau_{in}}\right)\rho_{ab}^{(0)}, \tag{8.20}$$

resulting in

$$\rho_{ab}^{(0)}(t) = \rho_{ab}^{(0)}(0)e^{(j\omega_{ba}-1/\tau_{in})\,t} \longrightarrow 0 \text{ for } t \gg \tau_{in}. \tag{8.21}$$

Similarly,

$$\rho_{ba}^{(0)}(t) = 0 \text{ for } t \gg \tau_{in}. \tag{8.22}$$

Thus, we can regard the *zeroth-order terms as consisting of only diagonal elements.*

The diagonal elements of the i-th elements ($i \geq 1$) are

$$\frac{d\rho_{bb}^{(i)}}{dt} = \frac{j}{\hbar}\left(R_{ba}\rho_{ab}^{(i-1)} - \rho_{ba}^{(i-1)}R_{ab}\right)E - \left(\frac{1}{\tau_b} + \frac{1}{\tau_s}\right)\rho_{bb}^{(i)}, \tag{8.23}$$

$$\frac{d\rho_{aa}^{(i)}}{dt} = \frac{j}{\hbar}\left(R_{ab}\rho_{ba}^{(i-1)} - \rho_{ab}^{(i-1)}R_{ba}\right)E - \left(\frac{1}{\tau_a} + \frac{1}{\tau_s}\right)\rho_{aa}^{(i)}. \tag{8.24}$$

The *off-diagonal elements* are

$$\frac{d\rho_{ab}^{(i)}}{dt} = \left(j\omega_{ba} - \frac{1}{\tau_{in}}\right)\rho_{ab}^{(i)} + \frac{j}{\hbar}R_{ab}\left(\rho_{bb}^{(i-1)} - \rho_{aa}^{(i-1)}\right)E, \tag{8.25}$$

$$\rho_{ba}^{(i)} = \rho_{ab}^{(i)*}. \tag{8.26}$$

Whereas the zeroth-order terms consist of only diagonal elements, first-order terms consist of off-diagonal elements. Second-order terms are diagonal elements. By repeating the analysis, we find that the *diagonal elements consist of even-order terms* and the *off-diagonal elements consist of odd-order terms*:

$$\rho_{bb} = \rho_{bb}^{(0)} + \rho_{bb}^{(2)}(E^2) + \cdots + \rho_{bb}^{(2i)}(E^{2i}) + \cdots, \tag{8.27}$$

$$\rho_{aa} = \rho_{aa}^{(0)} + \rho_{aa}^{(2)}(E^2) + \cdots + \rho_{aa}^{(2i)}(E^{2i}) + \cdots, \tag{8.28}$$

$$\rho_{ab} = \rho_{ab}^{(1)}(E) + \rho_{ab}^{(3)}(E^3) + \cdots + \rho_{ab}^{(2i+1)}(E^{2i+1}) + \cdots, \tag{8.29}$$

$$\rho_{ba} = \rho_{ba}^{(1)}(E) + \rho_{ba}^{(3)}(E^3) + \cdots + \rho_{ba}^{(2i+1)}(E^{2i+1}) + \cdots. \tag{8.30}$$

Equations (8.23)–(8.26) can now be rewritten:

$$\frac{d\rho_{bb}^{(2i)}}{dt} = \frac{j}{\hbar}\left(R_{ba}\rho_{ab}^{(2i-1)} - \rho_{ba}^{(2i-1)}R_{ab}\right)E - \left(\frac{1}{\tau_b} + \frac{1}{\tau_s}\right)\rho_{bb}^{(2i)}, \tag{8.31}$$

$$\frac{d\rho_{aa}^{(2i)}}{dt} = \frac{j}{\hbar}\left(R_{ab}\rho_{ba}^{(2i-1)} - \rho_{ab}^{(2i-1)}R_{ba}\right)E - \left(\frac{1}{\tau_a} + \frac{1}{\tau_s}\right)\rho_{aa}^{(2i)}, \tag{8.32}$$

$$\frac{d\rho_{ab}^{(2i+1)}}{dt} = \left(j\omega_{ba} - \frac{1}{\tau_{in}}\right)\rho_{ab}^{(2i+1)} + \frac{j}{\hbar}R_{ab}\left(\rho_{bb}^{(2i)} - \rho_{aa}^{(2i)}\right)E, \tag{8.33}$$

$$\rho_{ba}^{(2i+1)} = \rho_{ab}^{(2i+1)^*}. \tag{8.34}$$

Here, one important point that allows our treatment to be expanded to more general situations is that we need to account for the possible existence of many lasing modes. The electric field E of the optical wave is expressed with a combination of cavity modes and their complex conjugates. We denote the *complex conjugate mode by a negative mode number*; hence

$$E = \sum_{p=0}^{\infty}\left\{E_p(t,\boldsymbol{r})e^{j\omega_p t} + E_p^*(t,\boldsymbol{r})e^{-j\omega_p t}\right\} = \sum_{p=-\infty}^{\infty} E_p(t,\boldsymbol{r})e^{j\omega_p t}, \tag{8.35}$$

where

$$E_{-p}(t,\boldsymbol{r}) = E_p^*(t,\boldsymbol{r}), \tag{8.36}$$

$$\omega_{-p} = -\omega_p. \tag{8.37}$$

The *zeroth-order terms* $\rho_{bb}^{(0)}$ and $\rho_{aa}^{(0)}$ are determined from the *current injection level* and in this section we suppose that the temporal variation of the zeroth-order terms is very small. We then represent the higher-order terms in terms of the zeroth-order terms. The *first-order term* is expressed as

$$\rho_{ab}^{(1)} = u^{(1)}(t)e^{(j\omega_{ba}t - 1/\tau_{in})t}. \tag{8.38}$$

By substituting (8.38) into (8.33) and setting $i = 0$, we obtain

$$\frac{du^{(1)}(t)}{dt} = \frac{j}{\hbar}R_{ab}\left(\rho_{bb}^{(0)} - \rho_{aa}^{(0)}\right)\sum_{p=-\infty}^{\infty} E_p e^{\{j(\omega_p - \omega_{ba}) + 1/\tau_{in}\}t}. \tag{8.39}$$

Integrating this equation gives

$$u^{(1)}(t) = \frac{j}{\hbar}R_{ab}\left(\rho_{bb}^{(0)} - \rho_{aa}^{(0)}\right)\sum_{p=-\infty}^{\infty} E_p \frac{e^{\{j(\omega_p - \omega_{ba}) + 1/\tau_{in}\}t} - 1}{j(\omega_p - \omega_{ba}) + 1/\tau_{in}}. \tag{8.40}$$

Substituting into (8.38) and taking the steady state solution for $t \gg \tau_{in}$ gives

$$\rho_{ab}^{(1)} = \frac{j}{\hbar}R_{ab}\left(\rho_{bb}^{(0)} - \rho_{aa}^{(0)}\right)\sum_{p=0}^{\infty}\frac{E_p e^{j\omega_p t}}{j(\omega_p - \omega_{ba}) + 1/\tau_{in}}, \tag{8.41}$$

$$\rho_{ba}^{(1)} = \rho_{ab}^{(1)^*} = -\frac{j}{\hbar}R_{ba}\left(\rho_{bb}^{(0)} - \rho_{aa}^{(0)}\right)\sum_{p=0}^{\infty}\frac{E_p^* e^{-j\omega_p t}}{j(\omega_{ba} - \omega_p) + 1/\tau_{in}}. \tag{8.42}$$

Here, we reduced the possible mode numbers to $p \geq 0$ by requiring $\omega_{ba} > 0$. The equation corresponding to the *second-order terms*, (8.31), is

$$\begin{aligned}
\frac{d\rho_{bb}^{(2)}}{dt} &= \frac{j}{\hbar}\left\{\frac{j|R_{ba}|^2\left(\rho_{bb}^{(0)} - \rho_{aa}^{(0)}\right)}{\hbar}\sum_{p=0}^{\infty}\frac{E_p e^{j\omega_p t}}{j(\omega_p - \omega_{ba}) + 1/\tau_{in}}\right.\\
&\quad \left. + \frac{j|R_{ba}|^2\left(\rho_{bb}^{(0)} - \rho_{aa}^{(0)}\right)}{\hbar}\sum_{p=0}^{\infty}\frac{E_p^* e^{-j\omega_p t}}{j(\omega_{ba} - \omega_p) + 1/\tau_{in}}\right\}E\\
&\quad - \left(\frac{1}{\tau_b} + \frac{1}{\tau_s}\right)\rho_{bb}^{(2)}\\
&= -\frac{|R_{ba}|^2\left(\rho_{bb}^{(0)} - \rho_{aa}^{(0)}\right)}{\hbar^2}\sum_{p=0}^{\infty}\sum_{q=-\infty}^{\infty}\left\{\frac{E_p E_q e^{j(\omega_p + \omega_q)t}}{j(\omega_p - \omega_{ba}) + 1/\tau_{in}} + \frac{E_p^* E_q e^{j(-\omega_p + \omega_q)t}}{j(\omega_{ba} - \omega_p) + 1/\tau_{in}}\right\}\\
&\quad - \left(\frac{1}{\tau_b} + \frac{1}{\tau_s}\right)\rho_{bb}^{(2)}.
\end{aligned} \tag{8.43}$$

By setting

$$\rho_{bb}^{(2)} = u^{(2)}(t)e^{-(1/\tau_b+1/\tau_s)}, \tag{8.44}$$

and substituting into (8.43), we can analyze $u^{(2)}(t)$ and substitute again $u^{(2)}(t)$ into (8.44). We then obtain

$$\begin{aligned}\rho_{bb}^{(2)} &= -\frac{|R_{ba}|^2(\rho_{bb}^{(0)} - \rho_{aa}^{(0)})}{\hbar^2} \\ &\times \sum_{p=0}^{\infty}\sum_{q=0}^{\infty}\left[\frac{E_pE_q^*e^{j(\omega_p-\omega_q)t}}{\{j(\omega_p-\omega_{ba})+1/\tau_{in}\}\{j(\omega_p-\omega_q)+1/\tau_b+1/\tau_s\}}\right. \\ &\qquad\left.+\frac{E_p^*E_qe^{j(-\omega_p+\omega_q)t}}{\{j(\omega_{ba}-\omega_p)+1/\tau_{in}\}\{j(-\omega_p+\omega_q)+1/\tau_b+1/\tau_s\}}\right].\end{aligned} \tag{8.45}$$

Here, the possible mode numbers are restricted to $p \geq 0$ and $q \geq 0$ through relations $E_{-q} = E_q^*, \omega_{-q} = -\omega_{-q}$, and $\omega_{ba} > 0$. Similarly, we find

$$\begin{aligned}\rho_{aa}^{(2)} &= \frac{|R_{ba}|^2(\rho_{bb}^{(0)} - \rho_{aa}^{(0)})}{\hbar^2} \\ &\times \sum_{p=0}^{\infty}\sum_{q=0}^{\infty}\left[\frac{E_pE_q^*e^{j(\omega_p-\omega_q)t}}{\{j(\omega_p-\omega_{ba})+1/\tau_{in}\}\{j(\omega_p-\omega_q)+1/\tau_a+1/\tau_s\}}\right. \\ &\qquad\left.+\frac{E_p^*E_qe^{j(-\omega_p+\omega_q)t}}{\{j(\omega_{ba}-\omega_p)+1/\tau_{in}\}\{j(-\omega_p+\omega_q)+1/\tau_a+1/\tau_s\}}\right].\end{aligned} \tag{8.46}$$

Substituting these results into (8.33) with $i = 1$ and by similar calculations, we have the *third-order terms*,

$$\begin{aligned}\rho_{ab}^{(3)} = &-\frac{jR_{ab}|R_{ba}|^2(\rho_{bb}^{(0)} - \rho_{aa}^{(0)})}{\hbar^3}\sum_{p=0}^{\infty}\sum_{q=0}^{\infty}\sum_{r=0}^{\infty} \\ &\times\left[\frac{E_pE_q^*E_re^{j(\omega_p-\omega_q+\omega_r)t}}{\{j(\omega_p-\omega_{ba})+1/\tau_{in}\}\{j(\omega_p-\omega_q+\omega_r-\omega_{ba})+1/\tau_{in}\}}\right. \\ &\quad\times\left\{\frac{1}{j(\omega_p-\omega_q)+1/\tau_b+1/\tau_s}+\frac{1}{j(\omega_p-\omega_q)+1/\tau_a+1/\tau_s}\right\} \\ &\quad+\frac{E_p^*E_qE_re^{j(-\omega_p+\omega_q+\omega_r)t}}{\{j(\omega_{ba}-\omega_p)+1/\tau_{in}\}\{j(-\omega_p+\omega_q+\omega_r-\omega_{ba})+1/\tau_{in}\}} \\ &\quad\left.\times\left\{\frac{1}{j(-\omega_p+\omega_q)+1/\tau_b+1/\tau_s}+\frac{1}{j(-\omega_p+\omega_q)+1/\tau_a+1/\tau_s}\right\}\right],\end{aligned} \tag{8.47}$$

$$\rho_{ba}^{(3)} = \rho_{ab}^{(3)*}. \tag{8.48}$$

In principle, we can extend the perturbation expansion to higher-order terms than third-order. In [2], an expression for the general term in the i-th-order perturbation and the summation of all i-th values have already been analyzed for a restricted condition. However, third-order expressions are sufficient to understand most nonlinear effects in the semiconductor laser.

8.3 Nonlinear Polarization and Gain Coefficient

Using the perturbation expansion of the density matrix, the *laser polarization* P_{st} is represented in the expanded form,

$$\begin{aligned} P_{st} &= N_t \mathrm{Tr}(\rho R) = N_t \sum_{b \text{ or } a} (\rho_{ab}R_{ba} + \rho_{ba}R_{ab}) \\ &= P^{(1)}(E) + P^{(3)}(E^3) + \cdots + P^{(2i+1)}(E^{2i+1}) + \cdots. \end{aligned} \tag{8.49}$$

Note that all even-order polarizations are zero in the semiconductor laser.

The *first-order polarization* term is obtained from the first-order density matrix,

$$\begin{aligned} P^{(1)} &= N_t \sum_{b\, or\, a} \left(\rho_{ab}^{(1)} R_{ba} + \rho_{ba}^{(1)} R_{ab}\right) \\ &= \frac{N_t}{\hbar} \sum_{b\, or\, a} |R_{ba}|^2 \left(\rho_{bb}^{(0)} - \rho_{aa}^{(0)}\right) \sum_{p=0}^{\infty} \left\{ \frac{jE_p e^{j\omega_p t}}{j(\omega_p - \omega_{ba}) + 1/\tau_{in}} + c.c. \right\} \\ &\equiv \varepsilon_o \sum_{p=-\infty}^{\infty} \chi^{(1)}(\omega_p) E_p e^{j\omega_p t} \\ &= \varepsilon_o \sum_{p=-\infty}^{\infty} \chi^{(1)}(\omega_p) \widehat{E}_p(t) \Phi_p(\mathbf{r}) e^{j\omega_p t}. \end{aligned} \tag{8.50}$$

This gives a *first-order susceptibility* of

$$\begin{aligned} \chi^{(1)}(\omega_p) &= j \frac{N_t}{\varepsilon_o \hbar} \sum_{b\, or\, a} \frac{|R_{ab}|^2 \left(\rho_{bb}^{(0)} - \rho_{aa}^{(0)}\right)}{j(\omega_p - \omega_{ba}) + 1/\tau_{in}} \\ &= \frac{1}{\varepsilon_o} \int_{W_g}^{\infty} \frac{|R_{ab}|^2 g_{cv}(W_{ba}) \{f_c(W_{ba}) - f_v(W_{ba})\} \{\hbar\omega_p - W_{ba} + j\hbar/\tau_{in}\}}{(\hbar\omega_p - W_{ba})^2 + (\hbar/\tau_{in})^2} dW_{ba}. \end{aligned} \tag{8.51}$$

If $\chi^{(1)}(\omega_p)$ has no temporal variation, we can define the *linear gain coefficient* as follows,

$$
\begin{aligned}
G_p^{(1)} &= \operatorname{Im}\left\{\frac{\omega_p}{\varepsilon_{eff}\widehat{E}_p(t)\Delta t}\int_{t-\Delta t}^{t}\int_{act} P^{(1)}\Phi_p^*(\boldsymbol{r})e^{-j\omega_p t}d^3\boldsymbol{r}\,dt\right\} \\
&= \operatorname{Im}\left\{\frac{\varepsilon_o\omega_p}{\varepsilon_{eff}}\int_{act}\chi^{(1)}(\omega_p)|\Phi_p(\boldsymbol{r})|^2 d^3\boldsymbol{r}\right\} \\
&= \frac{\omega_p}{n_r^2}\xi\operatorname{Im}\chi^{(1)}(\omega_p).
\end{aligned}
\tag{8.52}
$$

Here, ξ is the *field confinement factor* associated with the active region defined in (3.48). The linear gain coefficient $G_p^{(1)}$ is the same as G_p in (3.58) because we have not as yet taken into account saturation phenomena described by the higher-order terms.

Based on the numerical results presented in Sect. 6.3 and (6.71), the linear gain coefficient is approximately expressed as,

$$
G_p^{(1)} = \xi a\left\{N - N_g - b(\lambda_p - \lambda_o)^2\right\},
\tag{8.53}
$$

where a is the tangential coefficient, N_g the transparent electron density, b the coefficient characterizing the wavelength dispersion, and λ_o the wavelength at the gain peak. Numerical values of the GaAs laser are, $a = 2.75 \times 10^{-12}\,\mathrm{m^3}/s$, $b = 3 \times 10^{21}\,\mathrm{m^{-3}\,nm^{-2}}$, $N_g = 2.1 \times 10^{24}\,\mathrm{m^{-3}}$, and $\lambda_o \approx 850\,\mathrm{nm}$.

The third-order polarization $P^{(3)}$ is defined by the third-order density matrix elements, $\rho_{ab}^{(3)}$ and $\rho_{ba}^{(3)}$, and the third-order susceptibility $\chi^{(3)}(\omega_p, \omega_q, \omega_r)$ in the form

$$
\begin{aligned}
P^{(3)} &= N_t\sum_{b\,or\,a}\left(\rho_{ab}^{(3)}R_{ba} + \rho_{ba}^{(3)}R_{ab}\right) \\
&= -\frac{jN_t|R_{ba}|^4(\rho_{bb}^{(0)} - \rho_{aa}^{(0)})}{\hbar^3}\sum_{p=0}^{\infty}\sum_{q=0}^{\infty}\sum_{r=0}^{\infty} \\
&\quad\times\left[\frac{E_pE_q^*E_re^{j(\omega_p-\omega_q+\omega_r)t}}{\{j(\omega_p-\omega_{ba})+1/\tau_{in}\}\{j(\omega_p-\omega_q+\omega_r-\omega_{ba})+1/\tau_{in}\}}\right. \\
&\qquad\times\left\{\frac{1}{j(\omega_p-\omega_q)+1/\tau_b+1/\tau_s}+\frac{1}{j(\omega_p-\omega_q)+1/\tau_a+1/\tau_s}\right\} \\
&\qquad+\frac{E_p^*E_qE_re^{j(-\omega_p+\omega_q+\omega_r)t}}{\{j(\omega_{ba}-\omega_p)+1/\tau_{in}\}\{j(-\omega_p+\omega_q+\omega_r-\omega_{ba})+1/\tau_{in}\}} \\
&\qquad\left.\times\left\{\frac{1}{j(-\omega_p+\omega_q)+1/\tau_b+1/\tau_s}+\frac{1}{j(-\omega_p+\omega_q)+1/\tau_a+1/\tau_s}\right\}\right] + c.c. \\
&\equiv \varepsilon_o\sum_{p=-\infty}^{\infty}\sum_{q=-\infty}^{\infty}\sum_{r=-\infty}^{\infty}\chi^{(3)}(\omega_p,\omega_q,\omega_r)\widehat{E}_p\widehat{E}_q\widehat{E}_r\Phi_p(\boldsymbol{r})\Phi_q(\boldsymbol{r})\Phi_r(\boldsymbol{r})e^{j(\omega_p+\omega_q+\omega_r)t}.
\end{aligned}
\tag{8.54}
$$

The third-order susceptibility $\chi^{(3)}(\omega_p, \omega_q, \omega_r)$ is defined by mode numbers p, q, and r according to the order of the interaction for component with summed frequency $\omega_p + \omega_q + \omega_r$.

We now restrict the combination of the modes to specific modes and let $P_p^{(3)}$ denote the component for the summed frequency ω_p. Then $P_p^{(3)}$ is written as

$$P_p^{(3)} = \varepsilon_o \sum_{q=0}^{\infty} \sum{}' \chi^{(3)}(\omega_p, -\omega_q, \omega_q) \left| \widehat{E}_q \right|^2 \widehat{E}_p \left| \Phi_q(\mathbf{r}) \right|^2 \Phi_p(\mathbf{r}) e^{j\omega_p t}. \tag{8.55}$$

Here, the superscripted prime appended to the summation means to sum over all terms equivalent to $\chi^{(3)}(\omega_p, -\omega_q, \omega_q)$. If the modes are *identical*, i.e., $p = q$, then

$$\begin{aligned}
&\sum{}' \chi^{(3)}(\omega_p, -\omega_p, \omega_p) \\
&= \chi^{(3)}(\omega_p, -\omega_p, \omega_p) + \chi^{(3)}(-\omega_p, \omega_p, \omega_p) \\
&= -\frac{jN_t}{\varepsilon_o \hbar^3} \sum_{b \,\text{or}\, a} |R_{ba}|^4 \left(\rho_{bb}^{(0)} - \rho_{aa}^{(0)} \right) \times \frac{1}{j(\omega_p - \omega_{ba}) + 1/\tau_{in}} \\
&\quad \times \left[\frac{1}{j(\omega_p - \omega_{ba}) + 1/\tau_{in}} \left(\frac{1}{1/\tau_b + 1/\tau_s} + \frac{1}{1/\tau_a + 1/\tau_s} \right) \right. \\
&\quad \left. + \frac{1}{j(\omega_{ba} - \omega_p) + 1/\tau_{in}} \left(\frac{1}{1/\tau_b + 1/\tau_s} + \frac{1}{1/\tau_a + 1/\tau_s} \right) \right] \\
&= -\frac{jN_t}{\varepsilon_o \hbar^3} \sum_{b \,\text{or}\, a} |R_{ba}|^4 \left(\rho_{bb}^{(0)} - \rho_{aa}^{(0)} \right) \frac{1}{j(\omega_p - \omega_{ba}) + 1/\tau_{in}} \left[\frac{2(\tau_b + \tau_a)}{\left\{ (\omega_p - \omega_{ba})^2 + 1/\tau_{in}^2 \right\} \tau_{in}^2} \right], \\
&\quad \text{for } q = p,
\end{aligned} \tag{8.56}$$

where we assume $\tau_b, \tau_a \ll \tau_s$.

Another combination with *two different modes p* and *q* is

$$
\begin{aligned}
&\sum{}'\chi^{(3)}(\omega_p,-\omega_q,\omega_q)\\
&=\chi^{(3)}(\omega_p,-\omega_q,\omega_q)+\chi^{(3)}(-\omega_q,\omega_p,\omega_q)+\chi^{(3)}(\omega_q,-\omega_q,\omega_p)+\chi^{(3)}(-\omega_q,\omega_q,\omega_p)\\
&=-\frac{jN_t}{\varepsilon_o\hbar^3}\sum_{b\,\mathrm{or}\,a}|R_{ba}|^4\left(\rho_{bb}^{(0)}-\rho_{aa}^{(0)}\right)\times\frac{1}{j(\omega_p-\omega_{ba})+1/\tau_{in}}\\
&\times\left[\frac{1}{j(\omega_p-\omega_{ba})+1/\tau_{in}}\left\{\frac{1}{j(\omega_p-\omega_q)+1/\tau_b+1/\tau_s}+\frac{1}{j(\omega_p-\omega_q)+1/\tau_a+1/\tau_s}\right\}\right.\\
&+\frac{1}{j(\omega_{ba}-\omega_q)+1/\tau_{in}}\left\{\frac{1}{j(\omega_p-\omega_q)+1/\tau_b+1/\tau_s}+\frac{1}{j(\omega_p-\omega_q)+1/\tau_a+1/\tau_s}\right\}\\
&+\frac{1}{j(\omega_q-\omega_{ba})+1/\tau_{in}}\left\{\frac{1}{1/\tau_b+1/\tau_s}+\frac{1}{1/\tau_a+1/\tau_s}\right\}\\
&\left.+\frac{1}{j(\omega_{ba}-\omega_q)+1/\tau_{in}}\left\{\frac{1}{1/\tau_b+1/\tau_s}+\frac{1}{1/\tau_a+1/\tau_s}\right\}\right]\\
&=-\frac{jN_t}{\varepsilon_o\hbar^3}\sum_{b\,\mathrm{or}\,a}|R_{ba}|^4\left(\rho_{bb}^{(0)}-\rho_{aa}^{(0)}\right)\times\frac{1}{j(\omega_p-\omega_{ba})+1/\tau_{in}}\\
&\times\left[\left\{\frac{1}{j(\omega_p-\omega_{ba})+1/\tau_{in}}+\frac{1}{j(\omega_{ba}-\omega_q)+1/\tau_{in}}\right\}\right.\\
&\times\left\{\frac{1}{j(\omega_p-\omega_q)+1/\tau_b}+\frac{1}{j(\omega_p-\omega_q)+1/\tau_a}\right\}\\
&\left.+\frac{2(\tau_b+\tau_a)}{\left\{(\omega_q-\omega_{ba})^2+1/\tau_{in}^2\right\}\tau_{in}^2}\right],\\
&\text{for}\, q\neq p.
\end{aligned}
\tag{8.57}
$$

We find from (8.56) and (8.57) that the *third-order terms have twice the number of combinations for different modes than the combination with a single mode*. This difference originates from the second-order density matrix $\rho_{bb}^{(2)}$ and $\rho_{aa}^{(2)}$, which has not only DC terms for $q = p$ but also *beating terms* with exp $[\pm j(\omega_p - \omega_q)t]$ when there are more than two modes.

As found from (8.56) and (8.57), the imaginary part of the third-order susceptibility is negative, hence the third-order polarization acts to reduce the lasing gain. We now represent the term $|\hat{E}_q|^2$ in the third-order polarization $P_p^{(3)}$ in (8.55) by the photon number, as in (3.56) or (6.66),

$$
\left|\hat{E}_q\right|^2=\frac{\hbar\omega_q}{2\varepsilon_{eff}}S_q. \tag{8.58}
$$

The gain coefficient G_p expanded to include *third-order polarization* terms is defined as

$$G_p = G_p^{(1)} - \sum_q G_{p(q)}^{(3)} S_q. \tag{8.59}$$

Here, $G_{p(q)}^{(3)}$ is the *third-order nonlinear gain coefficient*, which accounts for the saturation effect from mode q to mode p. By substituting (8.55) into (8.12), we get

$$\begin{aligned} G_{p(q)}^{(3)} &= -\frac{\hbar\omega_p\omega_q}{2\varepsilon_{eff} n_r^2}\int_{act} \mathrm{Im}\left\{\sum{}'\chi^{(3)}(\omega_p, -\omega_q, \omega_q)\right\}\left|\Phi_p(\boldsymbol{r})\right|^2\left|\Phi_q(\boldsymbol{r})\right|^2 d^3\boldsymbol{r} \\ &= \frac{\hbar\omega_p\omega_q}{2\varepsilon_{eff}^2(1+\delta_{p,q})}\int_{act} \mathrm{Re}\left(\int_{W_g}^{\infty} |R_{ba}|^4 g_{cv}(W_{ba})[f_c(W_{ba}) - f_v(W_{ba})]\right. \\ &\quad \times \left[\left\{\frac{1}{j(\hbar\omega_p - W_{ba}) + \hbar/\tau_{in}} + \frac{1}{j(W_{ba} - \hbar\omega_q) + \hbar/\tau_{in}}\right\}\right. \\ &\quad \times \left\{\frac{1}{j(\hbar\omega_p - \hbar\omega_q) + \hbar/\tau_b} + \frac{1}{j(\hbar\omega_p - \hbar\omega_q) + \hbar/\tau_a}\right\} \\ &\quad \left. + \frac{2(\tau_b + \tau_a)}{\left\{(\hbar\omega_q - W_{ba})^2 + (\hbar/\tau_{in})^2\right\}\tau_{in}^2}\right] \\ &\quad \left.\times \frac{1}{j(\hbar\omega_p - W_{ba}) + \hbar/\tau_{in}} dW_{ba}\right) \times \left|\Phi_p(\boldsymbol{r})\right|^2\left|\Phi_q(\boldsymbol{r})\right|^2 d^3\boldsymbol{r} \end{aligned} \tag{8.60}$$

Equation (8.60) is applicable for both $q = p$ and $q \neq p$, because we have introduced the difference via the factor $1/(1 + \delta_{p,q})$. From a numerical calculation, the third-order gain saturation coefficient is approximately expressed as

$$G_{p(q)}^{(3)} = \frac{2}{1+\delta_{p,q}}\frac{\hbar\omega_p}{\varepsilon_o n_r^2}\left(\frac{\tau_{in}}{\hbar}\right)^2 \frac{|R_{ba}|^2 a}{1 + \tau_{in}^2(\omega_p - \omega_q)^2}\int_{act}(N - N_s)\left|\Phi_p(\boldsymbol{r})\Phi_q(\boldsymbol{r})\right|^2 d^3\boldsymbol{r}. \tag{8.61}$$

Care is needed in evaluating the 4th power spatial integration in (8.61). The form of the field distribution function $\Phi_p(\boldsymbol{r})$ has been given in (3.64). We suppose here that the field distribution in the width direction x is the transverse fundamental mode and well confined in the active region along the interval of width W, and that in the thickness direction y is the transverse fundamental mode but spread sufficiently into the cladding region from the active region along the interval of thickness d. The field distribution along the z direction has a different standing wave shape characterized by longitudinal mode number p or q. Hence, the *spatial integration* becomes

$$\int_{act} |\Phi_p(\boldsymbol{r})\Phi_q(\boldsymbol{r})|^2 d^3\boldsymbol{r} = \frac{3}{4}\frac{\xi^2}{V_a}(2 + \delta_{p,q}). \tag{8.62}$$

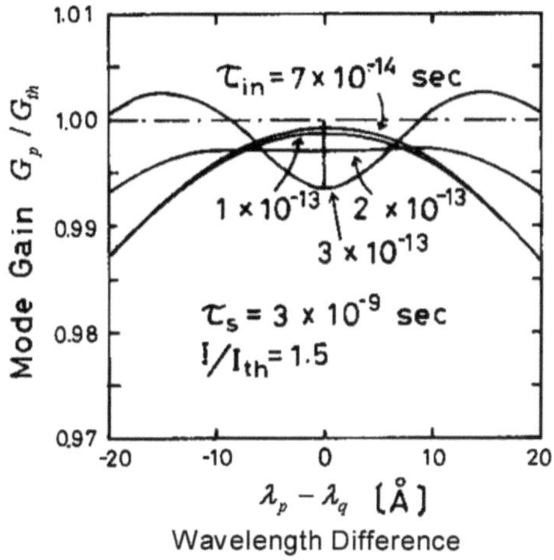

Fig. 8.1 Calculated example of the saturated gain in GaAs

Here, values of $3/2W$, ξ^2/d, and $(2 + \delta_{p,q}/2L$ come from the integrations along x, y, and z, respectively. Hence, the third-order saturation coefficient is approximately written as [1]

$$G^{(3)}_{p(p)} \equiv B = \frac{9\xi^2\hbar\omega_p a|R_{ba}|^2}{4V_a\varepsilon_o n_r^2}\left(\frac{\tau_{in}}{\hbar}\right)^2 (N - N_s), \tag{8.63}$$

$$\begin{aligned} G^{(3)}_{p(q)} \equiv D &= \frac{4}{3}\frac{B}{\{1 + \tau_{in}^2(\omega_p - \omega_q)^2\}} \\ &= \frac{4}{3}\frac{B}{1 + \left(\frac{\tau_{in}\omega_p}{\lambda_p}\right)^2 (\lambda_q - \lambda_p)^2}, \qquad \text{for } p \neq q \end{aligned} \tag{8.64}$$

where we have used the relation

$$\omega_p - \omega_q = 2\pi c\left(\frac{1}{\lambda_p} - \frac{1}{\lambda_q}\right) = \frac{\omega_p}{\lambda_p}(\lambda_q - \lambda_p). \tag{8.65}$$

We call B the self-saturation coefficient and D the symmetric (mutual) gain saturation coefficient because D has a symmetric profile with respect to dispersion $\pm(\lambda_p - \lambda_q)$. Numerical values for the GaAs lasers are $N_s = 1.7 \times 10^{24}\,\mathrm{m}^{-3}$, $|R_{ba}|^2 = 2.8 \times 20^{-57}\,\mathrm{C^2m^2}$, and $\tau_{in} = 1.0 \times 10^{-13}$ s. Other parameters are the same as those used for the linear gain coefficient.

A calculated example of the saturated gain for GaAs is shown in Fig. 8.1. Here, a single-mode oscillation with mode q is assumed with gain suppression from the mode q taken into account,

$$G_q = G_q^{(1)} - BS_q = G_{th}, \tag{8.66}$$

$$G_p = G_p^{(1)} - DS_q, \tag{8.67}$$

with

$$S_q = \frac{I - I_{th}}{e\,G_{th}}. \tag{8.68}$$

As found from Fig. 8.1, G_q of the oscillating mode is fixed to the threshold gain G_{th}, but gains of other modes are suppressed. The shapes of the saturated gain profile are determined by the dispersion coefficient b in the linear gain coefficient of (8.53) and the intraband relaxation time τ_{in}. If τ_{in} is long, the saturation becomes deep and narrow, a situation that is called *inhomogeneous gain saturation*. When τ_{in} is short, the saturation becomes shallow and wide, hence called *homogeneous gain saturation*. For a GaAs laser, the dispersions for both the linear gain and saturated gain are almost identical, as indicated by the lines $\tau_{in} = (0.7 - 1.0) \times 10^{-13}$ s. This type of saturation is called *partially homogeneous broadening*.

As found from Fig. 8.1, the difference between the modal gain G_p and threshold gain G_{th} is merely 1/1000. However, such a small difference in gain is very important in determining whether the laser operates in single mode or multi-mode.

The third-order gain saturation can show *stronger gain suppression* for the non-oscillating mode if several parameters are suitably selected. Such stronger gain reduction was observed in an experiment by Aiki et al. and theoretically explained by the present author, as seen in Fig. 8.2. By increasing the injection current I, the output power of the oscillating mode increases but the *output from the non-oscillating modes is reduced*. In the experiment, the oscillating mode itself jumped to another mode of longer wavelength thereby increasing the injection current because the temperature increased in the active region. The parameters used in the theoretical calculation of Fig. 8.2 differed slightly from those in other analyses. This reason is that previously we did not take into account the asymmetric gain saturation. Nonetheless, the presence of strong gain suppression was experimentally observed for the first time in Fig. 8.2.

8.4 Effect from Higher Order Transverse Mode

In the analyses up till now, we have assumed that the higher-order transverses modes do not exist and only the fundamental mode exists for all longitudinal modes. If we permit higher-order transverse modes, the discussion is different.

We suppose for the moment that a higher-order transverse mode exists in the width direction of the cavity. Then, in (8.62), the fourth-power spatial integration becomes

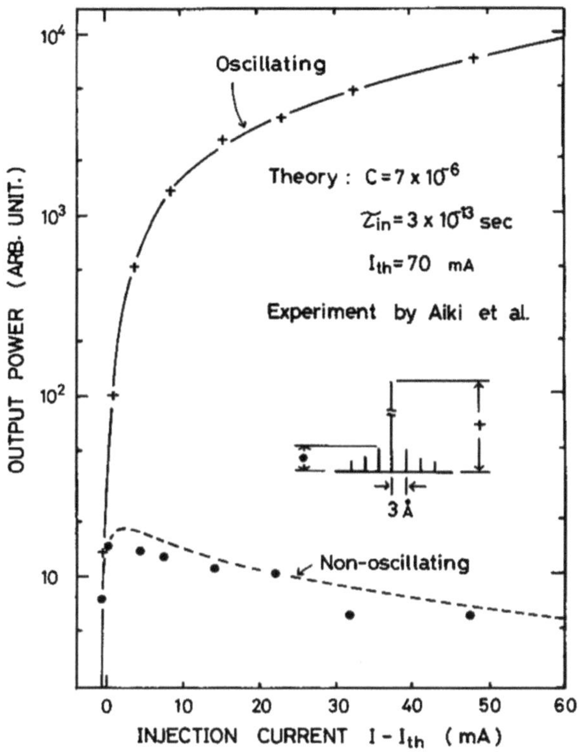

Fig. 8.2 Stronger gain suppression for non-oscillating modes

$$\int_{act} |\Phi_p(\boldsymbol{r})\Phi_q(\boldsymbol{r})|^2 d^3\boldsymbol{r} = \frac{1}{4}\frac{\xi^2}{V_a}(2+\delta_{p_x,q_x})(2+\delta_{p_z,q_z}). \tag{8.69}$$

Here, p_x and q_x are the transverse mode numbers, and p_z and q_z are the longitudinal mode numbers. This equation gives the integrated value of $(9/4)(\xi/V_a)$ for the same mode $p = q$, but ξ^2/V_a for distinct transverse and longitudinal modes, $p_x \neq q_x$ and $p_z \neq q_z$. Thus, the spatial integration becomes smaller for distinct modes by factor 4/9. The symmetric mutual saturation coefficient then becomes

$$D = \frac{8}{9}\frac{B}{\{1+\tau_{in}^2(\omega_p-\omega_q)^2\}} < B, \quad \text{for higher-order modes.} \tag{8.70}$$

This relation implies that *we are unable to produce the single longitudinal mode of operation in lasers which permits higher-order transverse modes to propagate*. Such modes can be excluded using a suitably designed *stripe structure*, the analysis of which was given in Chap. 4, from the perspective of the single longitudinal-mode operation.

8.5 Variation of Electron Density

With the diagonal elements of the density matrix given with even-order terms, the perturbation expansions for the *existing electron densities* are

$$N_c = N_t \sum_b \rho_{bb} = N_t \sum_b \left\{ \rho_{bb}^{(0)} + \rho_{bb}^{(2)}(E^2) + \cdots + \rho_{bb}^{(2i)}(E^{2i}) + \cdots \right\}, \tag{8.71}$$

$$N_v = N_t \sum_a \rho_{aa} = N_t \sum_a \left\{ \rho_{aa}^{(0)} + \rho_{aa}^{(2)}(E^2) + \cdots + \rho_{aa}^{(2i)}(E^{2i}) + \cdots \right\}. \tag{8.72}$$

However, as found from (8.45) and (8.46), the higher-order terms beyond the second-order term have *multi-solutions* with various combinations of modes. Let us show a counter example: We assume that a laser is in oscillation with an active single mode q. The electron density $N_c(\omega_q)$ contributing to the laser oscillation depends only on this q mode. However, if we measure the electron density with a probe light at ω_p which is generated from spontaneous emission, the measured density $N_c(\omega_q, \omega_p)$ must differ from $N_c(\omega_q)$. Therefore, *we cannot uniquely define* the term "*existing electron density*" for the laser.

Meanwhile, the zeroth-order electron density corresponds directly to the applied voltage or the potential needed to inject electrons and holes into the active region. The linear and nonlinear gain coefficients are expressed using the zeroth-order electron density. The *evaluation of the zeroth-order electron density is more essential* than describing the nature of the existing electron density.

Expressions for the zeroth-order terms are given in (8.18) and (8.19). Here we define the *injected electron density* N for the *zeroth-order terms* by the following:

$$\begin{aligned} N &= N_t \sum_b \left(\rho_{bb}^{(0)} - \tilde{\rho}_b \right) \\ &= \frac{N_t}{2} \left\{ \sum_b \left(\rho_{bb}^{(0)} - \tilde{\rho}_b \right) - \sum_a \left(\rho_{aa}^{(0)} - \tilde{\rho}_a \right) \right\} \\ &= \int_{W_g}^{\infty} g_{cv}(W_{ba}) \{ f_c(W_{ba}) - \tilde{f}_c(W_{ba}) \} dW_{ba}. \end{aligned} \tag{8.73}$$

Because the number of injected electrons and injected holes are identical owing to the neutral charge condition, this value is defined as the value of the average number of injected electrons and injected holes, that is, the average number of electron-hole pairs.

The temporal variation of the injected electron density is given by

$$\begin{aligned}\frac{dN}{dt} &= \frac{N_t}{2}\left\{\sum_b \frac{d\rho_{bb}^{(0)}}{dt} - \sum_a \frac{d\rho_{aa}^{(0)}}{dt}\right\} \\ &= -\frac{N_t}{2}\left\{\sum_b \frac{\rho_{bb}^{(0)} - \tilde{\tilde{\rho}}_b}{\tau_b} - \sum_a \frac{\rho_{aa}^{(0)} - \tilde{\tilde{\rho}}_a}{\tau_a}\right\} - \frac{N}{\tau_s} + \frac{I}{eV_a}.\end{aligned} \tag{8.74}$$

Here, injection current I is defined by

$$I = eN_t \int_{act} \sum_b \Lambda_b\, d^3\boldsymbol{r} = -eN_t \int_{act} \sum_a \Lambda_a\, d^3\boldsymbol{r}. \tag{8.75}$$

The quasi-equilibrium distribution function has been defined as

$$\int_{V_a} \sum_b \frac{\tilde{\tilde{\rho}}_b}{\tau_b} d^3\boldsymbol{r} = \int_{V_a} \sum_b \frac{\rho_{bb}}{\tau_b} d^3\boldsymbol{r}, \tag{8.76}$$

so that the electron number remains constant in each band in the active region. However, the electron density is not always spatially uniform in the active region. We introduce the diffusion constants D_c and D_v describing spatial diffusion as

$$N_t \sum_b \frac{\rho_{bb} - \tilde{\tilde{\rho}}_b}{\tau_b} = -D_c \nabla^2 (N_t \sum_b \rho_{bb}), \tag{8.77}$$

$$N_t \sum_a \frac{\rho_{aa} - \tilde{\tilde{\rho}}_a}{\tau_a} = -D_v \nabla^2 (N_t \sum_a \rho_{aa}). \tag{8.78}$$

With the diagonal elements ρ_{bb} and ρ_{aa} expanded to second-order, we get the following equations,

$$N_t \sum_b \frac{\rho_{bb}^{(0)} - \tilde{\tilde{\rho}}_b}{\tau_b} = -D_c \nabla^2 (N_t \sum_b \rho_{bb}) + N_t \sum_b \rho_{bb}^{(2)}, \tag{8.79}$$

$$N_t \sum_b \frac{\rho_{aa}^{(0)} - \tilde{\tilde{\rho}}_b}{\tau_b} = -D_v \nabla^2 (N_t \sum_a \rho_{aa}) + N_t \sum_a \rho_{aa}^{(2)}. \tag{8.80}$$

Here, we redefine the diffusion constant D_{diff} to treat simultaneously diffusion effects for electrons and holes,

$$
\begin{aligned}
D_{diff}\nabla^2 N &\equiv \frac{N_t}{2}\left\{D_c\nabla^2\Big(\sum_b \rho_{bb}^{(0)}\Big) - D_v\nabla^2\Big(\sum_a \rho_{aa}^{(0)}\Big)\right\} \\
&\approx \frac{N_t}{2}\left\{D_c\nabla^2\Big(\sum_b \rho_{bb}\Big) - D_v\nabla^2\Big(\sum_a \rho_{aa}\Big)\right\}.
\end{aligned}
\tag{8.81}
$$

Using the notation of (8.79)–(8.81), (8.74) is rearranged as

$$
\frac{dN}{dt} = \frac{N_t}{2}\left\{\sum_b \frac{\rho_{bb}^{(2)}}{\tau_b} - \sum_a \frac{\rho_{aa}^{(2)}}{\tau_a}\right\} + D_{diff}\nabla^2 N - \frac{N}{\tau_s} + \frac{I}{eV_a}. \tag{8.82}
$$

Next, we shall examine the second-order terms $\rho_{bb}^{(2)}$ and $\rho_{aa}^{(2)}$. The second-order density matrix has been derived as in (8.45) and (8.46). If the intraband relaxation is fast enough, $|\omega_p - \omega_q|\tau_b \ll 1$, the summed values of the second-order terms can be expressed approximately using the linear susceptibility,

$$
N_t \sum_b \frac{\rho_{bb}^{(2)}}{\tau_b} = j\sum_{p=0}^{\infty}\sum_{q=0}^{\infty}\frac{\varepsilon_o}{\hbar}\chi^{(1)}(\omega_p)E_pE_q^* e^{j(\omega_p-\omega_q)t} + c.c., \tag{8.83}
$$

$$
N_t \sum_a \frac{\rho_{aa}^{(2)}}{\tau_a} = -j\sum_{p=0}^{\infty}\sum_{q=0}^{\infty}\frac{\varepsilon_o}{\hbar}\chi^{(1)}(\omega_p)\,E_p\,E_q^* e^{j(\omega_p-\omega_q)\,t} + c.c.. \tag{8.84}
$$

Then (8.82) can be rewritten as

$$
\frac{dN}{dt} = \left\{j\sum_{p=0}^{\infty}\sum_{q=0}^{\infty}\frac{\varepsilon_o}{\hbar}\chi^{(1)}(\omega_p)E_pE_q^* e^{j(\omega_p-\omega_q)t} + c.c.\right\} + D_{diff}\nabla^2 N - \frac{N}{\tau_s} + \frac{I}{eV_a}. \tag{8.85}
$$

This equation states that the electron injection density N can be modulated with beat frequency of $\omega_p - \omega_q$ when the multi-modes p and q are taken into account. Frequency beating produces an *asymmetric gain saturation* effect to be described in a later section.

We focus for now on the linear susceptibility $\chi^{(1)}(\omega_p)$ which has both real and imaginary parts with both depending on the electron density. When the number of injected electron changes from N_o to $N_o + \Delta N$, the change to the susceptibility is given as

$$
\chi^{(1)}(\omega_p, N_o + \Delta N) = \chi^{(1)}(\omega_p, N_o) + \frac{\partial\chi^{(1)}(\omega_p)}{\partial N}\Delta N. \tag{8.86}
$$

We introduce here the variation ratio α of the real and imaginary parts of the first derivative of the susceptibility,

$$\alpha \equiv -\frac{\partial \mathrm{Re}\chi^{(1)}(\omega_p)/\partial N}{\partial \mathrm{Im}\chi^{(1)}(\omega_p)/\partial N}, \tag{8.87}$$

called the *linewidth enhancement factor* [3], whose idea was introduced to determine the spectral line width of the laser, as we shall show in a later chapter. The derivative is then written as

$$\begin{aligned}\frac{\partial \chi^{(1)}(\omega_p)}{\partial N} &= \frac{\partial \mathrm{Re}\chi^{(1)}(\omega_p)}{\partial N} + j\frac{\partial \mathrm{Im}\chi^{(1)}(\omega_p)}{\partial N} \\ &= j(1+j\alpha)\frac{\partial \mathrm{Im}\chi^{(1)}(\omega_p)}{\partial N}.\end{aligned} \tag{8.88}$$

We write the imaginary part of the linear susceptibility $\chi^{(1)}(\omega_p)$ in the approximate form

$$\mathrm{Im}\chi^{(1)}(\omega_p, N_o) = \kappa\left\{N_o - N_g - b(\lambda_p - \lambda_o)^2\right\} \tag{8.89}$$

with κ a coefficient of proportionality. The linear susceptibility $\chi^{(1)}(\omega_p)$ can be expressed as

$$\chi^{(1)}(\omega_p, N) = j\kappa\left\{(1+j\alpha)N - N_g - b(\lambda_p - \lambda_o)^2\right\}. \tag{8.90}$$

Therefore, the variation of the electron density in (8.85) is rewritten as

$$\begin{aligned}\frac{dN}{dt} = &-\sum_{p=0}^{\infty}\sum_{q=0}^{\infty}\frac{\varepsilon_o\kappa}{\hbar}\left[\left\{(1+j\alpha)N - N_g - b(\lambda_p - \lambda_o)^2\right\}E_pE_q^* e^{j(\omega_p-\omega_q)t} + c.c.\right] \\ &+ D_{diff}\nabla^2 N - \frac{N}{\tau_s} + \frac{I}{eV_a}.\end{aligned} \tag{8.91}$$

8.6 Frequency Beating of the Injected Electron Density with Lasing Modes

The mode numbers p and q appearing in (8.91) imply that all possible modes in the laser are to be scanned. However, to simplify the discussion, we restrict these modes to just two modes and set

$$\Omega \equiv \omega_q - \omega_p. \tag{8.92}$$

Summations over the combination of modes are then reduced resulting in

$$\begin{aligned}\frac{dN}{dt} = &-\frac{\varepsilon_o\kappa}{\hbar}\Big[\Big\{(1+j\alpha)N - N_g - b(\lambda_p-\lambda_o)^2\Big\}\Big\{|E_p|^2 + E_pE_q^*e^{-j\Omega t}\Big\}\\ &+\Big\{(1+j\alpha)N - N_g - b(\lambda_q-\lambda_o)^2\Big\}\Big\{|E_q|^2 + E_qE_p^*e^{j\Omega t}\Big\} + c.c.\Big]\\ &+ D_{diff}\nabla^2 N - \frac{N}{\tau_s} + \frac{I}{eV_a}.\end{aligned} \tag{8.93}$$

Here, we put the number of injected electron in the form

$$N = \widehat{N}(t) + N_\Omega e^{j\Omega t} + N_\Omega^* e^{-j\Omega t}, \tag{8.94}$$

where N_Ω is a complex amplitude of the vibrational component $\exp(j\Omega t)$, and $\widehat{N}(t)$ a slowly varying component. The amplitude of the vibration term is presumed much smaller than the slowly varying component, $|N_\Omega| \ll \widehat{N}(t)$. By substituting (8.94) into (8.93), we separate terms with $\widehat{N}(t)$ and N_Ω as

$$\begin{aligned}\frac{d\widehat{N}}{dt} = &-\frac{2\varepsilon_o\kappa}{\hbar}\Big[\Big\{\widehat{N} - N_g - b(\lambda_p-\lambda_o)^2\Big\}|E_p|^2\\ &+\Big\{\widehat{N} - N_g - b(\lambda_q-\lambda_o)^2\Big\}|E_q|^2\Big]\\ &+ D_{diff}\nabla^2\widehat{N} - \frac{\widehat{N}}{\tau_s} + \frac{I}{eV_a},\end{aligned} \tag{8.95}$$

$$\begin{aligned}&j\Omega N_\Omega + \frac{N_\Omega}{\tau_s} + \frac{2\varepsilon_o\kappa}{\hbar}\Big(|E_p|^2 + |E_q|^2\Big)N_\Omega - D_{diff}\nabla^2 N_\Omega\\ &= -\frac{2\varepsilon_o\kappa}{\hbar}\Big(\widehat{N} - N_g\Big)E_p^*E_q.\end{aligned} \tag{8.96}$$

Our next task is to determine the effect of spatial diffusion on N_Ω. As N_Ω is induced by $E_p^*E_q$, N_Ω might have a spatial distribution of the form $\Phi_p^*(\boldsymbol{r})\Phi_q(\boldsymbol{r})$ where

$$\Phi_p(\boldsymbol{r}) = \sqrt{\frac{2}{L}}T(x,y)\cos(\beta_p z). \tag{8.97}$$

Product functions such as $\cos(\beta_p z)\cos(\beta_q z)$ are expanded into two terms with spatial variations $\beta_q - \beta_p$ and $\beta_q + \beta_p$.

Setting N_Ω as

$$N_\Omega = N_1 \cos\{(\beta_q - \beta_p)z\} + N_2 \cos\{(\beta_q + \beta_p)z\}, \tag{8.98}$$

and substituting into (8.96), we obtain

$$\begin{aligned}
&\left[j\Omega + \frac{1}{\tau_s} + \frac{2\varepsilon_o\kappa}{\hbar}\frac{|T(x,y)|^2}{L}\left(|\widehat{E}_p|^2\{1+\cos(2\beta_p z)\} + |\widehat{E}_q|^2\{1+\cos(2\beta_q z)\}\right)\right.\\
&\quad \left. + (\beta_q - \beta_p)^2 D_{diff}\right] N_1 \cos\{(\beta_q - \beta_p)z\}\\
&\quad + \left[j\Omega + \frac{1}{\tau_s} + \frac{2\varepsilon_o\kappa}{\hbar}\frac{|T(x,y)|^2}{L}\left(|\widehat{E}_p|^2\{1+\cos(2\beta_p z)\} + |\widehat{E}_q|^2\{1+\cos(2\beta_q z)\}\right)\right.\\
&\qquad \left. + (\beta_q + \beta_p)^2 D_{diff}\right] N_2 \cos\{(\beta_q + \beta_p)z\}\\
&= -\frac{2\varepsilon_o\kappa}{\hbar}\left(\widehat{N} - N_g\right)\frac{|\widehat{E}_p^* \widehat{E}_q||T(x,y)|^2}{L}\left[\cos\{(\beta_q - \beta_p)z\} + \cos\{(\beta_q + \beta_p)z\}\right].
\end{aligned} \tag{8.99}$$

By multiplying by $\cos\{(\beta_q - \beta_p)z\}$ or $\cos\{(\beta_q + \beta_p)z\}$ and taking the spatial average over the laser cavity length L, we can separate terms with N_1 and N_2. Because the term $(\beta_q + \beta_p)^2 D_{diff}$ is much larger than the other bracketed terms [ ] on the left-hand side in (8.99), we find that $|N_1| \gg |N_2|$, and N_1 becomes

$$N_1 = -\frac{\frac{2\varepsilon_o\kappa}{\hbar}\left(\widehat{N} - N_g\right)\frac{|\widehat{E}_p^* \widehat{E}_q||T(x,y)|^2}{L}}{j\Omega + \frac{1}{\tau_s} + \frac{2\varepsilon_o\kappa}{\hbar}\frac{|T(x,y)|^2}{L}\left(|\widehat{E}_p|^2 + |\widehat{E}_q|^2\right) + (\beta_q - \beta_p)^2 D_{diff}}. \tag{8.100}$$

The diffusion term $(\beta_q - \beta_p)^2 D_{diff}$ can be neglected in many cases because $(\beta_q - \beta_p)^2 D_{diff}\tau_s \ll 1$. However, the effect of diffusion should not be overlooked in deriving (8.100), because frequency beating is affected by the character of the standing waves in the cavity. As the terms involving $\cos(\beta_q + \beta_p)z$ change rapidly, the terms have been dropped in the derivation of (8.100) from (8.99).

8.7 Asymmetric Gain Saturation Coefficient

The *asymmetric gain saturation* effect [4, 5] is caused by the beating vibration in the linear gain. We reexamine the linear gain by returning to the definition of the gain coefficient in (8.12) or (8.52):

$$
\begin{aligned}
G_p^{(1)} &= \mathrm{Im}\left\{ \frac{\omega_p}{\varepsilon_{eff}\widehat{E}_p(t)\Delta t} \int_{t-\Delta t}^{t} \int_{act} P^{(1)} \Phi_p^*(\boldsymbol{r}) e^{-j\omega_p t} d^3\boldsymbol{r}\, dt \right\} \\
&= \mathrm{Im}\left\{ \frac{\varepsilon_o \omega_p}{\varepsilon_{eff}\Delta t} \int_{t-\Delta t}^{t} \int_{act} \sum_q \chi^{(1)}(\omega_q) \frac{\widehat{E}_q}{\widehat{E}_p} \Phi_q(\boldsymbol{r}) \Phi_p^*(\boldsymbol{r}) e^{j(\omega_q-\omega_p)t} d^3\boldsymbol{r}\, dt \right\}.
\end{aligned}
\tag{8.101}
$$

One aim of this section is to take into account the effect of the other mode q to the linear gain coefficient through the beating vibration. We restrict the analysis to two modes, p and q, and express the linear susceptibility in the form of (8.90), thus

$$
\begin{aligned}
G_p^{(1)} =& \frac{\kappa\omega_p}{n_r^2 \Delta t} \mathrm{Re} \int_{t-\Delta t}^{t} \left[\int_{act} \left\{ (1+j\alpha)N - N_g - b(\lambda_p - \lambda_o)^2 \right\} |\Phi_p(\boldsymbol{r})|^2 d^3\boldsymbol{r} \right. \\
& \left. + \int_{act} \left\{ (1+j\alpha)N - N_g - b(\lambda_q - \lambda_o)^2 \right\} \Phi_q(\boldsymbol{r}) \Phi_p^*(\boldsymbol{r}) d^3\boldsymbol{r} \frac{\widehat{E}_q}{\widehat{E}_p} \right] e^{j\Omega t} dt.
\end{aligned}
\tag{8.102}
$$

We substitute into this equation the varying electron density

$$
N = \widehat{N} + \left\{ N_1(x,y)\, e^{j\Omega t} + N_1^*(x,y)\, e^{-j\Omega t} \right\} \cos\{\, (\beta_q - \beta_p) z \},
\tag{8.103}
$$

where $N_1(x, y)$ is given in (8.100) and $N_1^*(x, y)$ is its complex conjugate. By performing the spatial integration along the $z$direction and taking the time average with Δt, the linear gain becomes

$$
\begin{aligned}
G_p^{(1)} =& \xi \frac{\kappa\omega_p}{n_r^2} \left\{ \widehat{N} - N_g - b(\lambda_p - \lambda_o)^2 \right\} \\
& - \mathrm{Re}\left[\int_{-d/2}^{d/2} \int_{-W/2}^{W/2} \frac{(1+j\alpha) \frac{\kappa\omega_p}{n_{eff}^2} \frac{\varepsilon_o}{\hbar L} (\widehat{N} - N_g) |\widehat{E}_q|^2 |T(x,y)|^4}{-j\Omega + \frac{1}{\tau_s} + \frac{2\varepsilon_o \kappa}{\hbar L} \left(|\widehat{E}_p|^2 + |\widehat{E}_q|^2 \right) |T(x,y)|^2} dx\, dy \right].
\end{aligned}
\tag{8.104}
$$

The first term on the right-hand side is the already-derived conventional linear gain coefficient and the second term shows a reduction of the gain. We next define a linear gain coefficient A_p and a nonlinear gain coefficient $H_{p(q)}$ such that

$$
G_p^{(1)} = A_p - H_{p(q)} S_q,
\tag{8.105}
$$

where

$$\frac{\kappa\,\omega_p}{n_r^2} \equiv a, \tag{8.106}$$

$$A_p = \xi a\left\{\widehat{N} - N_g - b(\lambda_p - \lambda_o)^2\right\}. \tag{8.107}$$

To determine the form of the newly defined nonlinear gain coefficient, we need to perform some algebraic manipulations. The term with optical power $|\widehat{E}_p|^2 + |\widehat{E}_q|^2$ in the denominator of the second term can be expressed in terms of the photon number using (8.58) or (6.66), and is expressed in terms of the injection current at steady state using (8.95) or (6.72) as

$$\begin{aligned}\frac{2\varepsilon_o\kappa}{\hbar L}\left(|\widehat{E}_p|^2 + |\widehat{E}_q|^2\right)|T(x,y)|^2 &= \frac{a}{L}\left(S_p + S_q\right)|T(x,y)|^2\\ &= \frac{1}{\tau_s}\frac{I - I_{th}}{I_{th} - I_g}.\end{aligned} \tag{8.108}$$

This term is combined with the term $1/\tau_s$ as

$$\frac{1}{\tau_s} + \frac{2\varepsilon_o\kappa}{\hbar L}\left(|\widehat{E}_p|^2 + |\widehat{E}_q|^2\right)|T(x,y)|^2 = \frac{1}{\tau_s}\frac{I - I_g}{I_{th} - I_g}, \tag{8.109}$$

where I_g is the transparent injection current already introduced in (7.40). The numerator of the second term on the right-hand side of (8.104) is given in terms of the photon number S_q and a spatial integration. We assume similarly, as for the calculation of $G^{(3)}_{p(q)}$, that the field distribution in the width direction corresponds to the transverse fundamental mode and is well confined in the active region over width W. Moreover, in the thickness direction, the transverse fundamental mode spreads sufficiently into the cladding region from the active region of thickness d,

$$\int_{-d/2}^{d/2}\int_{-W/2}^{W/2} |T(x,y)|^4 dx\,dy \approx \frac{3\xi^2}{2dW}. \tag{8.110}$$

$H_{p(q)}$ then becomes [5]

$$H_{p(q)} = \frac{\dfrac{3\xi^2 a^2}{4V_a}\left(\widehat{N} - N_g\right)\left\{\dfrac{1}{\tau_s}\dfrac{I - I_g}{I_{th} - I_g} + \alpha(\omega_p - \omega_q)\right\}}{(\omega_p - \omega_q)^2 + \left(\dfrac{1}{\tau_s}\dfrac{I - I_g}{I_{th} - I_g}\right)^2}. \tag{8.111}$$

A calculated example of $H_{p(q)}$ for a GaAs laser is shown in Fig. 8.3. Although $H_{p(q)}$ is positive when the frequency difference between the two modes is small enough, $H_{p(q)}$ exhibits an asymmetric profile with respect to the frequency

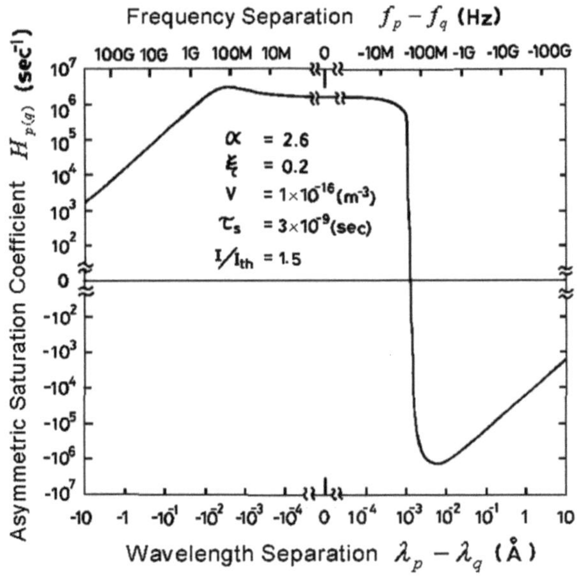

Fig. 8.3 Calculated example of asymmetric gain saturation coefficient

dispersion. Therefore, $H_{p(q)}$ is called the *asymmetric gain saturation coefficient*. The frequency difference between the two modes is converted to a wavelength difference using (8.65) and hence the asymmetric gain saturation is expressed as

$$H_{p(q)} \approx \frac{3\xi a G_{th} \lambda_p \alpha}{4 V_a \omega_p (\lambda_q - \lambda_p)}. \qquad (8.112)$$

A calculated example of the saturated gain and lasing modes are shown in Figs. 8.4 and 8.5, respectively. When the laser oscillates in mode q, the gain of the other mode is *suppressed if it has shorter wavelength* but is *increased if it has longer wavelength* owing to the asymmetric gain saturation effect. When the gain of the other mode increases above the threshold gain G_{th}, we can no longer establish a stable single-mode operation.

The asymmetric gain saturation is *proportional to the linewidth enhancement factor* α as indicated in (8.112). The linewidth enhancement factor tends to become larger in materials at longer wavelengths, such as InGaAsP, which is used in optical-fiber communications. We then need to use a *mode-selective laser cavity* such as a distributed feedback (DFB) or distributed Bragg reflector (DBR) structure *for the optical-fiber communications system*. Shorter wavelength lasers, such as those based on AlGaAs used in the optical disc system, show rather small linewidth enhancement factors enabling *stable single-mode operation*, even in the Fabry-Perot cavity, to be established.

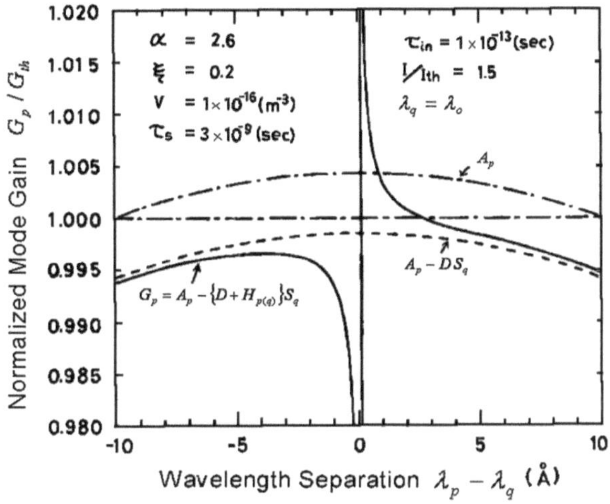

Fig. 8.4 Calculated example of the saturated gain

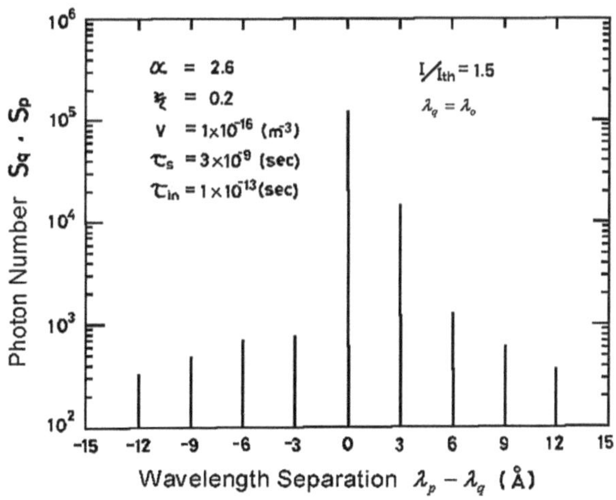

Fig. 8.5 Calculated example of the modal spectrum

8.8 Relation Among Nonlinear Gain Coefficients

A flow-chart of the perturbation expansions and derivations of the non-linear terms are summarized in Fig. 8.6. To begin, the *zeroth-order terms* of $\rho_{bb}^{(0)}$ in the conduction band and $\rho_{aa}^{(0)}$ in the valence band are formed from the *current injection* from the *external power source*. The *interaction* between the dipole (electron-hole pair) and the optical wave is introduced in the term *jRE*, where *R* is the *dipole moment* consisting of an electron-hole pair and *E* is the *electric field* component of the optical wave. The perturbation expansion is performed according to the interaction order with respect to jRE. The *first-order density matrix elements* $\rho_{ab}^{(1)}$ and $\rho_{ba}^{(1)}$ are constructed from the zeroth-order terms. The linear susceptibility

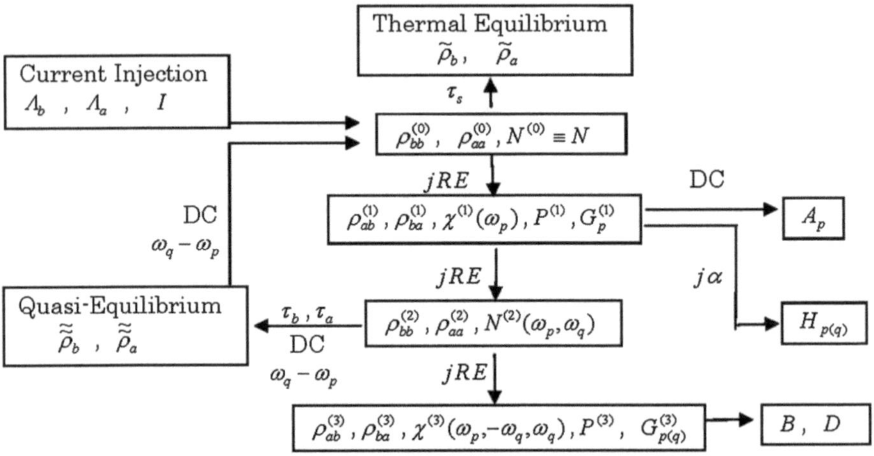

Fig. 8.6 Flowchart for perturbation expansions and derivations of the non-linear terms

$\chi^{(1)}(\omega_p)$, linear polarization $P^{(1)}$ and *linear gain coefficient* $G_p^{(1)}$ are formed from the first-order density matrix. As the interaction is *jRE*, the imaginary part of $\chi^{(1)}(\omega_p)$ determines the linear gain.

The electron transition up to first-order perturbation gives a reduction in electron number in the conduction band and an increase for the valence band. These variations are expressed by *second-order perturbation terms* of $\rho_{bb}^{(2)}$ and$\rho_{aa}^{(2)}$. With negative prefactor $(jRE)^2 = -R^2E^2$, the second-order terms contribute to a *reduced* electron injection density. We emphasize here that the second-order terms must include not only the identical frequency component but also the *mixed frequency components* such as $\exp[\pm j(\omega_p - \omega_q)t]$, because the electric field is applied twice in the form of $E^2 = \left\{\sum_p E_p \exp[j\omega_p t] + c.c.\right\}^2$. The two-fold interaction in the second-order terms directly affects the *third-order gain coefficients* associated with imaginary prefactor $(jRE)^3 = -jR^3E^3$ for B and D as the respective differences between the *self* and *mutual saturation* coefficients. The symmetric saturation coefficient D describes the beating vibration between the electron waves and the optical wave as given by $j(\omega_p - \omega_q + \omega_p - \omega_{ba}) + 1/\tau_{in}$ in (8.54). We recall that the coefficient $D_{p(q)}$ is characterized by the *intraband relaxation time* τ_{in}.

Meanwhile, the second-order terms associated with beating vibration reappear in the *quasi-equilibrium distribution functions* $\tilde{\tilde{\rho}}_b$ and $\tilde{\tilde{\rho}}_a$, which describe the dispersion of electrons after collisions and information loss of the vibration phase of each electron wave. The continual supply of electrons, however, maintains the *beating vibration*. The electron injection into the active region resulting from the applied voltage $eV = \mu_c - \mu_v$ depends on the injected density.value $\rho_{bb}^{(0)} - \rho_{aa}^{(0)}$. However, the injection impedance $\partial V/\partial I$ is very low in the actual operating state and is smaller than several tenths of Ω. Nevertheless, electrons and holes are applied by an external power source and pass through electrodes and cladding regions to reach the active region. The resistance or impedance for electron transportation is higher than several Ω. Therefore, the injected electrons are *driven by the current source I* and not with voltage source V.

As found from (8.74), *modulations* in the quasi-equilibrium functions $\tilde{\tilde{\rho}}_b$ and $\tilde{\tilde{\rho}}_a$ get relayed back to the zeroth-order terms $\rho_{bb}^{(0)}$ and $\rho_{aa}^{(0)}$, because the laser is driven with constant current I. The zeroth-order terms thus produce frequency beating by this *feedback-loop mechanism* via *the current injection*.

Subsequently, these zeroth-order terms reenter the first-order gain coefficient as *second-order terms*. Because the vibrating terms depend on phase $(1 + j\alpha)$ from other terms, the real part of the electron injection density generates a gain saturation term. *The asymmetric profile of the saturation coefficient* $H_{p(q)}$ comes from the variation of the refractive index based on the *real part of the linear susceptibility* $\mathrm{Re}\chi^{(1)}(\omega_p)$, as originally defined in (8.87).

8.9 Improved Rate Equations

As shown in the above sections, the semiconductor laser features many complicated nonlinear phenomena.

The electron density N consists of many terms as given in (8.94). The *vibrational terms* $N_\Omega \exp(j\Omega t)$ induced by the existing modes take account of *the asymmetric gain saturation effect*. The remaining term associated with the *slowly varying electron density* $\widehat{N}$ can *be regarded as the ordinary electron injection density* N separated from the asymmetric gain saturation effect.

The term C_p defined in (8.11) describes the contribution of spontaneous emission to the mode p. As discussed in Sect. 6.8 and given in (6.86), the term is written as $C_p = \xi a N$. In consequence, the following set of equations are listed as improved rate equations that include these nonlinear effects:

$$\frac{dS_p}{dt} = \left[A_p - BS_p - \sum_{q \neq p} D_{p(q)} S_q - G_{th}\right] S_p + \xi a N \tag{8.113}$$

$$A_p = \xi a \left\{N - N_g - b(\lambda_p - \lambda_o)^2\right\} \tag{8.114}$$

$$B = \frac{9\xi^2 \hbar \omega_p a |R_{ba}|^2}{4 V_a \varepsilon_o n_{eff}^2} \left(\frac{\tau_{in}}{\hbar}\right)^2 (N - N_s) \tag{8.115}$$

$$D_{p(q)} = D + H_{p(q)} \tag{8.116}$$

$$D = \frac{4}{3} \frac{B_p}{\{1 + \tau_{in}^2 (\omega_p - \omega_q)^2\}} \tag{8.117}$$

$$\begin{aligned} H_{p(q)} &= \frac{\frac{3\xi^2 a^2}{4V_a} \left(\widehat{N} - N_g\right) \left\{ \frac{1}{\tau_s} \frac{I - I_g}{I_{th} - I_g} + \alpha(\omega_p - \omega_q) \right\}}{(\omega_p - \omega_q)^2 + \left(\frac{1}{\tau_s} \frac{I - I_g}{I_{th} - I_g} \right)^2} \\ &\approx \frac{3\xi a G_{th} \lambda_p \alpha}{4 V_a \omega_p (\lambda_q - \lambda_p)} \end{aligned} \tag{8.118}$$

$$G_{th} = \frac{1}{\mathrm{v}} \left\{ \alpha_{loss} + \frac{1}{2L} \ln\left(\frac{1}{R_b R_f} \right) \right\} \tag{8.119}$$

$$\frac{dN}{dt} = -\sum_p \frac{A_p S_p}{V_a} - D_{diff} \nabla^2 N - \frac{N}{\tau_s} + \frac{I}{e V_a}, \tag{8.120}$$

where

$$V_a = W d L : \text{volume of the active region}, \tag{8.121}$$

$$\xi = \int_{-d/2}^{d/2} \int_{-W/2}^{W/2} |T(x, y)|^2 \, dx \, dy : \text{field confinement factor}. \tag{8.122}$$

Bibliography

1. M. Yamada, Transverse and longitudinal mode control in semiconductor injection lasers. IEEE J. Quantum Electron. QE-**19**(9), 1365–1380 (1983)
2. M. Ahmed, M. Yamada, An infinite order perturbation approach to gain calculation in injection semiconductor lasers. J. Appl. Phys. **84**(6), 3004–3015 (1998)
3. C.H. Henry, Theory of linewidth of semiconductor lasers. IEEE J. Quantum Electron. QE-**18**(9), 259–264 (1982)
4. N. Ogasawara, R. Ito, Longitudinal mode competition and asymmetric gain saturation in semiconductor injection lasers. II. Theory. Jpn. J. Appl. Phys. **27**(4), 615–626 (1988)
5. M. Yamada, Theoretical analysis of nonlinear optical phenomena taking into account the beating vibration of the electron density in semiconductor lasers. J. Appl. Phys. **66**(1), 81–89 (1989)

Chapter 9
Mode Competition

9.1 Two Modes Competition with Symmetric Gain Saturation

As the simplest case, we consider a laser operation with only two modes, p and q, both having positive-valued nonlinear gain coefficients, the values of which are as yet not fixed, and the contribution from spontaneous emission negligible. Under these conditions, the variation in photon numbers in each mode is given by

$$\frac{dS_p}{dt} = \left(A_p - BS_p - D_{p(q)}S_q - G_{th}\right)S_p, \tag{9.1}$$

$$\frac{dS_q}{dt} = \left(A_q - BS_q - D_{q(p)}S_p - G_{th}\right)S_q. \tag{9.2}$$

Let us examine the solutions of these simultaneous equations using a graphical analysis. Solutions of (9.1) are illustrated in Fig. 9.1a, where the steady-state condition $dS_p/dt = 0$ is given by the straight line $A_p - BS_p - D_{p(q)}S_q - G_{th} = 0$ drawn as line L_p. The region under this line corresponds to $dS_p/dt > 0$ and the motion of the operating point is indicated by the right-directed arrows. The region above this line corresponds to $dS_p/dt < 0$ and the motion is indicated by left-directed arrows. On the line L_p, horizontal motion does not exist. A similar analysis of (9.2) is illustrated in Fig. 9.1b, where the steady-state condition $dS_q/dt = 0$ produces the straight line $A_q - BS_q - D_{q(p)}S_q - G_{th} = 0$ drawn as line L_q. The direction of motion of the operating point is indicated by up and down arrows in the respective regions.

The solutions of these simultaneous equations (9.1) and (9.2) are given by superimposing Fig. 9.1a, b. They are classified into four types depending on the relative orientation of the lines, as shown in Fig. 9.2. The direction of motion between the lines depends on whether the line L_p is above or below L_q. In (a), L_p lies outside L_q with all motion directed towards point P, the point of intersection of L_p with the S_p axis. A *stable single-mode operation* with mode p is achieved with a

M. Yamada, *Theory of Semiconductor Lasers*,
Springer Series in Optical Sciences 185,
DOI: 10.1007/978-4-431-54889-8_9, © Springer Japan 2014

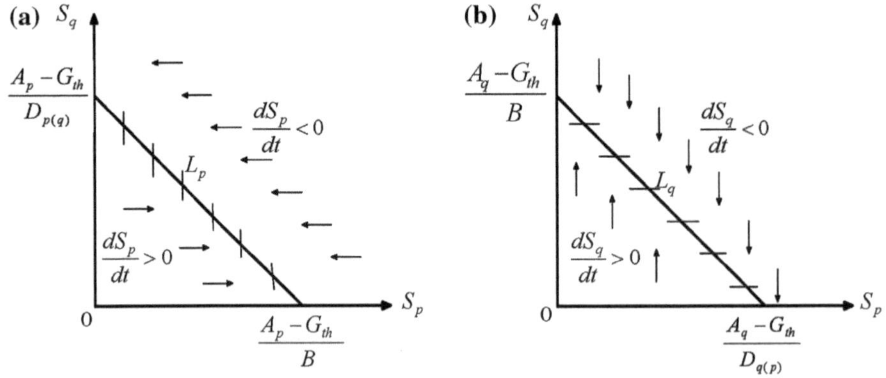

Fig. 9.1 Graphical solutions of the dynamic equations. **a** $\frac{dS_p}{dt} = \left(A_p - BS_p - D_{p(q)}S_q - G_{th}\right)S_p$. **b** $\frac{dS_q}{dt} = \left(A_q - BS_q - D_{q(p)}S_p - G_{th}\right)S_q$

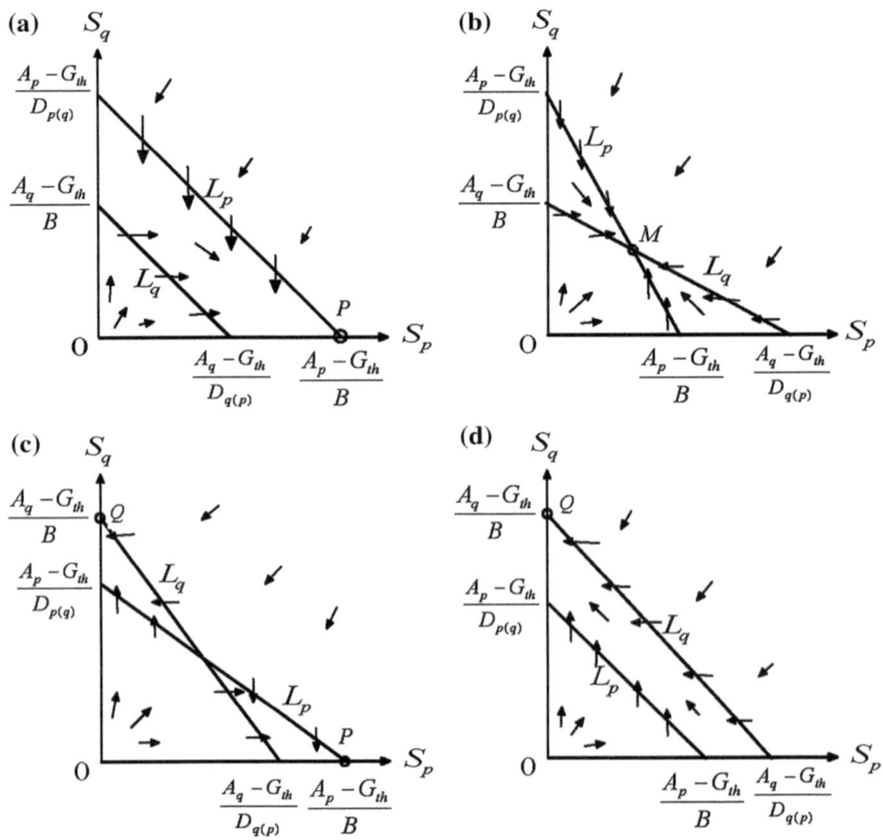

Fig. 9.2 Four types of solutions with two-mode competition. **a** p single. **b** p and q multi. **c** p or q bi-stable. **d** q single

photon number $S_p = (A_p - G_{th})/B$ as the steady state. In (b), L_p and L_q cross at point M, to which all motion is directed towards, thus producing a *stable multi-mode operation* with modes p and q. In (c), L_p and L_q also cross but the crossing point is a *saddle point* and all motion is directed away towards either point P on the S_p axis or point Q on the S_q axis. For this *bi-stable state*, the operation point at steady state can be either P or Q, which is determined by the *initial condition.* Finally, in (d), L_q lies outside L_p, and hence all motion is directed towards point Q, the intersection point between L_q and the S_q axis. A *stable single-mode operation* with mode q is achieved with photon number $S_q = (A_q - G_{th})/B$ as the steady state.

Thus, a single-mode operation with mode p is achieved in the set-up represented by Fig. 9.2a, c. The set-up requires that the intersection point P must lie outside the intersection point between L_q and the S_p axis, that is,

$$S_p = \frac{A_p - G_{th}}{B} > \frac{A_q - G_{th}}{D_{q(p)}}, \tag{9.3}$$

or equivalently

$$A_q - D_{q(p)}\, S_p < G_{th}, \tag{9.4}$$

which is commonly used in conventional analyses.

For a more precise analysis, we need to take into account the contribution from *spontaneous emission* which is added in (9.1) and (9.2) via terms C_p and C_q,

$$\frac{dS_p}{dt} = \left(A_p - BS_p - D_{p(q)}\, S_q - G_{th}\right)S_p + C_p, \tag{9.5}$$

$$\frac{dS_q}{dt} = \left(A_q - BS_q - D_{q(p)}S_p - G_{th}\right)S_q + C_q. \tag{9.6}$$

Their graphical solutions are sketched in Fig. 9.3. We have again indicated the steady-state solutions, $dS_p/dt = 0$ and $dS_q/dt = 0$, by lines L_p and L_q, respectively. When spontaneous emission is taken into account, L_p never crosses the S_q axis and asymptotes to $S_q \rightarrow \infty$ as $S_p \rightarrow 0$. Similarly, L_q never crosses the S_p axis and asymptotes to $S_p \rightarrow \infty$ as $S_q \rightarrow 0$.

Solutions of the simultaneous equations of (9.5) and (9.6) are again classified into four types, as shown in Fig. 9.4. The classification scheme is the same as for Fig. 9.2. However, the non-oscillating mode is not completely suppressed and produces a small output because of spontaneous emission. In (c), a *bi-stable state* still occurs if the contribution from *spontaneous emission is not large*.

Profiles of the linear gain, saturated gain, and the photon numbers are illustrated in Fig. 9.5 corresponding to Fig. 9.4. In (c), point P is chosen as the operating condition.

The classification of the four types (a)–(d) in Figs. 9.2 or 9.4 are determined not only by the non-linear gain coefficients B, $D_{p(q)}$, and $D_{q(p)}$, but also by the linear

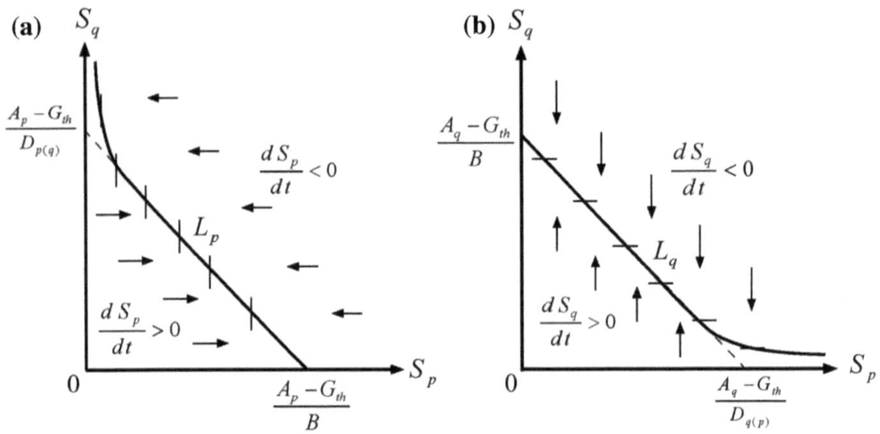

Fig. 9.3 Graphical solutions of the dynamic equations taking into account inclusion of the spontaneous emission. **a** $\frac{dS_p}{dt} = (A_p - BS_p - D_{p(q)}S_q - G_{th})S_p + C_p$. **b** $\frac{dS_q}{dt} = (A_q - BS_q - D_{q(p)}S_p - G_{th})S_q + C_q$

gain coefficients A_p and A_q. When the temperature of the laser is increased, the peak wavelength λ_o shifts to longer wavelengths because the band gap W_g diminishes with temperature. Because all of the injected electric power is not converted to optical power, the temperature of the laser rises as the injection current increases. The lasing mode then changes to another mode.

We suppose here that mode p is active in the single-mode operation given in Fig. 9.4a and changes to mode q in Fig. 9.4d through increases in injection current or temperature. There are two ways by which the change from (a) to (d) is performed. One involves steps (a)–(b)–(d) corresponding to "single-mode operation of p" to "multi-mode operation with p and q", and finally "single-mode operation of q". The other is (a)–(c)–(d) corresponding to "single-mode operation of p" to "single-mode operation of p" maintaining the same mode active, and finally "single-mode operation of q" where a jump in mode is made. However, when we reduce the injection current or the temperature from "single-mode operation of q" in (d) to "single-mode operation of p" in (a), the laser still exhibits "single-mode operation of q" in (c). Thus we can expect *hysteresis phenomena*[1, 2] associated with variations in lasing mode with changes in injection current or temperature. Figure 9.6 shows, together with a theoretical analysis, experimentally observed hysteresis phenomena related to the variation in lasing mode. These data are proof that the GaAs laser uses the bi-stable state, as in Fig. 9.4c, as the mutual saturation coefficient $G_{p(q)} \approx G_{q(p)}$is larger than the self-saturation coefficient B.

Some readers might have recognized that, as the curving in L_p and L_q becomes stronger, the situation given in Fig. 9.4c becomes similar to that in Fig. 9.4b when the contribution from spontaneous emission becomes larger than that for stimulated emission. Such a modification is expected. When injection current I is very

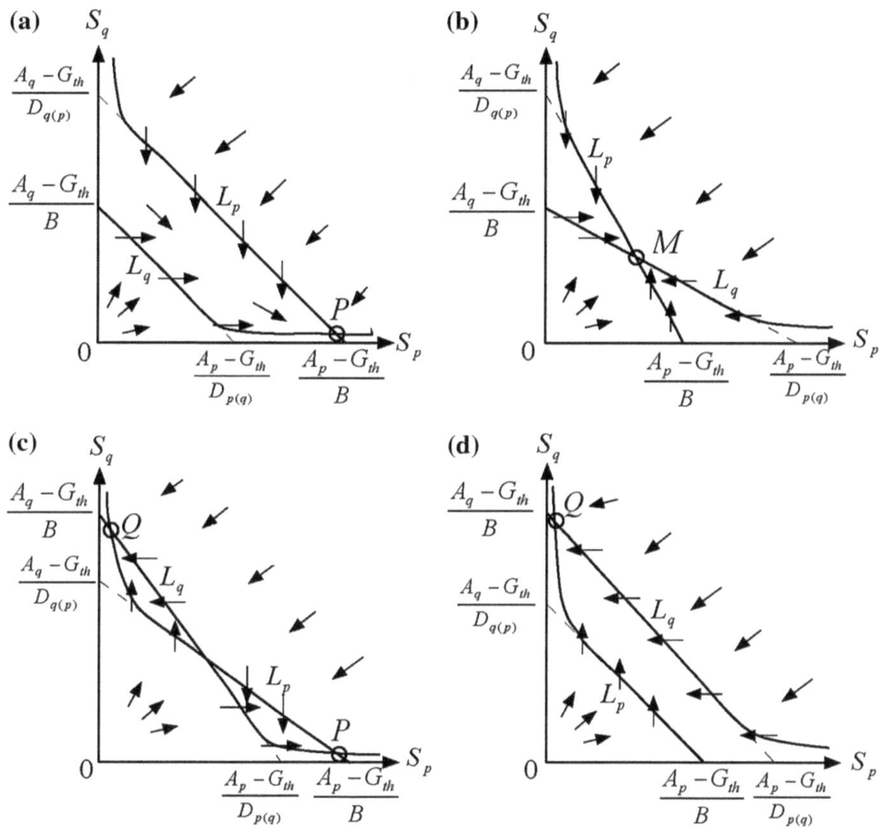

Fig. 9.4 Four types of solutions of two-mode competition taking into account spontaneous emission. **a** p single. **b** p and q multi. **c** p or q bi-stable. **d** single

near threshold current I_{th}, the laser shows *multi-mode operation* but not with the bi-stable state.

The presence of the bi-stable state is helpful in establishing single-mode operation, but is not always beneficial. *Noise from mode hopping* or *optical feedback* is generated aided by the *bi-stable state* as we shall discuss in the Chap. 10.

9.2 Mode Competition with Asymmetric Gain Saturation

The *asymmetric saturation coefficient* $H_{p(q)}$ changes sign depending on the difference in wavelength between the two modes p and q, as given in (8.118). We suppose here that mode q has a longer wavelength than mode p, $\lambda_q > \lambda_p$. If the

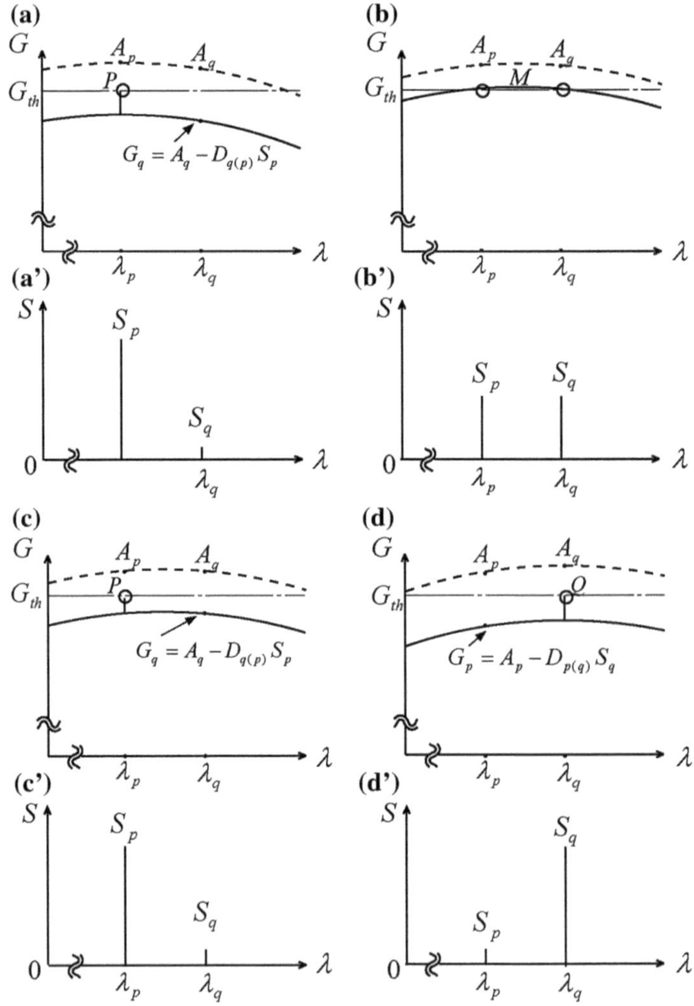

Fig. 9.5 Profiles of linear gain, saturated gain, and the photon numbers corresponding to Fig. 9.4. **a** and **a'** p single. **b** and **b'** p and q multi. **c** and **c'** p or q bi-stable. **d** and **d'** q single

separation between wavelengths is small or the α parameter is large enough, this results in

$$D_{p(q)} = D + H_{p(q)} > 0, \tag{9.7}$$

but also

$$D_{q(p)} = D + H_{q(p)} < 0. \tag{9.8}$$

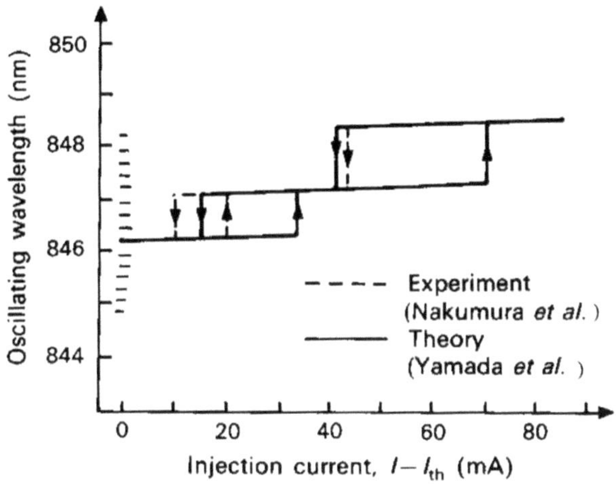

Fig. 9.6 Hysteresis phenomena in a GaAs laser associated with variation of the lasing mode with injection current

In this instance, the dynamic equations are

$$\frac{dS_p}{dt} = \left(A_p - BS_p - D_{p(q)}S_q - G_{th}\right)S_p + C_p, \tag{9.9}$$

$$\frac{dS_q}{dt} = \left(A_q - BS_q + |D_{q(p)}|S_p - G_{th}\right)S_q + C_q. \tag{9.10}$$

The graphical solution of (9.9) is the same as that in Fig. 9.3a. The solutions of (9.10) is sketched in Fig. 9.7, where in (a) the linear gain A_q is lower than the threshold gain level G_{th} and in (b) the linear gain A_q is higher than the threshold gain level G_{th}.

By superimposing these diagrams with Fig. 9.3a, we get four types of solutions, shown in Fig. 9.8, for the simultaneous equations (9.9) and (9.10) for two-mode competition. Figure 9.8a corresponds to $A_q < G_{th}$ and $(A_p - G_{th})/B > (G_{th} - A_q)/|D_{q(p)}|$, (b) corresponds to $A_q < G_{th}$ and $(A_p - G_{th})/B < (G_{th} - A_q)/|D_{q(p)}|$,(c) corresponds to $A_q > G_{th}$ and $(A_q - G_{th})/B < (A_p - G_{th})/D_{p(q)}$, and (d) corresponds to $A_q > G_{th}$ and $(A_q - G_{th})/B > (A_p - G_{th})/D_{p(q)}$.

Corresponding to these cases, profiles of the linear gain, saturated gain, and the photon numbers are illustrated in Fig. 9.9. In Fig. 9.8a, c, lines L_p and L_q cross in straight segments and the motion *rotates around crossing point M*, but barely converges on *M*. The photon numbers S_p and S_q should now show *temporal variations* or *fluctuations*, as illustrated in Fig. 9.9a', c'. Although (a) and (c) look like multi-mode operations, both are *unstable*. The phrase "stable multi-mode" should only be used when several modes are present with all producing constant optical power simultaneously.

Figure 9.8b, d depict instances in which the motion converges to points *P* and *Q*, respectively, and seem like stable single-mode operations. However, we must take care because we are analyzing here only two-mode operations. If we extend

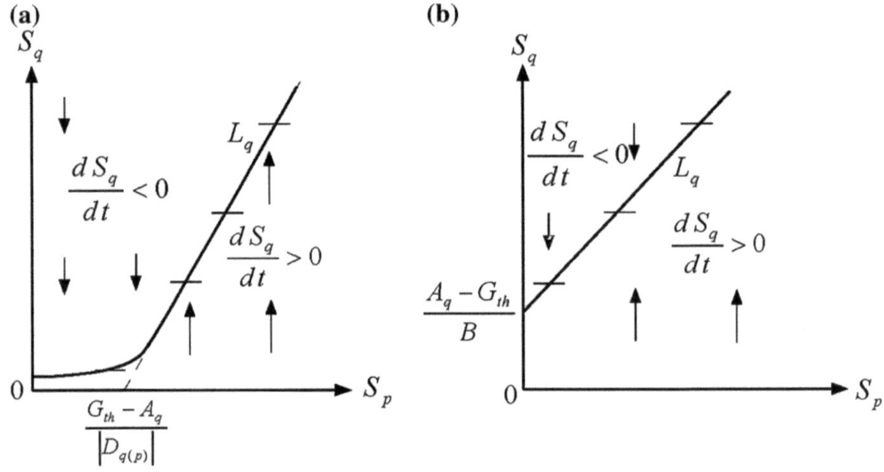

Fig. 9.7 Graphical solutions of the dynamic equation (9.10) for a mode of longer wavelength when the asymmetric gain saturation is dominant. **a** $A_q < G_{th}$. **b** $A_q > G_{th}$

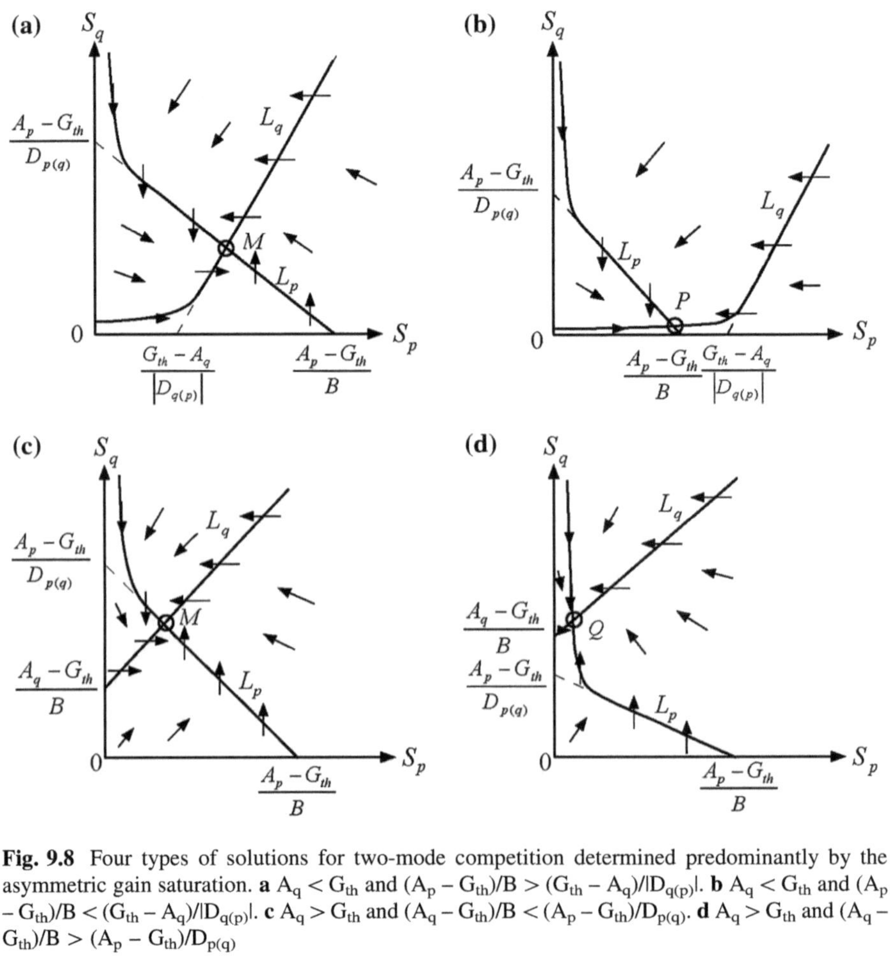

Fig. 9.8 Four types of solutions for two-mode competition determined predominantly by the asymmetric gain saturation. **a** $A_q < G_{th}$ and $(A_p - G_{th})/B > (G_{th} - A_q)/|D_{q(p)}|$. **b** $A_q < G_{th}$ and $(A_p - G_{th})/B < (G_{th} - A_q)/|D_{q(p)}|$. **c** $A_q > G_{th}$ and $(A_q - G_{th})/B < (A_p - G_{th})/D_{p(q)}$. **d** $A_q > G_{th}$ and $(A_q - G_{th})/B > (A_p - G_{th})/D_{p(q)}$

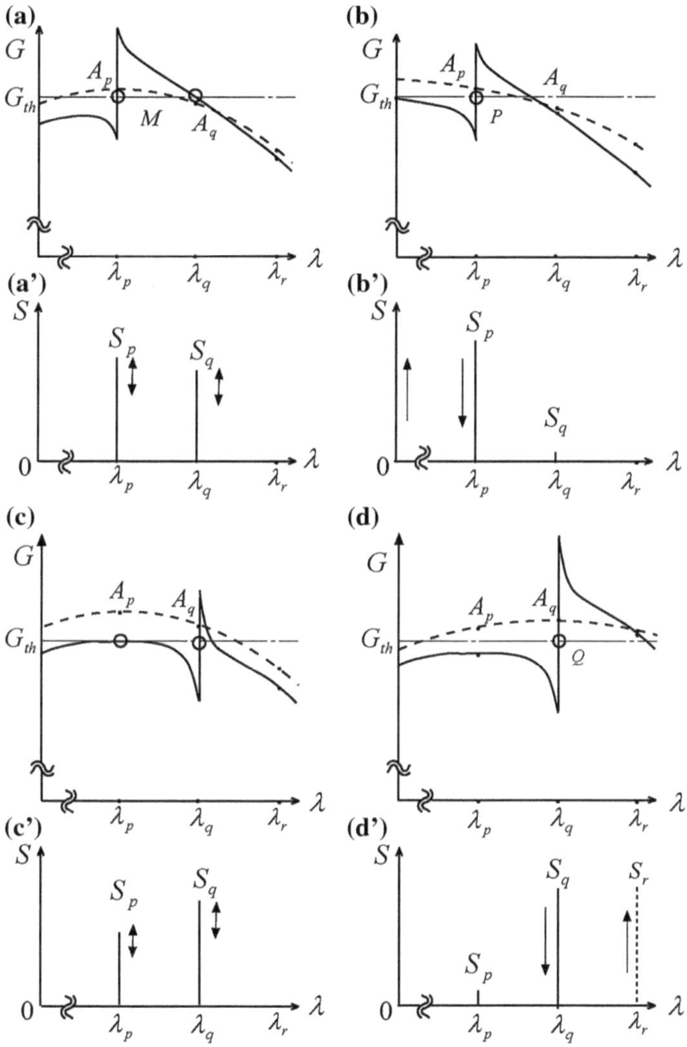

Fig. 9.9 Profiles of linear gain, saturated gain and photon numbers corresponding to Fig. 9.8. Classification of **a–d** is same as in Fig. 9.8

the gain profile to include other modes, as shown in Fig. 9.9b, the gain of the shorter wavelength mode can attain threshold gain level, hence the lasing mode can jump to the shorter wavelength mode.

A similar effect is expected in Fig. 9.9d for the longer wavelength mode *r*. As explained with regard to Fig. 9.9, the two-mode analysis of the lasing mode is not enough to determine the entire effect of asymmetric gain saturation. If the asymmetric gain saturation effect is strong, the lasing mode *jumps to a neighboring mode, then subsequently returns to the original mode* as observed for InGaAsP lasers. The mechanism for this *cyclic rotation* is illustrated in Fig. 9.10;

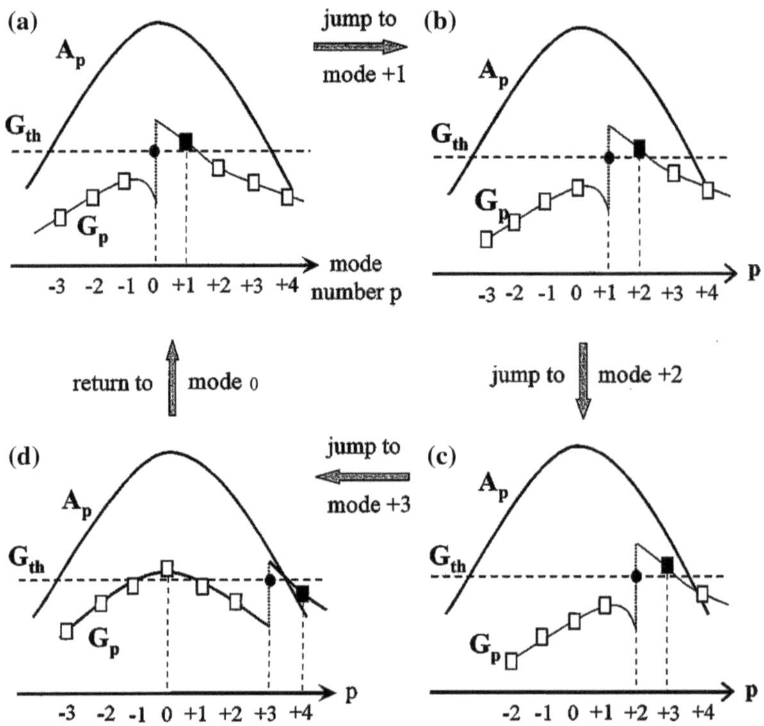

Fig. 9.10 Illustration of the mechanism for cyclic rotation of the lasing mode

there, A_p is the linear gain, G_p the saturated gain, and G_{th} the threshold gain level. Mode numbers are defined such that 0 is at the peak position of A_p, longer wavelengths correspond to positive numbers along the abscissa and shorter wavelengths correspond to negative numbers. Initially, laser oscillations start with mode 0, as shown in (a) because of its higher linear gain. However, when the saturated gain of mode +1 becomes higher than G_{th}, then, the oscillation mode jumps to mode +1, as shown in (b). When mode +1 is in oscillation, the gain of mode +2 rises above G_{th}, and the oscillation mode jumps to mode +2, as in (c). In a similar way, the oscillation mode jumps to mode +3 in (d). However, the gain of mode +4 is not high enough because the linear gain A_p for the mode +4 is very low. Meanwhile, the saturated gain of mode 0 recovers and exceeds G_{th}, because the linear gain of mode 0 is highest and the saturation becomes weaker owing to a sufficient wavelength separation $\lambda_{+3}-\lambda_0$. The laser oscillation returns to mode 0 and *repeats the rotation cycle* (a)→(b)→(c)→(d)→(a).

This cyclic rotation phenomenon of lasing modes, as seen in Fig. 9.11a, was experimentally observed for a $\lambda = 1.3$ μm InGaAsP laser. A simulation result is shown in Fig. 9.11b, where fluctuations caused by Langevin noise sources are included in the calculation to obtain a better agreement with experiment.

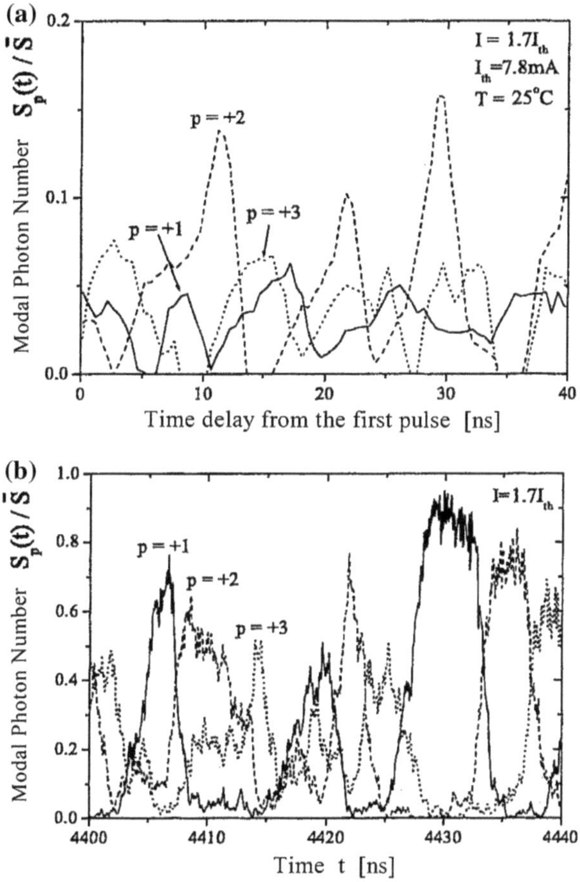

Fig. 9.11 Experimental observation and theoretical simulation of the cyclic rotation phenomena of lasing mode in a $\lambda = 1.3$ μm InGaAsP laser. **a** Experimental data. **b** Theoretical calculation

As shown in these figures, the lasing mode is *switching continually*. However, the time-averaged spectrum *looks much like a multi-mode operation*, as shown in Fig. 9.12 [3].

We have to distinguish between the *stable multi-mode operation* and the *unstable multi-mode operation* in which switching or jumping of the lasing mode is repeated [3].

9.3 Single Mode Operation by Help of Mode Selective Cavity

A laser having a *strong asymmetric gain saturation effect* is difficult to operate with a stable single longitudinal mode using a conventional Fabry-Perot cavity, which consists of simple two facet mirrors. For these lasers, we need to use the

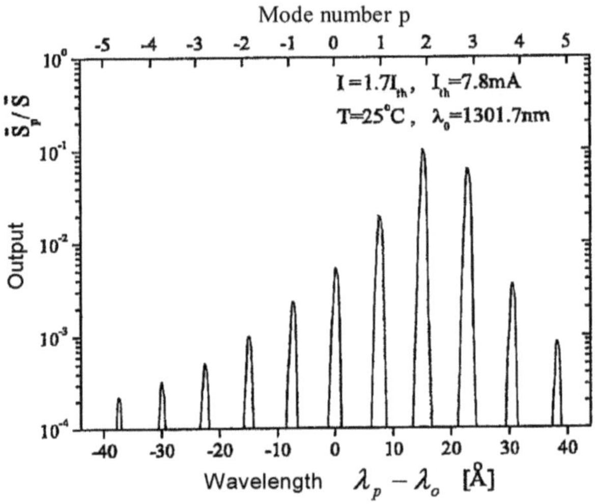

Fig. 9.12 Experimentally measured time-averaged spectrum in a laser operating with cyclic rotation of the lasing mode

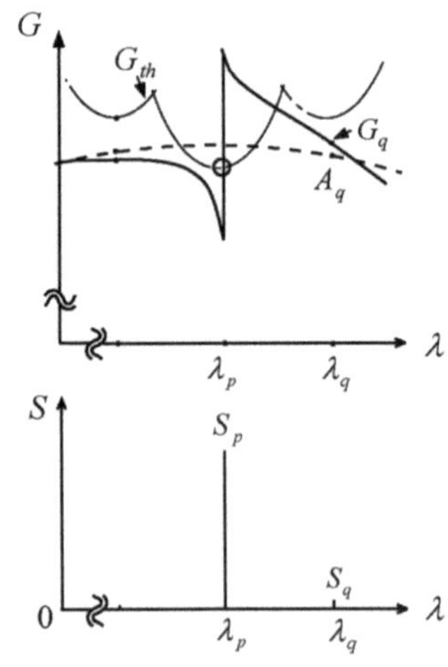

Fig. 9.13 Single-mode operation in a mode-selective cavity

mode-selective cavity structure such as the *distributed feedback (DFB)* structure or the *distributed Bragg reflector (DBR)*, where the threshold gain level G_{th} differs for each longitudinal mode. The laser oscillates with a mode having the lowest threshold gain level, as illustrated in Fig. 9.13. The operating state corresponds to Fig. 9.8b because gains of both the longer and shorter wavelength-side modes are lower than the threshold levels at those wavelengths. Several characteristics of the DFB and the DBR lasers will be given in Chap. 12.

Bibliography

1. M. Nakamura, K. Aiki, N. Chinone, R. Ito, J. Umeda, Longitudinal mode behaviors of mode-stabilized AlxGa1-xAs Injection lasers. J. Appl. Phys.**49**(9), 4644–4688 (1978)
2. M.Yamada, Y. Suematsu, Condition of single longitudinal mode operation in injection lasers with index-guiding structure. IEEE J. Quantum Electron. QE-**15**(8), 743–749 (1979)
3. M.Yamada, W. Ishimori, H. Sakaguchi, A. Moustafa, Time-dependent measurement of the mode-competition phenomena among longitudinal modes in long-wavelength lasers. IEEE J. Quantum Electron. **39**(12), 1548–1554 (2003)

Chapter 10
Noise

10.1 Classification of Noise

The term *noise* is defined with reference to the *signal*. Any phenomenon disturbing the signal is called noise. Therefore, the classification of noise, in a semiconductor laser, as suggested in Fig. 10.1, should be based on the light output as a carrier of transmitted information. One is *intensity noise* or *intensity-modulation (IM) noise*, where the intensity of the optical wave is used as the signal. Another is *frequency noise* or *frequency-modulation (FM) noise*, where the frequency of the optical wave is used to carry the signal. Frequency noise is directly related to *phase noise* as frequency is essentially the temporal derivative of the phase.

Another classification is based on the mechanism generating a disturbance in the optical wave. *Thermal noise* and *1/f noise* are major sources of noise in conventional electronics and are caused by electron transport phenomena. Semiconductor lasers also suffer from such noise, because the laser is driven by an electronic circuit. However, in a semiconductor laser, optical waves generate much higher levels of optical noise than electronic noise. In general, noise in semiconductor lasers are classified as in Fig. 10.2 according to the mechanism of generation.

As explained previously in Sects. 1.1 and 6.8 and Appendix 5, the transition of one electron from the conduction band to the valence band induces the emission of a single photon. This transition takes a finite interval of time. Both the photon number in the cavity and the electron number in the conduction band should be counted as discrete numbers but vary in time and hence are always accompanied by statistical fluctuations. This type of *intrinsic fluctuation* is called *quantum noise* or *shot noise*.

When the semiconductor laser is actually used as a light source connected with other optical devices, such as an optical lens, optical fiber or optical detector, the laser tends to reveal *extra noise*. This extra noise is called *optical feedback noise* because the extra noise is generated when the output light from the laser is reflected back from the optical device and re-injected into the laser cavity. There are two categories of mechanisms of optical feedback noise generation. The first is

M. Yamada, *Theory of Semiconductor Lasers*, 157
Springer Series in Optical Sciences 185,
DOI: 10.1007/978-4-431-54889-8_10, © Springer Japan 2014

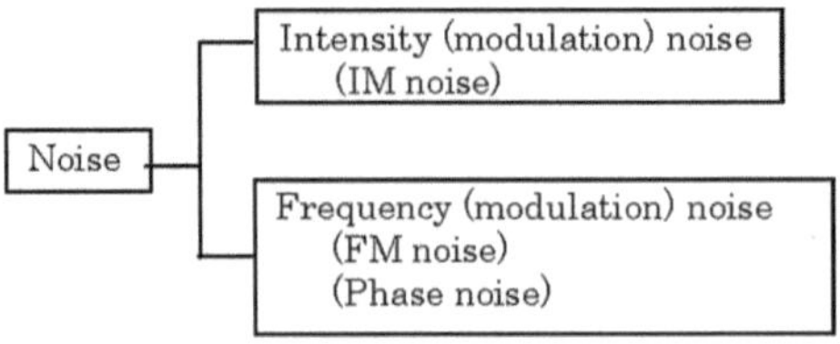

Fig. 10.1 Classification based on application of the optical wave

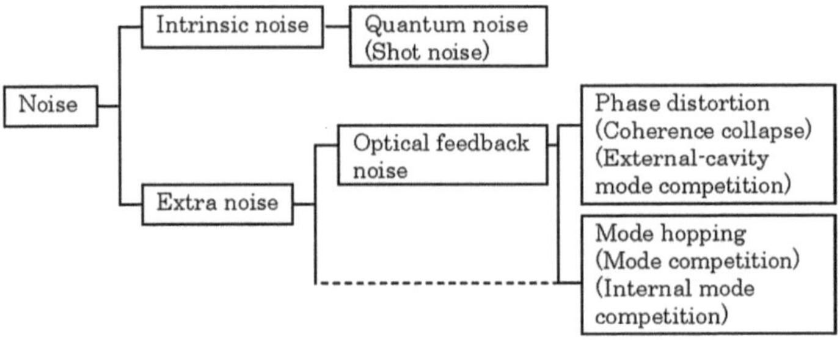

Fig. 10.2 Classification based on the mechanism generating the physical disturbance

due to phase distortion of the optical light called *coherence collapse*. The phase of lasing light is always fluctuating. Thus the phase of the re-injected light does not coincide with that of the oscillating light in the laser cavity. This difference in fluctuation of the optical phase enhances the fluctuation of the *optical intensity* through the relation between the real and imaginary parts of the *laser polarization*. This noise generating mechanism is alternatively explained in terms of *mode competition* phenomena associated with the *external cavity modes* formed in the space between the front facet and the reflecting point.

The second mechanism is due to *mode competition* or *mode hopping phenomena* of the *lasing mode*. Other names used here are the *internal mode* or *mode in a solitary laser* to distinguish them from the external-cavity mode. The re-injected light merges with the internally reflected light at the front facet. Because the re-injected light has a different phase from the internally reflected light, the equivalent reflection rate at the front facet has to have temporal variations, resulting in temporal variations in the threshold gain level G_{th} for each lasing mode. These variations subsequently cause *instantaneous shifts* in the lines L_p and L_q in the bi-stable operation given in Fig. 9.4c, resulting in *mode hopping* between the operating point P and Q.

As the total photon number $S_p + S_q$ is not constant during the mode hopping, the noise level of the laser output, called *mode hopping noise*, is large. This mode hopping noise can be induced *without optical feedback* in some lasers featuring *inferior transverse-mode control*.

We must confirm here that the *source of the extra noise is the intrinsic quantum noise*. When the intrinsic quantum fluctuation associated with an electron transition is amplified by the lasing mechanism, such as mode-competition phenomena, extra noise is generated.

10.2 Introduction of the Noise Sources

10.2.1 Noise Sources for Photon Number and Electric Phase

The origin of noise is the *fluctuation in the optical emission.* Let us take into account this fluctuation using Maxwell's classical wave equation. The optical polarization consists of the laser polarization P_{st} and spontaneous polarization P_{sp} as introduced in (8.1),

$$P = P_{st} + P_{sp}. \tag{10.1}$$

The electric field is represented as a sum of electric fields associated with each lasing mode p, as given in (8.3),

$$E = \sum_p \widehat{E}_p(t)\Phi_p(\boldsymbol{r})e^{j\omega_p t} + c.c. \tag{10.2}$$

By substituting (10.2) into the wave equation (8.2), the variation of the field amplitude is represented as

$$\frac{\partial \widehat{E}_p(t)}{\partial t} = -\frac{\sigma_{eff}}{2\varepsilon_{eff}}\widehat{E}_p(t) - \frac{j\omega_p}{2\varepsilon_{eff}\Delta t}\int_{t-\Delta t}^{t}\int_{act}(P_{st} + P_{sp})\,\Phi_p^*(\boldsymbol{r})e^{-j\omega_p t}d^3\boldsymbol{r}\,dt, \tag{10.3}$$

following the treatment given in Chap. 8. The laser polarization is expressed in terms of the laser susceptibility$\chi(\omega_p)$, in which both the linear and non-linear effects are included,

$$P_{st} = \varepsilon_o \sum_p \chi(\omega_p)\widehat{E}_p(t)\Phi_p(\boldsymbol{r})e^{j\omega_p t} + c.c. \tag{10.4}$$

Then, (10.3) is rewritten as

$$\frac{\partial \widehat{E}_p(t)}{\partial t} = -\frac{\sigma_{eff}}{2\varepsilon_{eff}}\widehat{E}_p(t) - \frac{j\omega_p}{2n_{eff}^2}\int_{act}\chi(\omega_p)\left|\Phi_p(\boldsymbol{r})\right|^2 d^3\boldsymbol{r}\widehat{E}_p(t) + U_p(t), \tag{10.5}$$

where $U_p(t)$, representing *contributions from spontaneously emitted light,* is defined as

$$U_p(t) = -\frac{j\omega_p}{2\varepsilon_{eff}\Delta t}\int_{t-\Delta t}^{t}\int_{act} P_{sp}\,\Phi_p^*(\boldsymbol{r})e^{-j\omega_p t}d^3\boldsymbol{r}\,dt. \tag{10.6}$$

We now factor the electric field amplitude as an absolute value and phase

$$\widehat{E}_p(t) = \left|\widehat{E}_p(t)\right| e^{j\theta_p(t)}, \tag{10.7}$$

and substitute it into (10.5) to obtain:

$$\begin{aligned}\frac{\partial \widehat{E}_p(t)}{\partial t} &= \frac{\partial|\widehat{E}_p(t)|}{\partial t} e^{j\theta_p(t)} + j\frac{\partial \theta_p(t)}{\partial t}|\widehat{E}_p(t)|e^{j\theta_p(t)} \\ &= -\frac{\sigma_{eff}}{2\varepsilon_{eff}}|\widehat{E}_p(t)|e^{j\theta_p(t)} - \frac{j\omega_p}{2n_{eff}^2}\int_{act}\chi(\omega_p)\left|\Phi_p(\boldsymbol{r})\right|^2 d\boldsymbol{r}|\widehat{E}_p(t)|e^{j\theta_p(t)} \\ &\quad + U_p(t).\end{aligned} \tag{10.8}$$

By multiplying by $e^{-j\theta_p(t)}$, we separate the real and imaginary parts, which correspond to the variation for the absolute value and the phase, respectively, of the amplitude:

$$\begin{aligned}\frac{\partial|\widehat{E}_p(t)|}{\partial t} &= -\frac{\sigma_{eff}}{2\varepsilon_{eff}}|\widehat{E}_p(t)| + \frac{\omega_p}{2n_{eff}^2}\mathrm{Im}\left\{\int_{act}\chi(\omega_p)\left|\Phi_p(\boldsymbol{r})\right|^2 d^3\boldsymbol{r}|\widehat{E}_p(t)|\right\} \\ &\quad + \mathrm{Re}\left\{U_p(t)e^{-j\theta_p(t)}\right\},\end{aligned} \tag{10.9}$$

$$\frac{\partial \theta_p(t)}{\partial t} = -\frac{\omega_p}{2n_{eff}^2}\mathrm{Re}\left\{\int_{act}\chi(\omega_p)\left|\Phi_p(\boldsymbol{r})\right|^2 d^3\boldsymbol{r}\right\} + \frac{1}{|\widehat{E}_p(t)|}\mathrm{Im}\left\{U_p(t)^{-j\theta_p(t)}\right\}. \tag{10.10}$$

Equation (10.9) is rewritten as a variation in the square of the amplitude as

$$\begin{aligned}\frac{d|\widehat{E}_p(t)|^2}{dt} &= 2|\widehat{E}_p(t)|\frac{d|\widehat{E}_p(t)|}{dt} \\ &= -\frac{\sigma_{eff}}{\varepsilon_{eff}}|\widehat{E}_p(t)|^2 + \frac{\omega_p}{n_{eff}^2}\mathrm{Im}\left\{\int_{act}\chi(\omega_p)\left|\Phi_p(\boldsymbol{r})\right|^2 d^3\boldsymbol{r}|\widehat{E}_p(t)|^2\right\} \\ &\quad + 2\left|\widehat{E}_p(t)\right|\mathrm{Re}\left\{U_p(t)e^{-j\theta_p(t)}\right\}.\end{aligned} \tag{10.11}$$

Inclusion of spontaneous emission is given by the last term for this relation, but the derivation using only a classical treatment of the detailed properties of spontaneous emission is difficult. One way is to introduce a suitable modification based on a *quantum mechanical treatment*. We rewrite the relation between the square of the electric field amplitude and photon number in the form

$$\frac{2\varepsilon_{eff}}{\hbar\omega_p}\left|\widehat{E}_p(t)\right|^2 = \begin{cases} S_p + 1 & \text{for optical emission} \\ S_p & \text{for optical absorption}\end{cases}. \tag{10.12}$$

Then, (10.11) is rewritten as a rate equation by adding a *source of fluctuation* $F_p(t)$ to (8.113) as

$$\frac{dS_p}{dt} = \left[A_p - BS_p - \sum_{q \neq p} D_{p(q)} S_q - G_{th}\right] S_p + \xi a N + F_p(t), \tag{10.13}$$

where

$$F_p(t) = \frac{4\varepsilon_{eff}}{\hbar \omega_p} \left|\widehat{E}_p(t)\right| \mathrm{Re}\left\{U_p(t) e^{-j\theta_p(t)}\right\}. \tag{10.14}$$

For the variation of the phase θ_p in (10.10), we include only the first-order polarization terms and contributions from spontaneous emission, by neglecting effects from higher-order polarization terms, to simply the calculation. With (8.90) and (8.106), the first term on the right-hand side of (10.10) is written as

$$-\frac{\omega_p}{2n_{eff}^2} \mathrm{Re}\left\{\int_{cav} \chi(\omega_p) \left|\Phi_p(\boldsymbol{r})\right|^2 d^3\boldsymbol{r}\right\} = \frac{\alpha \xi a}{2}(N - \bar{N}), \tag{10.15}$$

where $\bar{N}$ is the *time-averaged value* of the injected electron density. As a determination of the phase change is required to evaluate the phase noise or the frequency noise, the determination is performed for variations around the averaged value $\bar{N}$. Then, (10.10) is rewritten as

$$\frac{d\theta_p}{dt} = \frac{\alpha \xi a}{2}(N - \bar{N}) + T_p(t), \tag{10.16}$$

where

$$T_p(t) = \frac{1}{\left|\widehat{E}_p(t)\right|} \mathrm{Im}\left\{U_p(t)^{-j\theta_p(t)}\right\}. \tag{10.17}$$

Here, we determine the properties of $F_p(t)$ and $T_p(t)$. The time-averaged values of $F_p(t)$ and $T_p(t)$ must be zero; that is,

$$\overline{F_p(t)} = 0, \tag{10.18}$$

$$\overline{T_p(t)} = 0. \tag{10.19}$$

The *auto-correlation* or *mutual correlation* between *noise sources* of the photon numbers for modes p and q are

$$\left\langle F_p(t)F_q(t')\right\rangle = \left(\frac{4\varepsilon_{eff}}{\hbar\omega_p}\right)^2 \left|\widehat{E}_p(t)\right|^2 \left\langle \mathrm{Re}\left\{U_p(t)e^{-j\theta_p(t)}\right\}\mathrm{Re}\left\{U_q(t')e^{-j\theta_q(t')}\right\}\right\rangle. \tag{10.20}$$

Next, we establish a quantum mechanical criterion. The origin of the fluctuations is based on the fact that both the photon and electron numbers depend on *discrete values* and the presence of the zero-point energy. When electrons make transitions, photon numbers increase or decrease. During the time interval of a transition, quantum numbers associated with both the electron and photon should show *an uncertainty* of one quantum. Then, the total fluctuation depends on the number of transitions, because each transition gives an *independent contribution to the fluctuation*. As each transition gives either zero or one as an expected quantum number, the statistical distribution of the total fluctuation follows the *Poisson distribution*. The amount of *dispersion* for a Poisson distribution is the same as the *mean value* of the distribution. Then, the *correlated value* for optical emission is the same as the *total transition probability at steady state* given in (10.13). Hence we have

$$\left\langle F_p(t)F_q(t')\right\rangle = \left[\left\{\xi a(N+N_g)+G_{th}\right\}S_p+\xi aN\right]\delta_{p,q}\delta(t-t'). \tag{10.21}$$

Here, the linear gain coefficient follows from (8.114). We find that the optical absorptions from terms aN_g and G_{th} are non-negative. This is because the generating phases for the fluctuations from emission and absorption are independent and both mechanisms *increase these fluctuations*. The modes p and q should have a slight correlation through the nonlinear gain coefficient $\mathrm{D}_{p(q)}$, but is ignored because the nonlinear gain is much smaller than the other terms. Note that $\delta_{p,q}$ is the *Krönecker delta*, signifying that modes p and q are orthogonal as noise sources, and $\delta(t-t\prime)$ is the *Dirac delta-function*.

The auto-correlation and mutual correlation between the *noise sources for the phase variation* of modes p and q are

$$\left\langle T_p(t)T_q(t')\right\rangle = \frac{1}{\left|\widehat{E}_p(t)\right|^2}\left\langle \mathrm{Im}\left\{U_p(t)^{-j\theta_p(t)}\right\}\mathrm{Im}\left\{U_q(t')^{-j\theta_q(t')}\right\}\right\rangle. \tag{10.22}$$

As (10.16) is not an equation indicating the change of the photon number, we *cannot evaluate* (10.22) from (10.16) directly. However, we can connect this equation to (10.21) in the following way. The correlation between the photon number fluctuation and phase fluctuation should be *orthogonal*, hence

$$\left\langle \mathrm{Re}\left\{U_p(t)e^{-j\theta_p(t)}\right\}\mathrm{Im}\left\{U_p(t)e^{-j\theta_p(t)}\right\}\right\rangle = 0. \tag{10.23}$$

However, the *auto-correlations* of the real and imaginary parts must be *identical*, because the fluctuation is caused by random vibrations,

$$\left\langle \mathrm{Re}^2\left\{U_p(t)e^{-j\theta_p(t)}\right\}\right\rangle = \left\langle \mathrm{Im}^2\left\{U_p(t)e^{-j\theta_p(t)}\right\}\right\rangle. \tag{10.24}$$

Using this relation and (10.12), we obtain a relation for the noise sources between the photon number and phase, (10.20) and (10.22), as

$$\begin{aligned}\left\langle T_p(t)T_q(t')\right\rangle &= \left(\frac{\hbar\omega_p}{4\varepsilon_{eff}}\right)^2 \frac{1}{\left|\widehat{E}_p(t)\right|^4}\left\langle F_p(t)F_q(t')\right\rangle \\ &= \frac{1}{4\left(S_p+1\right)^2}\left\langle F_p(t)F_q(t')\right\rangle.\end{aligned} \tag{10.25}$$

The cross-correlation of the fluctuation sources between photon number and phase is obviously zero as given by (10.23):

$$\left\langle F_p(t)T_q(t')\right\rangle = 0. \tag{10.26}$$

The noise sources $F_p(t)$ and $T_p(t)$ after the introduction of quantum mechanical properties are of *Langevin type*.

10.2.2 Noise Source for the Electron Density

The variation of the electron density N has been derived as a variation in the diagonal elements of the density matrix, as discussed in Chaps. 6 and 8, and is formulated in a rate equation given as (8.120). It seems sufficient then to add a fluctuation term to (8.120). However, here we need to include the electron number $\int_{act} N d^3\mathbf{r} \approx NV_a$ as a quantum number given by an integer; unfortunately, $N\,[\mathrm{m}^{-3}]$ is not a quantum number. If the electron distribution in the active region is spatially uniform, the variation of the electron number is given by modifying (8.120) and by adding a *Langevin noise source* $W(t)$; that is, we have

$$\frac{d(NV_a)}{dt} = -\sum_p A_p S_p - \frac{NV_a}{\tau_s} + \frac{I}{e} + W(t). \tag{10.27}$$

Evaluating the noise source is given by summing up the probabilities for electron transitions,

$$\langle W(t)W(t')\rangle = \left\{\xi a \sum_p (N+N_g)S_p + \frac{NV_a}{\tau_s} + \frac{I}{e}\right\}\delta(t-t'). \tag{10.28}$$

If the distribution of the electron density is not uniform in the active region, the noise source $W(t)$ adds further terms arising from spatial diffusion effects.

Given a single photon emission, the variations in the photon number and electron number should exhibit a cross-correlation. This type of cross-correlation is picked up by counting common terms in (10.13) and (10.27), where we need to select those spontaneous emissions NV_a/τ_s caused by the radiation to all possible modes for both the guided modes in the laser cavity and radiation modes outside it, as was discussed in Sect. 6.8. Then, only a fraction $CNV_a/\tau_s = \xi aN$ corresponds to spontaneous emission for mode p. Therefore, the *cross-correlation* is given by

$$\langle F_p(t)W(t')\rangle = -\xi a\{(N+N_g)S_p + N\}\delta(t-t'). \tag{10.29}$$

The fluctuating source $T_p(t)$ of the optical phase θ_p introduced in (10.16) does not have any specified relationship with changes in the electron number given in (10.27). We then set

$$\langle T_p(t)W(t')\rangle = 0. \tag{10.30}$$

10.3 Frequency Components of Noise Sources

The temporal variation of any physical quantity can be expanded in terms of its frequency components by the *Fourier transformation* as

$$F_p(t) = \int_{-\infty}^{\infty} F_{p\Omega} e^{j\Omega t} d\Omega, \tag{10.31}$$

$$T_p(t) = \int_{-\infty}^{\infty} T_{p\Omega} e^{j\Omega t} d\Omega, \tag{10.32}$$

$$W(t) = \int_{-\infty}^{\infty} W_{\Omega} e^{j\Omega t} d\Omega. \tag{10.33}$$

The correlated values of noise sources in the frequency domain are given as Fourier transforms of the temporally correlated values as demonstrated in Appendix 11:

$$\begin{aligned}\left\langle F_{p\Omega}F_{q\Omega}^{*}\right\rangle &= \frac{1}{T}\int_0^T \int_{-\infty}^{\infty} F_p(t)F_q(t+\tau)e^{j\Omega t} d\tau dt \\ &= \left[\{\xi a(N+N_g) + G_{th}\}S_p + \xi aN\right]\delta_{p,q},\end{aligned} \tag{10.34}$$

$$\left\langle T_{p\Omega}T_{q\Omega}^{*}\right\rangle = \frac{1}{4(S_p+1)^2}\left\langle F_{p\Omega}F_{q\Omega}^{*}\right\rangle, \tag{10.35}$$

$$\langle W_\Omega W_\Omega^* \rangle = \xi a \sum_p (N + N_g) S_p + \frac{NV_a}{\tau_s} + \frac{I}{e}, \tag{10.36}$$

$$\langle F_{p\Omega} W_\Omega^* \rangle = \left\langle F_{p\Omega}^* W_\Omega \right\rangle = -\xi a \{ (N + N_g) S_p + N \}, \tag{10.37}$$

$$\langle T_{p\Omega} W_\Omega^* \rangle = 0. \tag{10.38}$$

$$\langle T_{p\Omega} F_\Omega^* \rangle = 0. \tag{10.39}$$

Note that the complex conjugate notation can be dropped in many cases when the value obtained is real, i.e.,

$$\left\langle F_{p\Omega} F_{p\Omega}^* \right\rangle = \left\langle F_{p\Omega}^2 \right\rangle. \tag{10.40}$$

10.4 Noise Analysis Under Stable Single Mode Operation

10.4.1 Using Equations

The more intrinsic properties of noise are obtained when the laser is operating in a stable single mode. We suppose in this section that only one mode is active in laser oscillations and drop the mode-number suffix. The equations to be used are [1]

$$\frac{dS}{dt} = (A - G_{th}) S + \xi a N + F(t), \tag{10.41}$$

$$\frac{d\theta}{dt} = \frac{\alpha \xi a}{2} (N - \bar{N}) + T(t), \tag{10.42}$$

$$\frac{dN}{dt} = -A \frac{S}{V_a} - \frac{N}{\tau_s} + \frac{I}{eV_a} + \frac{W(t)}{V_a}, \tag{10.43}$$

$$A = \xi a (N - N_g). \tag{10.44}$$

We also ignore the terms involving gain saturation to simply the discussion. The noise sources are expanded in their frequency components

$$F(t) = \int_{-\infty}^{\infty} F_\Omega e^{j\Omega t} d\Omega, \tag{10.45}$$

$$T(t) = \int_{-\infty}^{\infty} T_\Omega e^{j\Omega t} d\Omega, \tag{10.46}$$

$$W(t) = \int_{-\infty}^{\infty} W_\Omega e^{j\Omega t} d\Omega. \tag{10.47}$$

Next, the photon number, optical phase, and carrier density are separated into DC terms and fluctuation terms:

$$S = \bar{S} + \int_{-\infty}^{\infty} S_\Omega e^{j\Omega t} d\Omega. \tag{10.48}$$

$$\frac{d\theta}{dt} = 2\pi\left(\bar{f} + \int_{-\infty}^{\infty} f_\Omega e^{j\Omega t} d\Omega\right). \tag{10.49}$$

$$N = \bar{N} + \int_{-\infty}^{\infty} N_\Omega e^{j\Omega t} d\Omega, \tag{10.50}$$

where we assume that $\bar{S} \gg |S_\Omega| d\Omega$ and $\bar{N} \gg |N_\Omega| d\Omega$ hold. The intensity noise is evaluated with $\langle S_\Omega^2 \rangle$. As the shift in the optical frequency is obtained from $d\theta/dt$, the frequency noise is given by$\langle f_\Omega^2 \rangle$.

10.4.2 Intensity Noise

We substitute (10.45), (10.48), and (10.50) into (10.41) with (10.44), and separate the DC and fluctuation terms:

$$\bar{N} = \frac{\bar{S}}{\bar{S}+1} N_{th}, \tag{10.51}$$

$$\{j\Omega + \xi a(N_{th} - \bar{N})\} S_\Omega = \xi a(\bar{S} + 1) N_\Omega + F_\Omega. \tag{10.52}$$

Here, N_{th} is the threshold electron density related to the threshold gain level and the threshold current by

$$G_{th} = \xi a(N_{th} - N_g), \tag{10.53}$$

$$I_{th} = \frac{e N_{th} V_a}{\tau_s}. \tag{10.54}$$

We note that $\bar{N}$ is very close to N_{th}, but analyze $I \approx I_{th}$, we need to maintain $\bar{N}$ and N_{th} as distinct. We substitute (10.47), (10.48), and (10.50) into (10.43) with (10.44), and separate the DC and fluctuations terms, obtaining

$$\frac{\xi a}{V_a}(\bar{N} - N_g)\bar{S} + \frac{\bar{N}}{\tau_s} - \frac{I}{eV_a} = 0, \tag{10.55}$$

$$\left\{ j\Omega + \frac{1}{\tau_s} + \frac{\xi a}{V_a}\bar{S} \right\} N_\Omega = -\frac{\xi a}{V_a}(\bar{N} - N_g)S_\Omega + \frac{W_\Omega}{V_a}. \tag{10.56}$$

By substituting (10.51) into (10.55), $\bar{S}$ is expressed in terms of the injection current,

$$\bar{S} = \frac{I - I_{th} + \frac{eG_{th}I_g}{I_{th} - I_g} + \sqrt{\left(I - I_{th} + \frac{eG_{th}I_g}{I_{th} - I_g}\right)^2 + 4eG_{th}I}}{2eG_{th}}, \tag{10.57}$$

where I_g is the current level corresponding to N_g,

$$I_g = \frac{eN_g V_a}{\tau_s}. \tag{10.58}$$

We substitute N_Ω of (10.56) into (10.52), then S_Ω is given as

$$S_\Omega = \frac{\frac{\xi a(\bar{S}+1)}{V_a} W_\Omega + \left(j\Omega + \frac{1}{\tau_s} + \frac{\xi a\bar{S}}{V_a}\right) F_\Omega}{Y}, \tag{10.59}$$

with

$$Y = \frac{(\xi a)^2(\bar{N} - N_g)(\bar{S} + 1)}{V_a} + \left\{ j\Omega + \frac{1}{\tau_s} + \frac{\xi a\bar{S}}{V_a} \right\} \left\{ j\Omega + \frac{1}{\tau_s} + \xi a(N_{th} - \bar{N}) \right\}. \tag{10.60}$$

The correlated value of the photon fluctuation is then given with the noise sources as

$$\begin{aligned} \langle S_\Omega^2 \rangle = \Bigg[& \left\{ \frac{\xi a(\bar{S}+1)}{V_a} \right\}^2 \langle W_\Omega^2 \rangle + \left\{ \Omega^2 + \left(\frac{1}{\tau_s} + \frac{\xi a\bar{S}}{V_a} \right) \right\} \langle F_\Omega^2 \rangle \\ & + \frac{2\xi a(\bar{S}+1)}{V_a} \left(\frac{1}{\tau_s} + \frac{\xi a\bar{S}}{V_a} \right) \langle F_\Omega W_\Omega \rangle \Bigg] \div |Y|^2. \end{aligned} \tag{10.61}$$

The correlated values of the noise sources have been given in (10.34), (10.36), and (10.37), but have a simplified form when developed for a single-mode operation. Specifically, using (10.51) and (10.53), (10.34) becomes

$$\langle F_\Omega^2 \rangle = 2\xi a N_{th} \bar{S}, \tag{10.62}$$

using (10.55), (10.36) becomes

$$\langle W_\Omega^2 \rangle = 2\bar{N}\left(\xi a \bar{S} + \frac{V_a}{\tau_s}\right). \tag{10.63}$$

Substitution of (10.51) into (10.37) gives

$$\langle F_\Omega W_\Omega \rangle = -\xi a \bar{S}(N_{th} + N_g). \tag{10.64}$$

By substituting (10.62)–(10.64) into (10.61) and by suitable manipulations, rearranging the right-hand side of (10.61) gives

$$\langle S_\Omega^2 \rangle = \frac{2\xi a \bar{S}\left[\Omega^2 N_{th} + \left(\frac{1}{\tau_s} + \frac{\xi a \bar{S}}{V_a}\right)\left\{\frac{\xi a \bar{S}}{V_a}(N_{th} - N_g) + \frac{N_{th}}{\tau_s} - \frac{\xi a N_g}{V_a}\right\}\right]}{|Y|^2}. \tag{10.65}$$

We perform other reformulations. The coefficient $a\xi$ is given in terms of threshold gain and current as

$$\xi a = \frac{G_{th}}{N_{th} - N_g} = \frac{G_{th} V_a e}{(I_{th} - I_g)\tau_s}. \tag{10.66}$$

By substituting this relation into (10.65) and using relation $N_{th}/\tau_s \gg \xi a N_g / V_a$, (10.65) becomes

$$\langle S_\Omega^2 \rangle = \frac{2G_{th} I_{th} \bar{S}}{I_{th} - I_g}\left[\Omega^2 + \frac{1}{\tau_s^2}\left(1 + \frac{eG_{th}\bar{S}}{I_{th} - I_g}\right)\left(1 + \frac{eG_{th}\bar{S}}{I_{th}}\right)\right] \div |Y|^2, \tag{10.67}$$

with

$$\begin{aligned} Y = & \frac{eG_{th}^2}{(I_{th} - I_g)\tau_s}\left(\bar{S} - \frac{1}{I_{th}/I_g - 1}\right) + \frac{1}{\tau_s}\left(1 + \frac{eG_{th}\bar{S}}{I_{th} - I_g}\right)\frac{G_{th} I_{th}}{(I_{th} - I_g)(\bar{S} + 1)} - \Omega^2 \\ & + j\Omega\left\{\frac{1}{\tau_s} + \frac{eG_{th}\bar{S}}{\tau_s(I_{th} - I_g)} + \frac{G_{th} I_{th}}{(I_{th} - I_g)(\bar{S} + 1)}\right\}. \end{aligned} \tag{10.68}$$

With the intensity noise evaluated in terms of the *relative intensity noise* (RIN), we have [1]

$$\mathrm{RIN} = \frac{\langle S_\Omega^2 \rangle}{\bar{S}^2} = \frac{2G_{th}I_{th}}{(I_{th} - I_g)\bar{S}} \left\{ \Omega^2 + \frac{1}{\tau_s^2}\left(1 + \frac{eG_{th}\bar{S}}{I_{th} - I_g}\right)\left(1 + \frac{eG_{th}\bar{S}}{I_{th}}\right) \right\} \div |Y|^2. \tag{10.69}$$

A calculated example of the frequency spectrum with noise frequency

$$f = \Omega/2\pi \tag{10.70}$$

is shown in Fig. 10.3 and was obtained using a model for the AlGaAs laser with parameter values of $\tau_s = 3.0 \times 10^{-9}$ s, $G_{th} = 5.03 \times 10^{11}\ \mathrm{s}^{-1}$, $I_{th} = 16$ mA and $I_g/I_{th} = 0.7$. Other more detailed examples include parameter values $L = 300$ μm, $R_f = 0.2$, $R_b = 0.7$, $n_{eff} = 3.5$, $\alpha_{loss} = 20\ \mathrm{cm}^{-1}$, $g_{th} = 52.8\ \mathrm{cm}^{-1}$, $a = 2.75 \times 10^{-12}\ \mathrm{m}^3/\mathrm{s}$, $\xi = 0.2$, $N_g = 2.1 \times 10^{24}\ \mathrm{m}^{-3}$, $N_{th} = 3.01 \times 10^{24}\ \mathrm{m}^{-3}$, and $V_a = 3 \times 0.11 \times 300\ \mathrm{m}^3 = 100\ \mathrm{m}^3 = 1.0 \times 10^{-16}\ \mathrm{m}^3$. The frequency spectrum shows an almost flat profile at lower frequencies and a peak at the relaxation frequency.

The variation of RIN with injection current I at low frequencies, $\Omega = 0$, is shown as the line marked "single mode" in Fig. 10.4. RIN is *high at the threshold current*. The explanation is that fluctuations induced by both spontaneous and stimulated emissions are independent, and the sum of these two contributions is greatest around the threshold level. The line marked "multi-mode" is obtained by taking into account, using a method in [1], the resulting multi-mode behavior around the threshold current. As spontaneous emission contributes to each lasing mode, the fluctuation based on spontaneous emission *is reduced* by including all lasing modes. Figure 10.3 is actually calculated by considering this type of multi-mode behavior.

If the injection current I is larger than the threshold current I_{th}, the RIN expressed in (10.69) can be simplified. The photon number $\bar{S}$ in (10.57) is simply

$$\bar{S} = \frac{I - I_{th}}{eG_{th}}, \quad \text{for } I > I_{th}, \tag{10.71}$$

and becomes large enough, $\bar{S} \gg 1$, that the electron number density can be set to its threshold value,

$$\bar{N} = N_{th}, \quad \text{for } I > I_{th}. \tag{10.72}$$

The denominator function Y is then rewritten as

$$Y = \Omega_R^2 - \Omega^2 + j\frac{\Omega}{\tau_s}\frac{I - I_g}{I_{th} - I_g}, \quad \text{for } I > I_{th}, \tag{10.73}$$

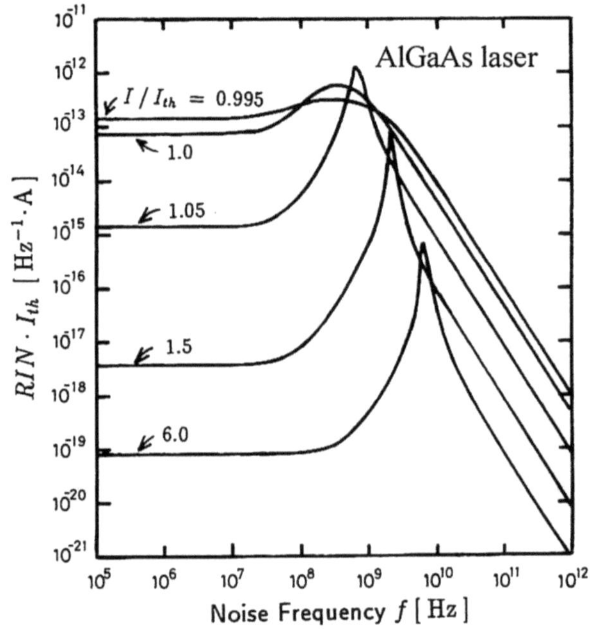

Fig. 10.3 Injection current dependence of RIN spectra against noise frequency for a AlGaAs laser

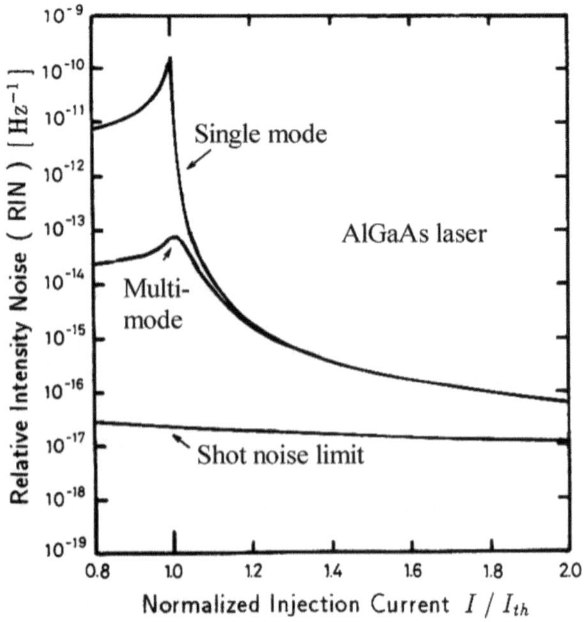

Fig. 10.4 Variation of RIN with injection current at low noise frequency

where Ω_R is the *relaxation oscillation angular frequency* which was introduced in (7.39),

$$\Omega_R^2 = \frac{G_{th}}{\tau_s}\frac{I - I_{th}}{I_{th} - I_g}. \tag{10.74}$$

Then we have

$$\mathrm{RIN} = \frac{2eG_{th}^2 I_{th}}{(I_{th} - I_g)(I - I_{th})} \cdot \frac{\dfrac{1}{\tau_s^2}\dfrac{I(I - I_g)}{I_{th}(I_{th} - I_g)} + \Omega^2}{\left(\dfrac{G_{th}}{\tau_s}\dfrac{I - I_{th}}{I_{th} - I_g} - \Omega^2\right)^2 + \left(\dfrac{\Omega}{\tau_s}\dfrac{I - I_g}{I_{th} - I_g}\right)^2}, \quad \text{for } I > I_{th}. \tag{10.75}$$

For low frequencies, this expression can be simplified further:

$$\mathrm{RIN}|_{\Omega=0} = \frac{2eI(I - I_g)}{(I - I_{th})^3}, \quad \text{for } I > I_{th}. \tag{10.76}$$

If the injection current becomes much higher, RIN approaches

$$RIN \to \frac{2e}{I}, \quad \text{for } I \gg I_{th}, \tag{10.77}$$

called the *shot noise limit*, the lower curve indicated in Fig. 10.4.

10.4.3 Frequency Noise and Spectrum Linewidth

From (10.16), (10.32), (10.49), and (10.50), the fluctuation term of the optical frequency is given as

$$f_\Omega = \frac{1}{2\pi}\left(T_\Omega + \frac{\alpha\xi a}{2}N_\Omega\right). \tag{10.78}$$

The frequency fluctuation has two contributions, one depending on the *phase fluctuation* T_Ω *at the optical emission* and the other on the *fluctuation of the electron density* N_Ω. The latter originally comes from the intensity fluctuation but effects on the frequency fluctuation through a change in *refractive index* as indicated by parameter α, which is defined in (8.87).

By substituting (10.56) into (10.52), the *electron density fluctuation* N_Ω is given as

$$N_\Omega = \frac{\{j\Omega + a\xi(N_{th} - \bar{N})\}W_\Omega - a\xi(\bar{N} - N_g)F_\Omega}{YV_a}. \tag{10.79}$$

Because there are no correlations between T_Ω and F_Ω and W_Ω, as already stated in (10.38) and (10.39), the fluctuation for T_Ω has no correlation with N_Ω. Thus the *frequency noise* is given by

$$\langle f_\Omega^2 \rangle = \frac{1}{4\pi^2}\left\{\langle T_\Omega^2 \rangle + \left(\frac{\alpha a \xi}{2}\right)^2 \langle N_\Omega^2 \rangle\right\}. \tag{10.80}$$

As frequency noise is present when the laser is oscillating in single mode with higher injection current than the threshold level, $I > I_{th}$, we can employ (10.71) and (10.72). We then have fluctuations

$$\langle T_\Omega^2 \rangle = \frac{eG_{th}^2 I_{th}}{2(I_{th} - I_g)(I - I_{th})}, \tag{10.81}$$

$$\langle N_\Omega^2 \rangle = \frac{2I_{th}\{G_{th}^2(I - I_{th}) + \Omega^2(I - I_g)\}}{e(I_{th} - I_g)|Y|^2 V_a^2}, \tag{10.82}$$

which substituted along with (10.66), (10.73), and (10.74) into (10.80), yields frequency noise [1]

$$\langle f_\Omega^2 \rangle = \frac{eG_{th}^2 I_{th}}{8\pi^2(I_{th} - I_g)} \times \left[\frac{1}{I - I_{th}} + \frac{\alpha^2}{(I_{th} - I_g)^2 \tau_s^2} \cdot \frac{G_{th}^2(I - I_{th}) + \Omega^2(I - I_g)}{\left(\dfrac{G_{th}}{\tau_s}\dfrac{I - I_{th}}{I_{th} - I_g} - \Omega^2\right)^2 + \left(\dfrac{\Omega}{\tau_s}\dfrac{I - I_g}{I_{th} - I_g}\right)^2}\right]. \tag{10.83}$$

A calculated example of the frequency noise is shown in Fig. 10.5. We have again assumed our model for a AlGaAs laser with identical parameter values as those used for Figs. 10.3 and 10.4. The calculations near the threshold current, $I \approx I_{th}$, are performed based on a more accurate equation using (10.57).

Because the sources of fluctuation have *Poisson (Gaussian) distribution profiles*, the spectral *linewidth* (full width at half maximum) Δf is evaluated from the frequency noise as

$$\Delta f = 4\pi\langle f_0^2 \rangle. \tag{10.84}$$

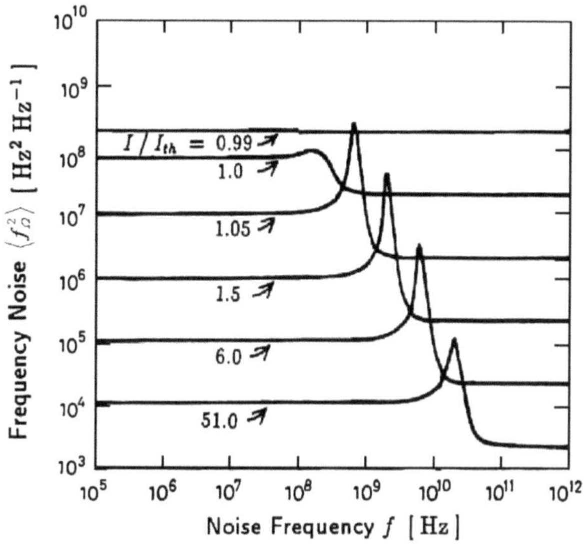

Fig. 10.5 Frequency noise

By putting $\Omega = 0$ in (10.83), we obtain the famous form

$$\Delta f = \frac{1+\alpha^2}{2\pi} \frac{eG_{th}^2}{(1 - I_g/I_{th})(I - I_{th})}. \tag{10.85}$$

As the parameter α plays an important role in determining the spectral linewidth, α is called the *linewidth enhancement factor* [2].

The generating mechanism of both intensity and frequency noise under a stable single-mode operation is summarized in Fig. 10.6. The source of the noise is the intrinsic fluctuation in spontaneous emission and contains both intensity and phase fluctuations. The intensity fluctuation is amplified by lasing gain, thus revealing the intensity noise. The phase fluctuation of the noise source gives one contribution to the frequency noise. Another contribution comes from the intensity fluctuation through a change in the refractive index.

10.5 Mode Competition Noise

10.5.1 Mode Competition due to Symmetric Gain Saturation

As discussed in Chap. 9, semiconductor lasers exhibit several types of *mode-competition phenomena*. Noise associated with such phenomena *is enhanced* in these lasers and is called *mode competition noise*. One typical noise is *mode hopping noise*.

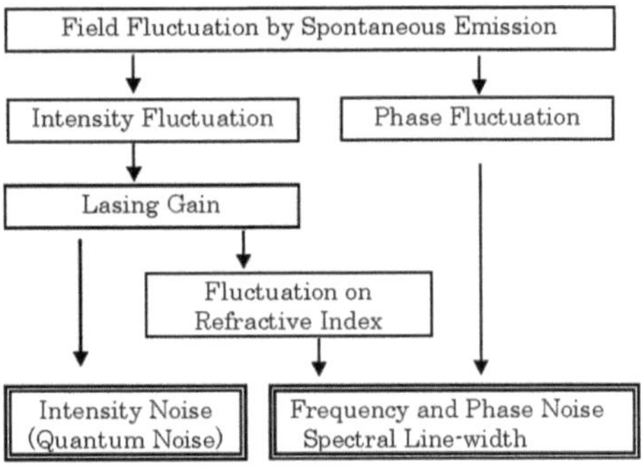

Fig. 10.6 Generating mechanism of intensity and frequency noise under stable single-mode operation

If the symmetric gain saturation is more dominant than asymmetric gain saturation, the lasing operation should be in the *bi-stable state*, as shown in Fig. 9.2c or 9.4c. In Fig. 10.7a, we suppose that the operating position at the steady state is P, assuming two-mode competition between modes p and q. The *Langevin noise source*, added in (10.13), gives a small instantaneous *random walk* around the position P as illustrated in Fig. 10.7a. Recall that lines L_p and L_q represent the steady state solution to $dS_p/dt = 0$ and $dS_q/dt = 0$, respectively.

If L_q shifts to the right and upwards either through increasing the value $(A_q - G_{th})/B$ or $(A_p - G_{th})/D_{p(q)}$ and makes contact with L_p at a single point, as shown in Fig. 10.7b, the operating point begins to fluctuate *moving in an extensive fashion to the other operating position Q*, but returning back to the original operating position P. Because the return distance is *longer*, returning back to the original position P takes a longer time. This effect gives rise to an *increased noise in the lower frequency range*. If the fluctuating operating point cannot return back to the original position P, the operating point reaches the position Q. This is called *mode hopping*. As the total photon number $S_p + S_q$ is not constant during mode hopping, optical power fluctuates even more. Mode hopping must be observed in the operating range showing *hysteresis* in Fig. 9.6. This type of mode hopping is caused by changes in the linear gain coefficient A_p or threshold gain G_{th}.

Data analyzed using an approximate two-mode competition model are shown in Fig. 10.8. In using the equations

$$\frac{dS_p}{dt} = \left[A_p - BS_p - DS_q - G_{th(p)}\right]S_p + \xi aN + F_p(t), \tag{10.86}$$

$$\frac{dS_q}{dt} = \left[A_q - BS_q - DS_p - G_{th(q)}\right]S_q + \xi aN + F_q(t), \tag{10.87}$$

$$\frac{dN}{dt} = -\frac{A_pS_p + A_qS_q}{V_a} - \frac{N}{\tau_s} + \frac{I}{eV_a} + \frac{W(t)}{V_a}, \tag{10.88}$$

$$A_p = \xi a\left\{N - N_g - b(\lambda_p - \lambda_o)^2\right\}, \tag{10.89}$$

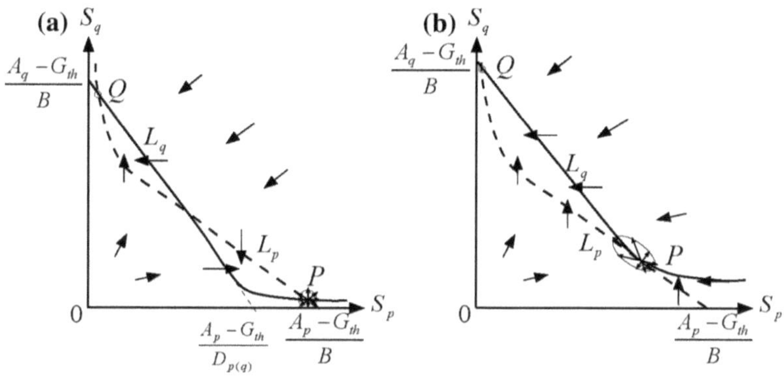

Fig. 10.7 Operating point fluctuation under symmetric gain saturation. **a** Bi-stable. **b** Critical setting for mode jump

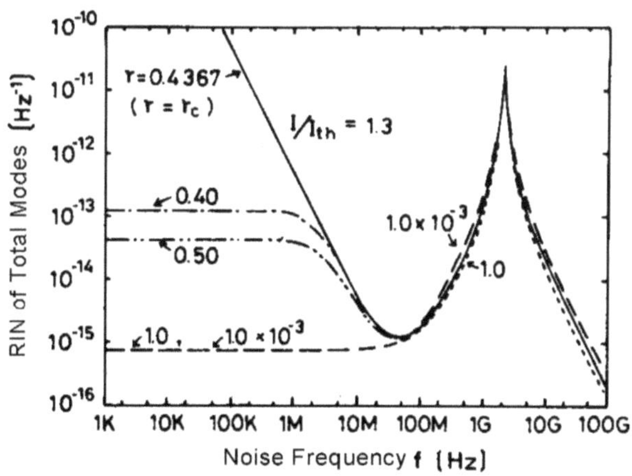

Fig. 10.8 Analyzed data based on small fluctuation two-mode competition model

$$A_q = \xi a\left\{N - N_g - b(\lambda_q - \lambda_o)^2\right\}, \tag{10.90}$$

the threshold gains, $G_{th(p)}$ and $G_{th(q)}$ of the two modes are assumed to vary. With noise sources $F_p(t)$, $F_q(t)$, and $W(t)$ expanded into their frequency components, (10.31) and (10.33), the photon numbers and the electron density can also be expanded similarly:

$$S_p(t) = \bar{S}_p + \int_{-\infty}^{\infty} S_{p\Omega} e^{i\Omega t} d\Omega, \tag{10.91}$$

$$S_q(t) = \bar{S}_q + \int_{-\infty}^{\infty} S_{q\Omega} e^{i\Omega t} d\Omega, \tag{10.92}$$

$$N(t) = \bar{N} + \int_{-\infty}^{\infty} N_{\Omega} e^{j\Omega t} d\Omega. \tag{10.93}$$

Equations (10.91)–(10.93) are substituted into (10.86)–(10.90) together with (10.31) and (10.33). Again the fluctuation terms are supposed to be much smaller than the DC terms. Given the numerical values of the coefficients for an AlGaAs laser, the RIN value is estimated as

$$\mathrm{RIN} = \frac{\left\langle (S_{p\Omega} + S_{q\Omega})(S_{p\Omega}^{*} + S_{q\Omega}^{*}) \right\rangle}{(\bar{S}_p + \bar{S}_q)^2}. \tag{10.94}$$

The parameter r in Fig. 10.8 is defined as the ratio of the DC terms of the photon number

$$r = \frac{\bar{S}_q}{\bar{S}_p}. \tag{10.95}$$

A value of $r = 10^{-3}$ indicates single-mode operation with mode p, $r = 1.0$ multi-mode operation, and $r = r_c = 0.4367$ a critical setting with the operating point jumping from P to the another point Q, as illustrated in Fig. 10.7b. Noise is sufficiently low when the laser operates in a stable single- or multi-mode, and highest in the lower frequency range operating at the critical point $r = r_c$. However, experimental data and more recently numerical simulations (see data in Figs. 10.9, 10.12, and 10.13), show that noise does not reach excessively high values at zero frequency. As this analysis was performed assuming fluctuation values are much smaller than DC values, the large noise values predicted are over-estimates and that in reality noise should exhibit saturation. Nevertheless, mode competition or hopping within the lasing mode produces extra noise which becomes higher at low frequencies [3]. We call this type of noise *low-frequency-type noise.*

Examples of experimentally observed noise characteristics are shown in Fig. 10.9. These data were recorded from an AlGaAs laser without optical feedback with careful tuning of the operating current and control of the cavity temperature. Presence of mode hopping was confirmed by optical spectrometer as noise was increased. Mode hopping noise tends to be generated when operating at higher current than threshold current because of the higher rate of stimulated emission than spontaneous emission [4].

Mode competition noise is enhanced with the optical feedback as we will show in Sect. 10.6.

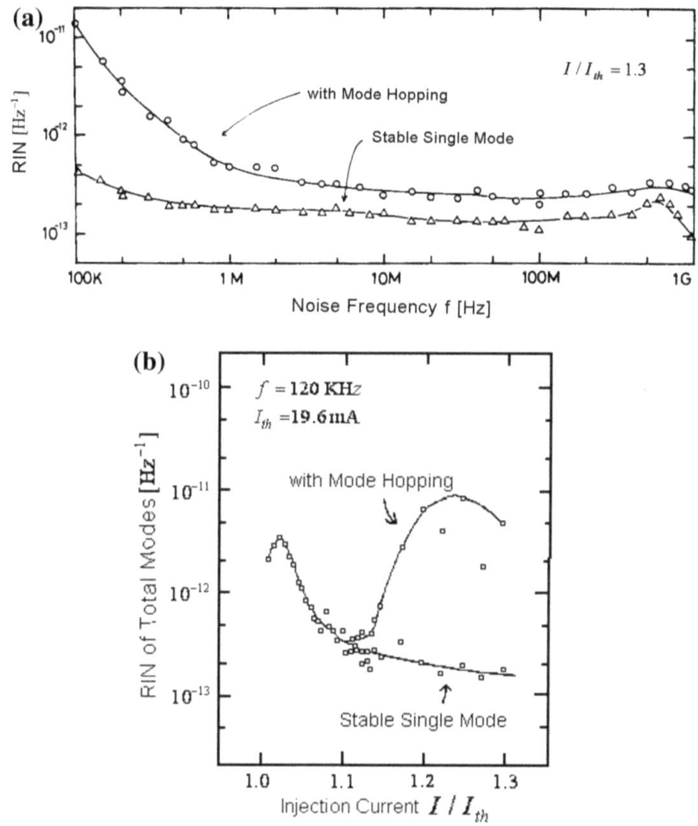

Fig. 10.9 Example of experimentally measured intensity noise in an AlGaAs laser. **a** Frequency spectrum. **b** Variation with injection current

10.5.2 *Mode Competition Including Asymmetric Gain Saturation*

As discussed in Sect. 9.2, mode competition becomes more complicated by introducing asymmetric gain saturation. Approximations with only two modes are not sufficient, because cyclic rotation with the lasing mode can be generated by asymmetric gain saturation, as evident in Figs. 9.10 and 9.11. Here, we show numerically simulated noise characteristics. The equations used are

$$\frac{dS_p}{dt} = \left[A_p - BS_p - \sum_{q \neq p} (D + H_{p(q)})S_q - G_{th}\right] S_p + C_p \frac{NV_a}{\tau_s} + F_p(t), \qquad (10.96)$$

$$A_p = \xi a \left\{ N - N_g - b(\lambda_p - \lambda_o)^2 \right\}, \qquad (10.97)$$

$$B = \frac{9\xi^2 \hbar \omega_p a |R_{ba}|^2}{4V_a \varepsilon_o n_{eff}^2} \left(\frac{\tau_{in}}{\hbar}\right)^2 (N - N_s), \tag{10.98}$$

$$D = \frac{4}{3} \frac{B}{\{1 + \tau_{in}^2 (\omega_p - \omega_q)^2\}}, \tag{10.99}$$

$$H_{p(q)} = \frac{3\xi a G_{th} \lambda_p \alpha}{4V_a \omega_p (\lambda_q - \lambda_p)}, \tag{10.100}$$

$$C_p = \frac{\xi a \tau_s}{V_a [\{2(\lambda_p - \lambda_o)/\delta\lambda\}^2 + 1]}, \tag{10.101}$$

$$\frac{dN}{dt} = -\sum_p \frac{A_p S_p}{V_a} - \frac{N}{\tau_s} + \frac{I}{eV_a} + \frac{W(t)}{V_a}, \tag{10.102}$$

where C_p is the spontaneous emission coefficient in which the wavelength profile is taken into account via the half width of $\delta\lambda$.

We shall provide numerical calculations based on (10.96)–(10.102). The sources of fluctuation can be generated using random number generators in a computer. Because noise sources of the photon number $F_p(t)$ have no correlation among the lasing modes, these fluctuations can be given by independent random number generators. However, the noise source of the electron number $W(t)$ is not independent from noise sources of the photons $F_p(t)$ displaying mutual correlations presented in (10.29). We adopt then a method to orthogonalize fluctuation sources from photon generation and electron reduction in the following [5].

We suppose a new transformed quantity $(NV_a + \sum_p k_p S_p)$, instead of the simple electron density N in (10.102), where k_p is a coefficient to be suitably determined. The associated fluctuation of this quantity is $\left\{W(t) + \sum_p k_p F_p(t)\right\}$, which needs to be orthogonal to the sources of the photon fluctuation by satisfying

$$\left\langle \left\{ W(t) + \sum_p k_p F_p(t) \right\} F_q(t') \right\rangle = 0. \tag{10.103}$$

Given the fluctuation sources of the photon mode satisfy the orthogonal condition of (10.21), the coefficient k_p is found to be

$$k_p = -\frac{\langle W(t) F_p(t') \rangle}{\langle F_p(t) F_p(t') \rangle}. \tag{10.104}$$

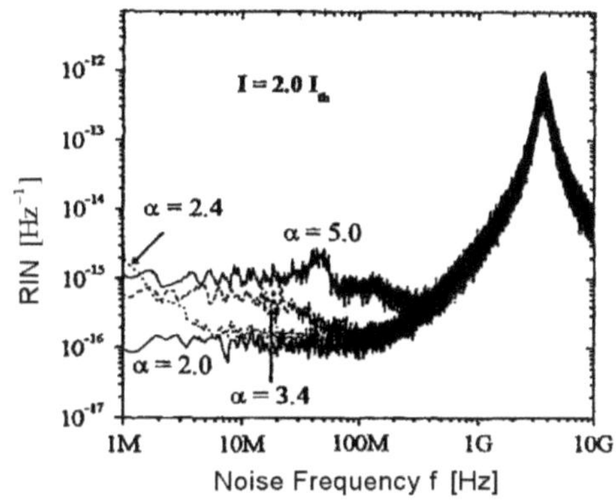

Fig. 10.10 Numerical simulated example of the mode-competition noise

The dynamic equation of the transformed quantity is then

$$\begin{aligned}&\frac{d(NV_a + \sum_p k_p S_p)}{dt}\\&= -\sum_p A_p S_p - \frac{NV_a}{\tau_s} + \frac{I}{e}\\&\quad + \sum_p k_p \left\{ \left[A_p - BS_p - \sum_{q \neq p} (D + H_{p(q)}) S_q - G_{th} \right] S_p + C_p \frac{NV_a}{\tau_s} \right\}\\&\quad + \left\{ W(t) + \sum_p k_p F_p(t) \right\}.\end{aligned} \quad (10.105)$$

We use this equation in the numerical calculation instead of (10.102) and employ a random number generator for $\left\{ W(t) + \sum_p k_p F_p(t) \right\}$. Equations for the other photon numbers are the same as in (10.96).

The numerical calculation is performed by integrating the simultaneous dynamic equations. The time step for the integration $\Delta t = t_i - t_{i-1}$ is 10 ps for the plot in Fig. 10.10. The source of the fluctuation at t_i is given from values at t_{i-1},

$$F_p(t_i) = \sqrt{\frac{\{\xi a(N(t_{i-1}) + N_g) + G_{th}\} S_p(t_{i-1}) + C_p \frac{N(t_{i-1})V_a}{\tau_s}}{\Delta t}} g_p, \quad (10.106)$$

where g_p is a random number generating source featuring a Gaussian distribution with $\bar{g}_p = 0$ and $\left\langle g_p^2 \right\rangle = 1$. Other noise sources are similarly defined while maintaining non-correlated relations among noise sources.

The RIN value is evaluated by taking the Fourier transformation from the temporal variation of total photon numbers obtained

$$\mathrm{RIN} = \frac{\frac{1}{T}\left|\frac{1}{2\pi}\int_0^T \sum_p S_p(t)e^{-j\Omega t}dt\right|^2}{\left\{\frac{1}{T}\int_o^T \sum_p S_p(t)dt\right\}^2}. \tag{10.107}$$

Numerical simulation examples of mode-competition noise are shown in Fig. 10.10. The α parameter, called the line enhancement factor, represents the contribution of asymmetric gain saturation and is used as a parameter in the calculation. Numerical simulations were made for the AlGaAs laser with parameter settings of $\lambda = 850\,\text{nm}$ for $\alpha = 2.0$ and 2.4, and for the InGaAsP laser with $\lambda = 1.3\ \mu\text{m}$ for $\alpha = 3.4$ and 5. A set of 15 modes ($p = -7$ to 7) were used in the calculation. The wavelength α at the peak of the linear gain and spontaneous emission is set for the central mode of $p = 0$. In this simulation, optical feedback was not taken into account.

With $\alpha = 2.0$, mode-competition noise was well suppressed. For $\alpha = 2.4$ and 3.4, low-frequency-type noise appeared.

At $\alpha = 5.0$, noise has increased over a wide frequency range with a small peak at around$f \approx 60$ MHz. In this instance, the asymmetric saturation seems strong enough, resulting in the cyclic rotation of the lasing mode, as illustrated in Figs. 9. 10 and 9.11. The small peak around $f \approx 60$ MHz must be the rotation period. Even when the rotation of the lasing mode is not generated with stable two-mode operation, as indicated in Fig. 9.8a, c, the operating point barely converges to crossing point M. Therefore, fluctuations continue. More generally though, we call such noise *flat-type noise* as the extra noise extends over a wide frequency range.

10.6 Optical Feedback Noise

10.6.1 Model of Optical Feedback

The conceptual model of optical feedback is illustrated in Fig. 10.11, where an object providing *optical reflection* is located at distance ℓ away from the laser. The distance is assumed much longer than the laser cavity, $\ell \gg L$. The fraction of reflected light returning to the surface of the front facet is denoted by Γ. $(1 - R_f)\eta\Gamma$ is the power rate coupling to the lasing field in the active region, where η is the coupling coefficient for the returning reflected light in the active region. The *time delay* for the light output to return is

$$\tau = \frac{2\ell}{c}. \tag{10.108}$$

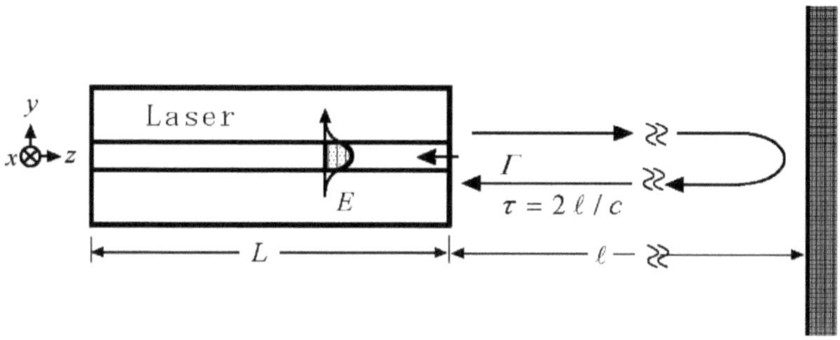

Fig. 10.11 Re-injection of the output light

The electric field in the laser is represented as a traveling wave,

$$E(t,\mathbf{r}) = \sum_p \left\{ E_p^{(+)}(z,t)T(x,y)e^{-j\beta_p z} + E_p^{(-)}(z,t)T(x,y)e^{-j\beta_p(L-z)} \right\} e^{j\omega_p t} + c.c., \tag{10.109}$$

where we have taken into account the temporal variation of the amplitude. The backward propagating component is given with summed values of the directly reflected wave at the front facet and the re-injected wave, which exist at time interval τ previous,

$$\begin{aligned} E_p^{(-)}(L,t) =& \sqrt{R_f} E_p^{(+)}(L,t) \\ &+ (1-R_f)\sqrt{\eta\Gamma} e^{-j\omega_p\tau} E_p^{(+)}(L,t-\tau), \end{aligned} \tag{10.110}$$

where $E_p^{(+)}(L,t-\tau)$ is the amplitude of the forward wave at $t-\tau$ and $\exp(-j\omega_p\tau)$ is the phase delay of the one-time round trip. The amplitude of the electric field passing through the front facet is $\sqrt{1-R_f}$. Because the re-injected wave passes the front facet twice, the variation becomes $\left(\sqrt{1-R_f}\right)^2 = 1-R_f$. We introduce here a complex function $\Theta_p(t)$ to account for optical feedback,

$$E_p^{(-)}(L,t) = \Theta_p(\tau)\sqrt{R_f} E_p^{(+)}(L,t), \tag{10.111}$$

with

$$\begin{aligned} \Theta_p(\tau) \equiv& 1 + (1-R_f)\sqrt{\frac{\eta\Gamma}{R_f}} e^{-j\omega_p\tau} \frac{E_p^{(+)}(L,t-\tau)}{E_p^{(+)}(L,t)} \\ \equiv& |\Theta_p(\tau)| e^{-j\phi_p(\tau)}. \end{aligned} \tag{10.112}$$

The oscillation conditions given by (3.29), (3.30), and (3.33) then become

$$e^{(g_p-\alpha_{loss})L}\sqrt{R_f R_b}|\Theta_p|e^{-2j\beta_p L-j\phi_p}=1, \tag{10.113}$$

$$2\beta_p L+\phi_p=2p\pi, \tag{10.114}$$

$$g_p=\alpha_{loss}+\frac{1}{2L}\ln\frac{1}{R_f R_b}-\frac{1}{L}\ln|\Theta_p|. \tag{10.115}$$

As shown by these equations, we have accounted for the effect of optical feedback with changes in the threshold gain and phase.

Next, we represent the optical field in the cavity as a standing wave

$$E(t,\boldsymbol{r})=\sum_p \widehat{E}_p(t)\Phi_p(\boldsymbol{r})e^{j\omega_p t}+c.c. \tag{10.116}$$

With optical feedback taken into account, we get a dynamic equation for the change in the complex amplitude $\widehat{E}_p(t)$ corresponding to (10.5),

$$\frac{\partial\widehat{E}_p(t)}{\partial t}=\frac{1}{2}\left\{G_p-G_{tho}+j\alpha\xi a(N-\bar{N})+\frac{c}{n_{eff}}\ln\Theta_p\right\}\widehat{E}_p(t)+U_p(t), \tag{10.117}$$

where G_p is the mode gain including gain saturation, and G_{tho} the threshold gain of a solitary laser without optical feedback,

$$G_{tho}=\frac{c}{n_{eff}}\left\{\alpha_{loss}+\frac{1}{2L}\ln\frac{1}{R_b R_f}\right\}. \tag{10.118}$$

We separate the amplitude and phase terms,

$$\widehat{E}_p(t)=|\widehat{E}_p(t)|e^{j\theta_p(t)}, \tag{10.119}$$

and substitute this equation into (10.117), perform manipulations similar as in (10.8)–(10.11), and change the absolute value to the photon number as given by (10.12). We finally derive a set of dynamic equations [6] for $S_p(t)$, $\Theta(t)$, and $N(t)$:

$$\frac{dS_p}{dt}=\left\{G_p-G_{tho}+\frac{c}{n_{eff}L}\ln|\Theta_p(\tau)|\right\}S_p+C_p\frac{\xi aNV_a}{\tau_s}+F_p(t), \tag{10.120}$$

$$G_p=A_p-BS_p-\sum_q D_{p(q)}S_q, \tag{10.121}$$

$$\Theta_p(\tau) \equiv 1 + (1 - R_f)\sqrt{\frac{\eta\Gamma}{R_f}} e^{j\{-\omega_p\tau + \theta_p(t-\tau) - \theta_p(t)\}}\sqrt{\frac{S_p(t-\tau)}{S_p(t)}} \\ \equiv |\Theta_p(\tau)| e^{-j\phi_p(\tau)} \tag{10.122}$$

$$\frac{d\theta_p}{dt} = \frac{\alpha\xi a}{2}(N - \bar{N}) - \frac{c}{2n_{eff}L}(\phi_p - \bar{\phi}_p) + T_p(t), \tag{10.123}$$

$$\frac{dN}{dt} = -\sum_p \frac{A_p S_p}{V_a} - \frac{N}{\tau_s} + \frac{I}{eV_a} + \frac{W(t)}{V_a}. \tag{10.124}$$

The gain and other coefficients are the same as in (10.97)–(10.101). The correlation relations for Langevin noise sources are as given in (10.21), (10.25), (10.28), and (10.29). The RIN level is evaluated by (10.107).

10.6.2 Calculated Examples

Numerical simulation data of noise under operations with optical feedback are shown in Fig. 10.12, where (a) shows the noise spectrum and (b) shows the dependence on the optical feedback ratio. The simulation uses the model of an AlGaAs laser with a line enhancement factor of $\alpha = 2.4$. In the numerical calculation, 13 longitudinal modes are used, but the noise source is ignored by putting $W(t) = 0$ because orthogonalization given by (10.103) requires extensive computing time in actual calculations. Fluctuation of the electron density N induced by photon fluctuations and optical feedback were of course included. The time step of the integration was $\Delta t = 5$ ps. The calculation of a single line in Fig. 10.12a took one and half days on the author's computer, which is a conventional personal computer running Microsoft's Windows XP operating system.

In Fig. 10.12a, the line with $\eta\Gamma = 0$ corresponds to free operations without optical feedback. With increasing optical feedback, noise rises in the low frequency range below 10 MHz corresponding to *low-frequency-type noise*. A further increase in feedback reduces this low-frequency-type noise and an excess noise extending over the high-frequency range above 100 MHz appears corresponding to *flat-type noise*.

Variations of the noise at fixed frequencies with feedback are shown in Fig. 10.12b. Low-frequency-type noise peaks at $\eta\Gamma \approx 5.28 \times 10^{-4}$, whereas flat-type noise increases gradually with feedback ratio. Experimentally measured data corresponding to Fig. 10.12 are shown in Fig. 10.13 [7]. Because we cannot measure η from the experiment, experimental data are given with Γ and not $\eta\Gamma$. Although the experimental data give higher noise than those from numerical simulations, the simulations well fit the experimental data [6, 7].

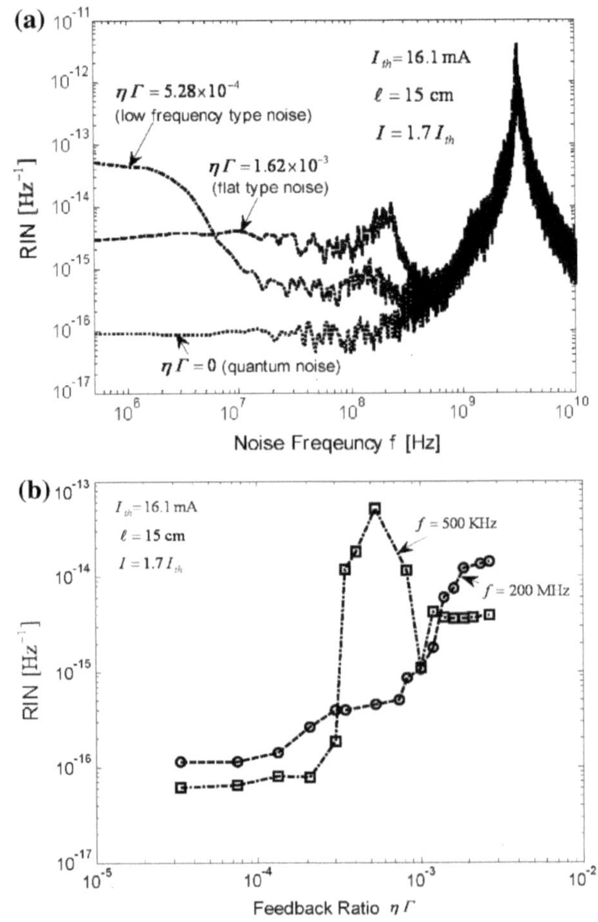

Fig. 10.12 Example of a numerical simulation of optical feedback noise modeled for an AlGaAs laser. **a** Frequency spectrum **b** Variation with feedback ratio

10.6.3 Mechanism of Noise Generation

The generating mechanism for optical feedback noise is summarized in Fig. 10.14. The source of noise is the intrinsic fluctuations associated with optical emission similar to quantum noise. This noise source is amplified with the lasing operation, but suffers additional instabilities due to optical feedback. There are two mechanisms causing such noise. One is by *mode hopping among the internal cavity modes*, which means the laser modes of a solitary laser.

The threshold gain for mode p changes to

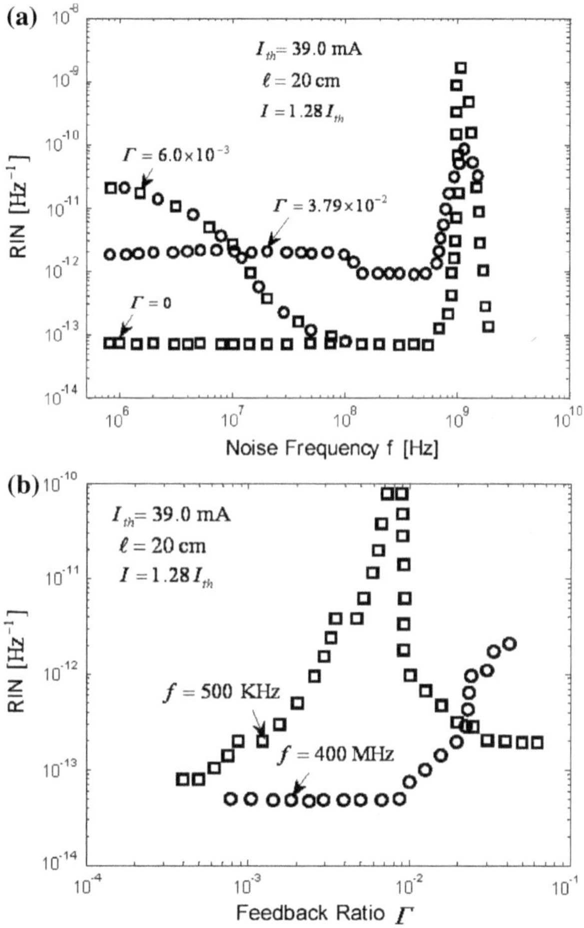

Fig. 10.13 Experimentally measured data corresponding to the numerical simulation modeled after an AlGaAs laser given in Fig. 10.12. **a** Frequency spectrum **b** Variation with feedback ratio

$$G_{th(p)} = G_{tho} - \frac{c}{n_{eff}L} \ln|\Theta_p(\tau)|, \tag{10.125}$$

as given by (10.120). When feedback $\eta\Gamma$ is not so large, its effect can be more simply written using $\ln \Theta_p = \ln|\Theta_p| - j\phi_p$ and $\ln(1 + x) \approx x$. We have

$$\ln|\Theta_p(\tau)| \approx (1 - R_f)\sqrt{\frac{\eta\Gamma}{R_f}}\sqrt{\frac{S_p(t-\tau)}{S_p(t)}}\cos\{\omega_p\tau - \theta_p(t-\tau) + \theta_p(t)\}, \tag{10.126}$$

$$\phi_p(\tau) \approx (1 - R_f)\sqrt{\frac{\eta\Gamma}{R_f}}\sqrt{\frac{S_p(t-\tau)}{S_p(t)}}\sin\{\omega_p\tau - \theta_p(t-\tau) + \theta_p(t)\}. \tag{10.127}$$

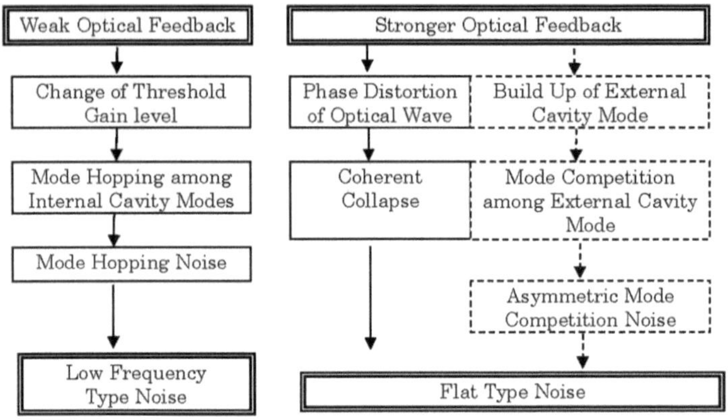

Fig. 10.14 Mechanisms of the optical feedback noise

As found from (10.125) and (10.126), the threshold gain $G_{th(p)}$ is changed with optical feedback as given by the phase difference $\{\theta_p(t) - \theta_p(t-\tau)\}$ or frequency difference $\{d\theta_p(t)/dt - d\theta_p(t-\tau)/dt\}$. This effect is understood as follows. When the output light is reflected back towards the laser, the lasing optical frequency in the cavity already differs from that of the output light because the lasing optical frequency always has an intrinsic fluctuation. The effective threshold gain $G_{th(p)}$, which is given by the superposition of reflected and re-injected light at the front facet, then incurs fluctuation. This change in $G_{th(p)}$ induces *mode hopping*, as well as *mode-hopping noise*, which was explained in Sect. 10.5.

Figure 10.15 shows the time-traced variations of the lasing modes under optical feedback in a numerical calculation for the model used to obtain Fig. 10.12: (a) corresponds to operations when the laser exhibits *low-frequency-type noise* with feedback ratio of $\eta\Gamma = 5.28 \times 10^{-4}$. Mode hopping is clearly seen in this figure, where mode numbers are indicated. Similar to that shown in Fig. 9.10, the mode at the linear gain peak is denoted $p = 0$; the longer wavelength side corresponds to positive numbers whereas the shorter wavelength side corresponds to negative numbers.

Figure 10.15b shows data for a laser exhibiting *flat-type noise* with feedback ratio of $\eta\Gamma = 1.63 \times 10^{-3}$. The operation becomes *multi-mode* for the internal cavity modes. Mode hopping is stabilized resulting in a *reduction in low-frequency-type noise*. However, each mode shows rather large fluctuation values, which must come from fluctuations of $\ln|\Theta_p|$ and φ_p described by (10.126) and (10.127).

This type of fluctuation is associated with distortions in the optical phase $\theta_p(t)$, called *coherence collapse*, as found from (10.123). A further increase in optical feedback tends to bring about unstable operations, called *chaos*.

An alternative understanding of the generating mechanism for flat-type noise is to estimate the number of the *external cavity modes* [8]. The phase delay $\omega_p\,\tau$ in

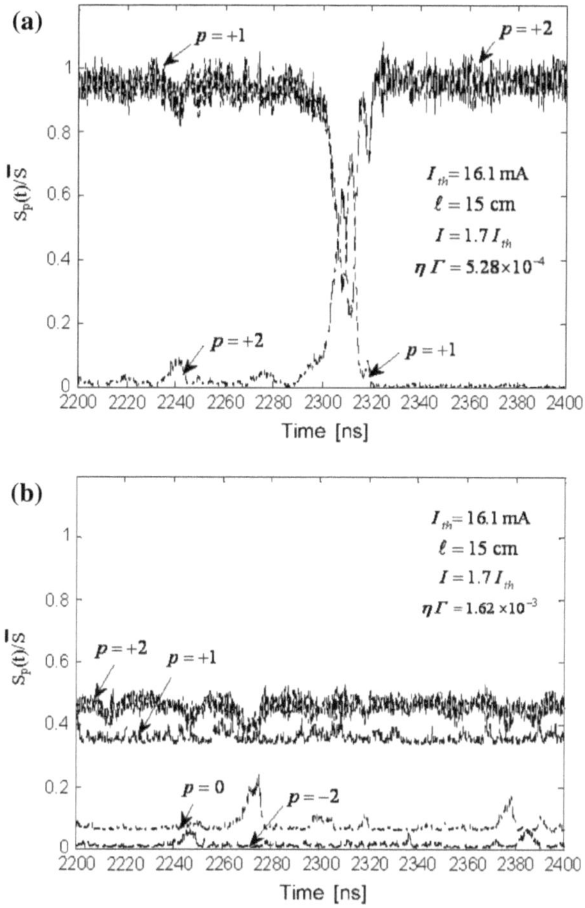

Fig. 10.15 Temporal variations of the internal cavity modes under optical feedback. **a** When low-frequency-type noise appears. **b** When flat-type noise appears

(10.126) and (10.127) is very large leading to other possible solutions. We denote a neighboring solution by p' such as

$$(\omega_p - \omega_{p'})\tau = \left(\frac{2\pi}{\lambda_p} - \frac{2\pi}{\lambda_{p'}}\right)2\ell = 2\pi. \tag{10.128}$$

The wavelength separation $\Delta\lambda$ is then

$$\Delta\lambda = \lambda_{p'} - \lambda_p = \frac{\lambda^2}{2\ell}. \tag{10.129}$$

The appearance of a new solution corresponds to forming a standing wave in the space between the front facet of the laser and the reflecting point. We call this new solution an external cavity mode, as illustrated in Fig. 10.16.

Fig. 10.16 Appearance of external cavity modes

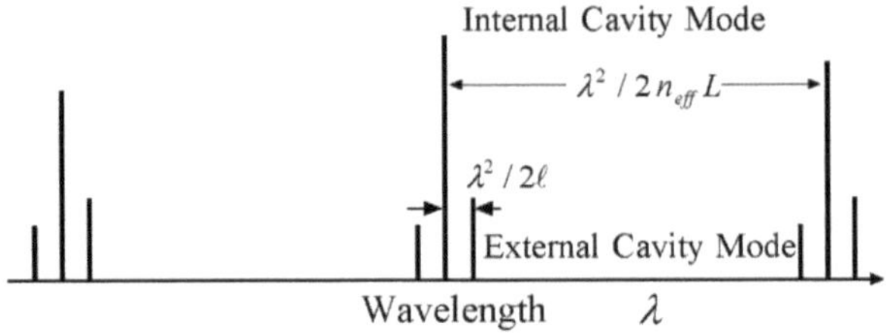

The subject of phase distortion can be discussed in terms of mode competition among the external modes. With such competition, the asymmetric gain saturation $H_{p'(p)}$ must be very strong because the wavelength difference $\Delta\lambda$ is very small. The increase in flat-type noise is explained by asymmetric gain saturation as presented in Sect. 10.5.2.

We need to reconfirm here that counting the external cavity modes is an alternative method to evaluate the time-delayed phase change in terms of $\{\omega_p\tau + \theta_p(t) - \theta_p(t-\tau)\}$.

10.7 Noise Reduction

Optical feedback noise generated when using semiconductor lasers is a difficult problem to solve. Inserting an optical isolator is the simplest means to mitigate optical feedback. However, because optical isolators are expensive, several other techniques have been developed to reduce noise using cheaper methods.

These techniques include

- *superposition of high-frequency current* [9],
- use of *self-pulsation lasers* [10],
- *electric negative feedback* [4].

The techniques used to reduce noise with each of these methods are explained below.

10.7.1 Superposition of High Frequency Current

The operating method is rather simple. We superimpose a *high-frequency current* with a frequency of 100 MHz to 1 GHz on the *injection current* of the semiconductor laser, as illustrated in Fig. 10.17. In most cases, the low-frequency-type optical feedback noise can then be reduced.

We examine the reduction mechanism of noise based on two-mode competition. We assume an injection current of the form

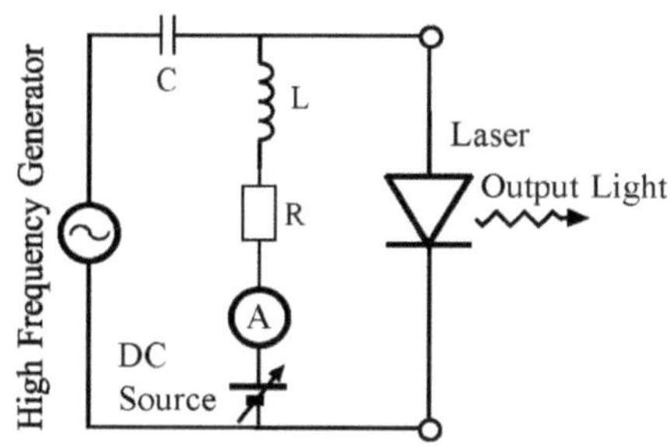

Fig. 10.17 Superposition of high-frequency current on the injection current

$$I = \bar{I} + I_M \cos(\Omega_M t) = \bar{I} + \frac{I_M}{2}\left(e^{j\Omega_M t} + e^{-j\Omega_M t}\right), \tag{10.130}$$

where $\bar{I}$ is the DC current, and I_M the amplitude of the superposing high-frequency current with Ω_M its angular frequency. Under this modulation, the photon numbers and the electron density are modulated and written as

$$S_p = \tilde{S}_p + S_{p\Omega_M} e^{j\Omega_M t} + S^*_{p\Omega_M} e^{-j\Omega_M t}, \tag{10.131}$$

$$S_q = \tilde{S}_q + S_{q\Omega_M} e^{j\Omega_M t} + S^*_{q\Omega_M} e^{-j\Omega_M t}, \tag{10.132}$$

$$N = \tilde{N} + N_{\Omega_M} e^{j\Omega_M t} + N^*_{\Omega_M} e^{j\Omega_M t}, \tag{10.133}$$

where terms with suffix Ω_M are modulation components and terms with symbol ~ are other components including fluctuation terms. By substituting these equations into (10.86) and (10.87), together with (10.89) and (10.90), we obtain

$$\frac{d\tilde{S}_p}{dt} = \left[\tilde{A}_p - B\tilde{S}_p - D\tilde{S}_q - G_{th(p)}\right]\tilde{S}_p + K_p + F_p(t), \tag{10.134}$$

$$\frac{d\tilde{S}_q}{dt} = \left[\tilde{A}_q - B\tilde{S}_q - D\tilde{S}_p - G_{th(q)}\right]\tilde{S}_q + K_q + F_q(t), \tag{10.135}$$

$$\tilde{A}_p = \xi a\left\{\tilde{N} - N_g - b(\lambda_p - \lambda_o)^2\right\}, \tag{10.136}$$

$$\tilde{A}_q = \xi a\left\{\tilde{N} - N_g - b(\lambda_q - \lambda_o)^2\right\}, \tag{10.137}$$

where

$$K_p = \xi a(\tilde{N} + N_{\Omega_M} S^*_{p\Omega_M} + N^*_{\Omega_M} S_{p\Omega_M}), \tag{10.138}$$

$$K_q = \xi a(\tilde{N} + N_{\Omega_M} S^*_{q\Omega_M} + N^*_{\Omega_M} S_{q\Omega_M}). \tag{10.139}$$

The effects from the superposition of high-frequency current is included in the newly introduced terms K_p and K_q.

Flow diagrams of the two-mode competition operating with the superposition of high-frequency current are shown in Fig. 10.18. The axes in these figures represent $\tilde{S}_p$ and $\tilde{S}_q$. The lines L_p and L_q are the respective steady-state trajectories for $d\tilde{S}_p/dt = 0$ and $d\tilde{S}_q/dt = 0$ in (10.134) and (10.135). If K_p and K_q are small, the flow diagram of Fig. 10.18a is the same as those in Figs. 9.4c and 10.7a, indicating an active bi-stable state for the operating points that tends to give mode hopping through optical feedback by instantaneous changes in $G_{\mathrm{th}(p)}$ and $G_{\mathrm{th}(q)}$. However, if the terms K_p and K_q are larger, the lines L_p and L_q are more concave and the operating points P and Q approach each other and ultimately overlap to form the point M, as depicted in Fig. 10.18b. This instance represents a *mono-stable state* of a *stable multimode operation* with modes p and q [11]. Although the new point M can fluctuate owing to instantaneous changes in $G_{\mathrm{th}(p)}$ and $G_{\mathrm{th}(q)}$, this fluctuation is much smaller than that occurring in modehopping. Thus *mode hopping noise is suppressed* under these circumstances [12].

Numerically calculated examples of optical feedback noise and its suppression by the superposition of the high-frequency current are shown in Fig. 10.19, where (a) gives the frequency spectrum and (b) the variation of the suppressed noise with amplitude I_M of the high-frequency current. The equations used are (10.120)–(10.124) and (10.130) but setting $W(t) = 0$ to simplify the treatment. We used the same model for the AlGaAs laser as that for Fig. 10.12, again employing 13 internal cavity modes. Suppression of optical feedback noise is well demonstrated in the RIN plots.

Here, we summarize the use of noise suppression by high-frequency [13]:

(1) The *most efficient frequency* for current superposition to use is the *relaxation oscillation frequency* f_R, because the photon numbers and the electron density are mainly modulated at this frequency. However, conventional semiconductor lasers, used for example in optical pick up, are not designed for operations with the higher GHz frequencies. Direct excitation with f_R is thus difficult. However, with the lower frequency of f_M driving the current, vibrations with *higher harmonic components* reach f_R and noise suppression becomes viable.

(2) If the modulation frequency f_M is related to the *external cavity frequency* $f_{ex} = 1/\tau = 2\ell/\lambda^2$, that is, $mf_M = nf_{ex}$ where m and n are positive integers, *noise suppression does not work*. This unusual phenomenon is caused by phase locking from the term $\{\theta_p(t-\tau) - \theta_p(t)\}$ for feedback light in (10.122), (10.126), and (10.127), resulting in a very small modulation of the electron density N_{Ω_M}.

(3) Noise reduction by high-frequency superposition is very *effective for low-frequency-type noise*, which is caused by mode-hopping among the internal cavity modes, but is *not so effective for flat-type noise*, which is caused by phase distortion.

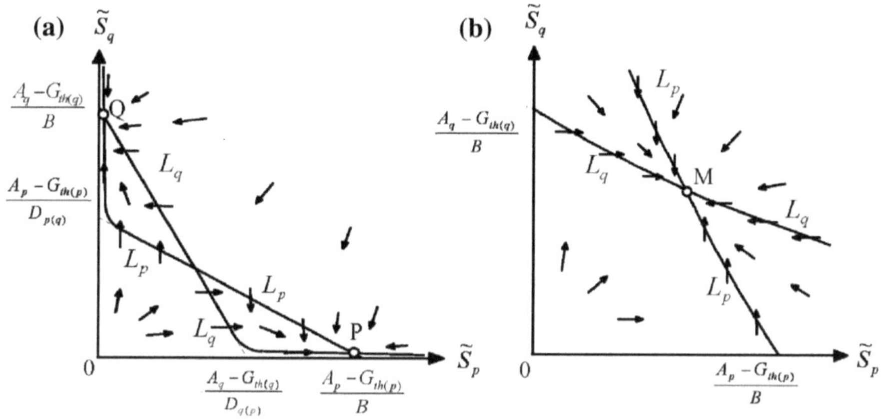

Fig. 10.18 Two-mode competition under the superposition of high-frequency current. **a** For small K_p and K_q, **b** For large K_p and K_q

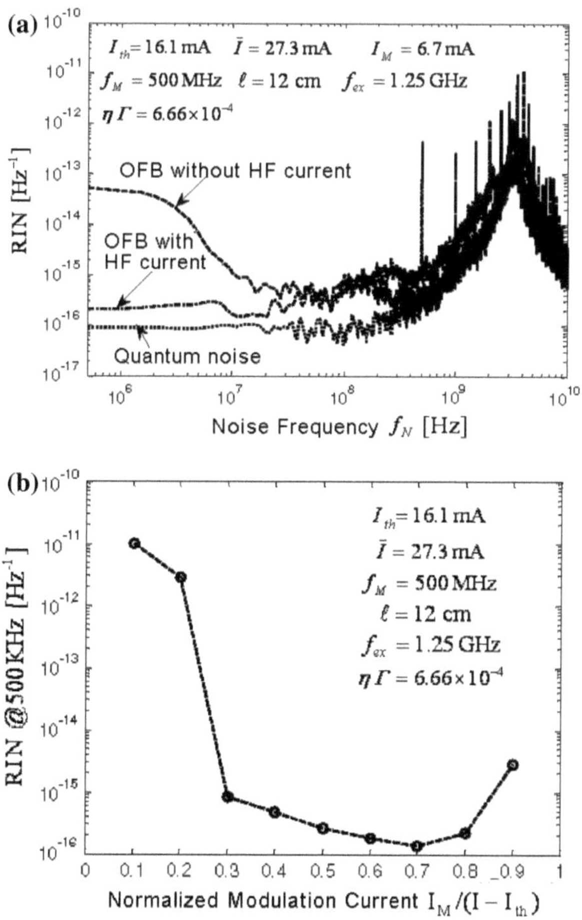

Fig. 10.19 Example from numerical simulation of suppression of optical feedback (OFB) noise by superposition of high-frequency (HF) current. The same model for an AlGaAs laser is used as in Fig. 10.12. **a** Noise frequency spectrum, **b** Variation with amplitude of suppressed noise by modulated current superposition

10.7.2 Use of Self-Pulsation Lasers

Self-pulsation lasers are lasers in which the output light automatically produces a pulsating periodic variation of several hundred MHz, even when the driving current is a regular DC current [10]. Cross-sectional structures of the self-pulsation lasers are illustrated in Fig. 10.20. The self-pulsation lasers consist of an *active region (Act.)* and *saturable absorbing regions (Sat. Ab.)*.

The operating mechanism of these self-pulsation lasers is similar to the relaxation oscillation in conventional semiconductor laser, which has been introduced for direct modulation in Sect. 7.3 and noise characteristics in Sect. 10.4. The relaxation oscillation of frequency f_R in conventional semiconductor lasers is caused by periodic variations of photon and electron numbers, but is relaxed because of the slight delay in their mutual responses.

In the self-pulsation lasers, the saturable absorbing region is placed around the active region, which speeds up the temporal variation in photon number. Hence, the resonance dynamics of frequency f_R is continually maintained giving rise to self-pulsation.

As the electron density N and the photon numbers $\sum_p S_p$are modulated by a high frequency f_R, self-pulsation lasers mitigate optical feedback noise by almost the same mechanism that underlies the superposition of high-frequency current, explained in the last section. As noise reduction is obtained automatically without any additional device such as the high-frequency generator, the self-pulsation laser is quite suited in applications such as optical pick-up in optical disc systems.

However, the self-pulsation laser is not without drawbacks. We offer the following comments.

(1) In a certain range of application distance ℓ to the reflecting point, the oscillation frequency of the self-pulsation is locked with $f_{ex} = c/2\ell$. The phase relation between the modulation terms of the electron density N_Ω and the photon number $S_{p\Omega}$ then changes resulting in a diminishing of noise reduction [14].

(2) The threshold current level in a self-pulsation laser has to be higher than that in a conventional laser, because the saturable absorbing region consumes energy. Self-pulsation lasers in AlGaAs systems have been successfully produced. However, lasers based on these InGaN systems have not been used often because too much energy is lost to the saturable absorbing region.

10.7.3 Electric Negative Feedback

A third noise reduction method uses an *electric negative feedback* loop, as illustrated in Fig. 10.21. We assume that one part of the laser output S_Ω is detected and electrically amplified with negative phase and injected into the laser as I_Ω. We

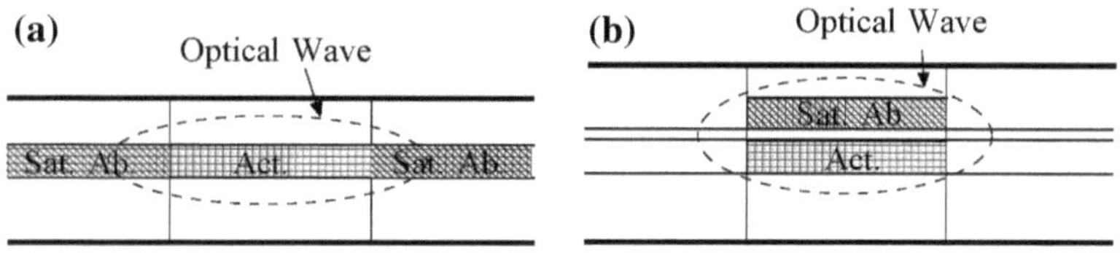

Fig. 10.20 Cross-sectional structures of the self-pulsation lasers. **a** Saturable absorbing regions locate in the lateral direction, **b** Saturable absorbing region locates in the thickness direction

ascribe a transfer function $T(\Omega)$ to the injected current I_Ω, which is added to the photon number along with generated noise $S_{i\Omega}$. Then,

$$S_\Omega = S_{i\Omega} + I_\Omega T(\Omega) \tag{10.140}$$

is obtained. From optical detection and amplification, the injection current component is given as

$$I_\Omega = -(N_{d\Omega} + \eta S_\Omega)A(\Omega), \tag{10.141}$$

where η is the detection efficiency of the photon number, $N_{d\Omega}$ the noise generated in the optical detector, and $A(\Omega)$ the amplification rate of the electric amplifier. By substituting (10.141) into (10.140), we get

$$S_\Omega = \frac{-N_{d\Omega}A(\Omega)T(\Omega) + S_{i\Omega}}{1 + A(\Omega)T(\Omega)}. \tag{10.142}$$

If we increase the amplification rate $A(\Omega)$ to a large value, then

$$S_\Omega = -N_{d\Omega}, \tag{10.143}$$

and the generated noise $S_{i\Omega}$ in the laser is completely suppressed. *Noise levels*, determined only with the *optical detector, can be reduced below quantum noise levels.*

However, this noise reduction method by the electric negative feedback also has several drawbacks.

(1) The electric negative feedback is not so effective in reducing mode-hopping noise (the low-frequency-type noise), because fluctuations in photon number S_Ω cannot be controlled by injection current I_Ω. Thus (10.140) does not hold for mode-hopping noise.
(2) The optical detection must be done using *light output from the front facet*, as shown in Fig. 10.21. In many cases, light output from the back facet is used in auto-power control. However, this is not suitable in reducing noise by electric negative feedback. The reason is that some lasers show a *negative correlation* between fluctuations in the output from the front facet and that from the back

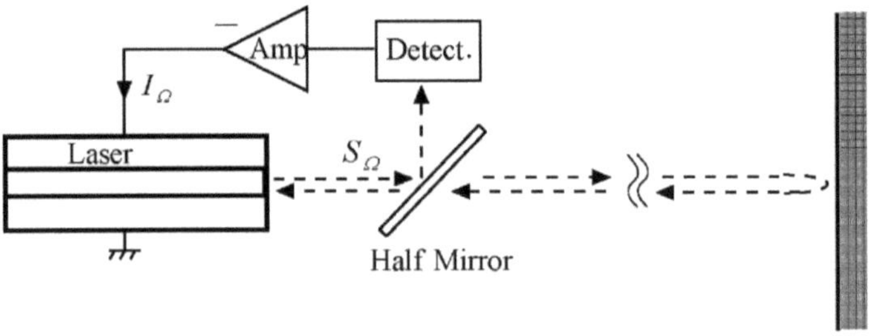

Fig. 10.21 Electric negative feedback

facet, because the spatial distribution of the optical wave along the cavity length differs for each longitudinal mode.

(3) The operating frequency range of the electric amplifier needs to be wide enough with the amplifier suitably designed with a cut-off frequency for the frequency range used. The amplification $A(\Omega)$ should have the same phase delay over this frequency range to maintain negative feedback. However, conventional electric amplifiers feature changes in phase around the cut-off frequency. Maintaining negative feedback over a wide frequency range is therefore not easy.

10.7.4 Hybrid Usage of the Noise Reduction Methods

The noise reduction methods mentioned above are capable of reducing noise but further reductions are required. We tried to use simultaneously two such methods, the high-frequency current super position and the electric negative feedback. The experimentally measured noise levels are shown in Fig. 10.22. The horizontal dashed line represents the level of quantum noise which is obtained without optical feedback under stable single-mode operation. This figure suggests that noise levels can be reduced below the *quantum noise level* even under optical feedback [15].

Another hybrid method is to combine positive and negative electric feedback. As mentioned in Sect. 10.7.3, the amplifier for negative feedback tends to change the phase of transmission in the higher frequency range and produces electric oscillation arising from positive feedback. We have exploited this property, that is, the electric feedback loop gives *positive feedback* revealing oscillation with frequency f_o, while the frequency range below f_o functions as *negative feedback*, with only *one chip of the electric amplifier*. Experimentally obtained data are shown in Fig. 10.23. The oscillation frequency produced from positive feedback is $f_o = 250$ MHz. Noise in the lower frequency range below 100 MHz is well suppressed below the *quantum noise level* [16].

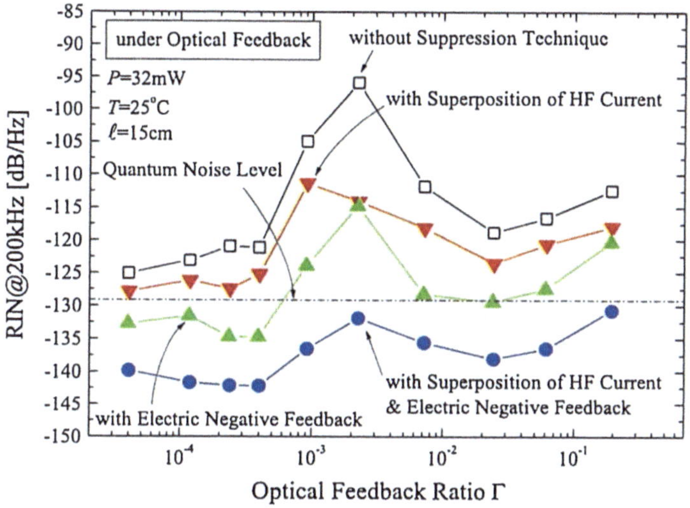

Fig. 10.22 Noise reduction by simultaneous use of high-frequency (HF) current superposition and electric negative feedback

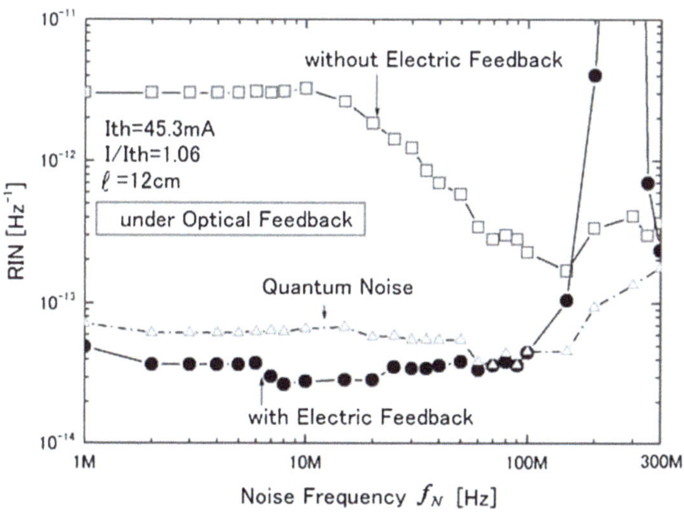

Fig. 10.23 Noise reduction by positive and negative electric feedback

Bibliography

1. M. Yamada, Variation of intensity noise and frequency noise with the spontaneous emission factor in semiconductor lasers. IEEE J. Quantum Electron. **30**(7), 1511–1519 (1994)
2. C.H. Henry, Theory of linewidth of semiconductor lasers. IEEE J. Quantum Electron. **QE-18**(9), 259–264 (1982)
3. M. Yamada, Theory of mode competition noise in semiconductor injection lasers. IEEE J. Quantum Electron. **QE-22**(7), 1052–1059 (1986)
4. M. Yamada, N. Nakaya, M. Funaki, Characteristics of mode-hopping noise and its suppression with the help of electric negative feedback in semiconductor lasers. IEEE J. Quantum Electron. **QE-23**(8), 1297–1302 (1987)
5. M. Ahmed, M. Yamada, Influence of instantaneous mode competition on the dynamics of semiconductor lasers. IEEE J. Quantum Electron. **38**(6), 682–693 (2002)
6. S.M.S. Imran, M. Yamada, Y. Kuwamura, Theoretical analysis of optical feedback noise based on multimode model of semiconductor lasers. IEEE J. Quantum Electron. **48**(4), 521–527 (2012)
7. M. Yamada, A. Kanamori, S. Takayama, Experimental evidence of mode competition phenomena on the feedback induced noise in semiconductor lasers. IEICE Trans. Electron. **E79-C**(12), 1766–1768 (1996)
8. M. Yamada, M. Suhara, Analysis of excess noise induced by optical feedback in semiconductor lasers based on mode competition theory. Trans. IEICE, **E-73**(1), 77–82 (1990)
9. A. Oishi, M. Chinone, M. Ojima, A. Arimoto, Noise characteristics of high-frequency superposed laser diodes for optical disc systems. Electron. Lett. **20**(20), 821–822 (1984)
10. S. Matsui, H. Takiguchi, H. Hayashi, S. Yamamoto, S. Yano, T. Higikata, Suppression of feedback-induced noise in short-cavity V-channeled substrate inner stripe lasers with self-oscillation. Appl. Phys. Lett. **43**(3), 219–221 (1983)
11. M. Yamada, T. Mizukami, Theoretical conditions for single longitudinal mode operation under direct modulation in injection lasers. Trans. IECE Jpn. **E65**(11), 614–620 (1982)
12. M. Yamada, T. Higashi, Mechanism of the noise reduction method by superposition of high-frequency current for semiconductor injection lasers. IEEE J. Quantum Electron. **27**(3), 380–388 (1991)
13. S.M.S. Imran, M. Yamada, Numerical analysis of suppression effects on optical feedback noise by superposition of high frequency current in semiconductor lasers. IEEE J. Quantum Electron. **49**(2), 196–204 (2013)
14. M. Yamada, Computer simulation of feedback induced noise in semiconductor lasers operating with self-sustained pulsation. IEICE Trans. Electron. **E81-C**(5), 768–780 (1998)
15. M. Yamada, I. Tera, K. Matsuoka, T. Hama, U. Kuwamura, Reduction of intensity noise in semiconductor lasers by simultaneous usage of the superposition of high frequency current and the electric negative feedback. IEICE Trans. Electron. **E95-C**(8), 1444–1446 (2012)
16. M. Yamada, M. Huda, E. Teraoka, Y. Kuwamura, Effective noise reduction by electric positive and negative feedback in semiconductor lasers. IEEE J. Quantum Electron. **45**(10), 1248–1254 (2009)

Part III
Structures for Superior Characteristics in the Semiconductor Lasers

Chapter 11
Quantum Well Structure

11.1 What is a Quantum Well?

The fundamental electron wave in a bulk semiconductor material is the *Bloch function*, defined in (5.16) and recalled here as

$$|b\rangle = |\varphi_b(\boldsymbol{r})\rangle = \frac{1}{\sqrt{V_e}} e^{j\boldsymbol{k}_b \boldsymbol{r}} u_b(\boldsymbol{r}), \quad \text{for bulk material,} \tag{11.1}$$

where $\boldsymbol{k}_b$ is the wave number, and $u_b(\boldsymbol{r})$ a periodic function having the same distribution at each lattice point and characterizing the effective mass in the band structure. V_e, the occupation volume of an electron, was introduced in (6.23) with a length of spread ℓ such that

$$V_e = \ell^3. \tag{11.2}$$

This length ℓ is regarded as a *coherence length* of the electron wave, which is determined by the *mutual scattering among electrons*. It is expected to be of order $\ell \approx 30$ nm in bulk semiconductor materials such as GaN, GaAs, and InP.

The purpose of a heterojunction structure is to confine electrons into an active region using cladding regions which act as potential barriers. These regions have higher band gaps than the active region, as illustrated in Fig. 1.10. When the thickness of the active region d is greater than the coherence length ℓ, the region is regarded as a *bulk semiconductor material*. However, if this thickness d is much smaller than the coherence length ℓ, $d \ll \ell$, the electron wave reflects at the boundaries and forms a *standing wave*, as illustrated in Fig. 11.1. The electron wave is confined to the lower energy gap region thus maintaining its coherence characteristic; we call the associated structure a *quantum well*. The center region having small band gap is called the *well region*, and the surrounding regions having large band gaps are called the *barrier regions*.

M. Yamada, *Theory of Semiconductor Lasers*,
Springer Series in Optical Sciences 185,
DOI: 10.1007/978-4-431-54889-8_11, © Springer Japan 2014

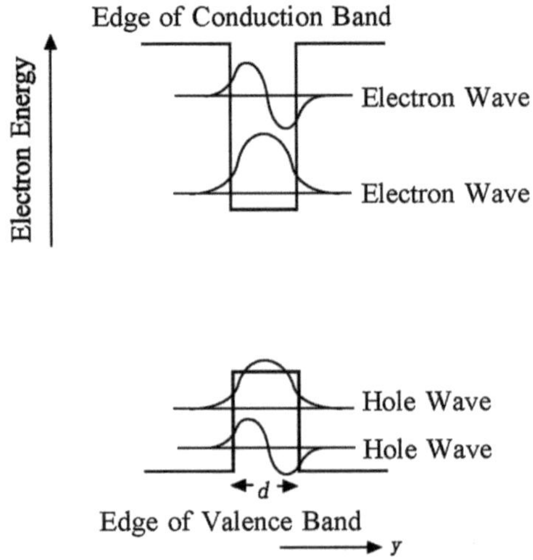

Fig. 11.1 Quantum well structure

The electron wave function in the quantum well structure is given as

$$|b\rangle = |\phi_b(\boldsymbol{r})\rangle = \psi_n(y)\,\frac{1}{\ell}\,e^{j(k_x x + k_z z)}u_b(\boldsymbol{r}) = \psi_n(y)\,\frac{1}{\ell}\,e^{j\boldsymbol{k}_{//}\boldsymbol{r}_{//}}u_b(\boldsymbol{r}), \tag{11.3}$$

where $\psi_n(y)$ is a *wave function* along the y direction and indexed by integer n which refers to its energy level. $\boldsymbol{k}_{//}$ and $\boldsymbol{r}_{//}$ are wave number and distance in the x–z plane given by the two-dimensional expressions

$$\boldsymbol{k}_{//} = \boldsymbol{e}_x\,k_x + \boldsymbol{e}_z\,k_z, \tag{11.4}$$

$$\boldsymbol{r}_{//} = \boldsymbol{e}_x\,x + \boldsymbol{e}_z\,z, \tag{11.5}$$

where $\boldsymbol{e}_x$ and $\boldsymbol{e}_z$ are dimensionless unit vectors for the x and z directions, respectively.

The wave function $\psi_n(y)$ is determined from its time-independent wave equation,

$$\left\{-\frac{\hbar^2}{2\,m_i}\frac{\partial^2}{\partial\,y^2} + U_i(y)\right\}\psi_n(y) = W_n\,\psi_n(y), \tag{11.6}$$

where m_i is the *effective mass* in the ith region, $U_i(y)$ the *height of the potential barrier* in the ith region measured with respect to the well region, and W_n the *energy component* for the y direction. Equation (11.6) is very similar to Maxwell's wave equation for the optical wave in a slab waveguide, discussed in Chap. 4. The solution in the various regions takes the form

$$\psi_n(y) = \begin{cases} A\cos\left(k_y y - \dfrac{(n-1)\pi}{2}\right) & \text{for } -\dfrac{d}{2} \le y \le \dfrac{d}{2} \\ A\cos\left(k_y \dfrac{d}{2} + \dfrac{(n-1)\pi}{2}\right) e^{\gamma(y+d/2)} & \text{for } y \le -\dfrac{d}{2} \\ A\cos\left(k_y \dfrac{d}{2} - \dfrac{(n-1)\pi}{2}\right) e^{-\gamma(y-d/2)} & \text{for } y \ge \dfrac{d}{2} \end{cases} \tag{11.7}$$

with energy value

$$W_n = \frac{\hbar^2 k_y^2}{2m_c} = U_B - \frac{\hbar^2 \gamma^2}{2m_B}, \tag{11.8}$$

where m_c and m_B are the effective masses of the conduction band in the well and the barrier regions, respectively, and U_B is the height of the potential barrier. The boundary conditions at $y = \pm d/2$ ensure the continuity of $\psi_n(y)$ and $(\partial\psi_n(y)/\partial y)/m_i$. The former requires the value of $W_n\,\psi_n(y)$ to be the same between the well and the potential barrier regions. The latter relates to the wave flow along the y direction, where $p_y/m_i = (-j\hbar\partial/\partial y)/m_i$ gives the velocity of the wave flow. The unknown parameters k_y and γ are solved from these boundary conditions. The amplitude A is determined from the normalization condition

$$\int_{-\infty}^{\infty} |\psi_n(y)|^2\, dy = 1. \tag{11.9}$$

By adding the energy components along the x and z directions, the energy level W_b of the conduction electron is given as

$$W_b = W_c + \frac{\hbar^2}{2m_c}(k_y^2 + \boldsymbol{k}_{//}^2) = W_c + \frac{\hbar^2}{2m_c}\left(k_x^2 + k_y^2 + k_z^2\right), \tag{11.10}$$

where W_c is the edge energy of the conduction band in the well region.

Equations (11.3) and (11.6) are defined for electrons in the conduction band. Similar equations are obtained for holes in the valence band by replacing the energy level index b for a and redefining the hole potential so that the downward direction is higher in energy, as depicted in Fig. 11.1.

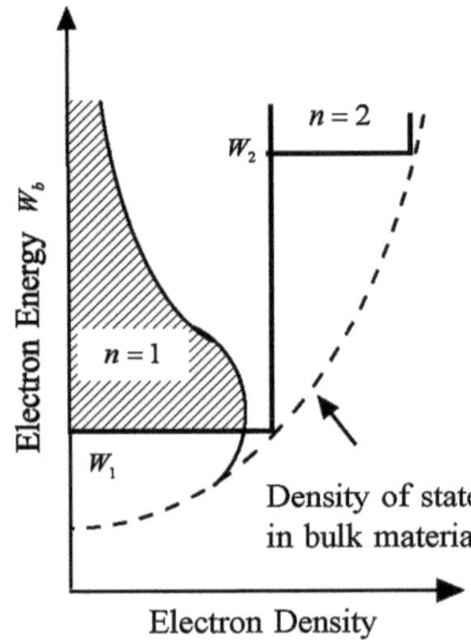

Fig. 11.2 Electron distribution in the conduction band

11.2 Advantage of Using Quantum Well Structures

The energy distribution of the electrons in the conduction band is illustrated in Fig. 11.2, where the broken line represents the density of states in the bulk semiconductor material given by a parabolic function as in (6.32). Note that the density of states in the quantum well structure form a rectangular shape for each energy level n.

In operation, the lasing gain is achieved by population inversion for a limited number of energy levels, which corresponds to the *height of the electron density labeled by the horizontal axis* in Fig. 11.2. Meanwhile, the *threshold current is determined from the total amount of injected electrons* indicated by the hatched area. As found from this figure, the quantum well structure gives the same population inversion with a smaller number of the injected electrons than for the bulk semiconductor. Therefore, we evaluate the density of states in the quantum well structure.

We consider a box in the well region of volume V_e

$$V_e = \ell \times d \times \ell. \tag{11.11}$$

The quantum number n for the electron wave along the y direction is restricted to integer values and hence wave number k_y in (11.7) and (11.10) is also restricted. The distributions along the x and z directions can assume periodic form depending on $\boldsymbol{k}_{//}$, as has been discussed for the bulk semiconductor treated in Sect. 6.3 using the wave number space representation depicted in Fig. 6.4.

The two-dimensional area of the wave number space is

$$\int d^2\boldsymbol{k}_{//} = 2\pi \int k_{//}\, dk_{//}. \tag{11.12}$$

Dividing this area by $(2\pi/\ell)^2$ gives the number of states with the same energy along the x and z directions. By including the spin degeneracy, the density of states $g_{nc}(W_b)\, dW_b$ of level n in the conduction band is defined as

$$2 \times \left(\frac{\ell}{2\pi}\right)^2 2\pi\, k_{//}\, dk_{//} = V_e\, g_{nc}(W_b)\, dW_b. \tag{11.13}$$

With energy W_b given in (11.10), we have

$$dW_b = \frac{\hbar^2\, k_{//}}{m_c}\, dk_{//}, \tag{11.14}$$

which, when substituted into (11.13), gives

$$g_{nc}(W_b) = \frac{m_c}{\pi\, \hbar^2\, d}\, u(W_b - W_n - W_c), \tag{11.15}$$

where $u(W_b{-}W_n{-}W_c)$ is a step function defined as $u(W_b - W_n - W_c) = 0$ if $W_b < W_n + W_c$ and $u(W_b - W_n - W_c) = 1$ if $W_b > W_n + W_c$. Thus, the density of the states obtained is as shown in Fig. 11.2.

11.3 Various Types of Quantum Well Structure

One aspect of quantum well structures in semiconductor lasers is that the well thickness is *too thin*. This is crucial as the area of the spatial cross-section enters in the *optical confinement factor* ξ in (3.14) and (3.16).

Several structures have been developed to resolve this issue. Examples are shown in Fig. 11.3. Figure 11.3a, there are a number of wells creating a structure called a *multi quantum well (MQW) structure*. One problem of this structure is that, if the thicknesses of the barrier layers are thin enough to allow coupling between electron waves in each well, the number of energy levels increases with the number of wells, as is shown in Appendix 12 [1]. That is, the MQW structures tend to lose the advantage that single quantum well structures have. Additionally, the injection of electrons and holes into the wells is harder when barrier thicknesses are rather large, because then electrons and holes have to pass through many barrier layers to reach the more distant wells.

Figure 11.3b, intermediate layers are placed between the well and the cladding layers; this is called the *separate confinement hetero (SCH)-structure*. Because the

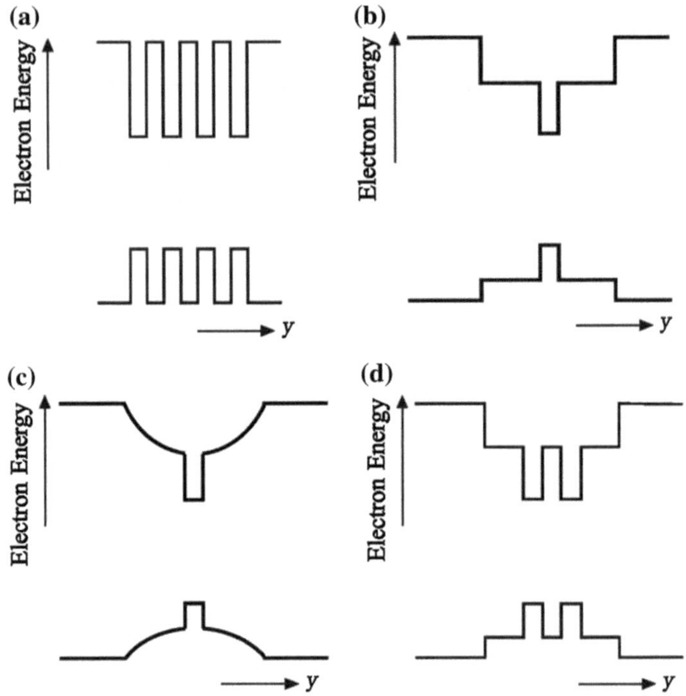

Fig. 11.3 Various types of quantum well structure. **a** Multi Quantum Well (MQW), **b** Separate Confinement Hetero (SCH), **c** Graded-Index SCH (GRIN-SCH), **d** Double Well SCH

refractive index becomes higher in materials having smaller band gaps, the intermediate layers help to concentrate the optical wave in the well. The main problem with this structure is that the injected electrons and holes tend to flow over into the intermediate layers.

Figure 11.3c, the band gaps of the intermediate layers gradually increase with distance from the well. This structure is called a *graded-index separate-confinement hetero* (*GRIN-SCH*)*-structure*. The design aim of this structure is to halt the rapid increase of electron and hole overflow in the structure (b). The problem with this structure is the required precision in fabrication technology.

Figure 11.3d combines the MQW structure in (a) with the SCH-structure in (b). As a general rule, the threshold current density J_{th} increases with the total thickness of the wells and decreases with confinement factor ξ defined as the cross-sectional area between electron wave and optical wave. Which structure is more suitable and how many wells are needed depends on materials used.

11.4 Polarization Dependent Gain

11.4.1 Dipole Moment

One feature of quantum well structures is that the *lasing gain differs depending on the optical polarization* of the laser light. Most quantum well lasers are designed to oscillate with the TE mode. We begin our discussion by determining properties of the dipole moment.

As has been introduced in (5.23)–(5.28), the interaction between the electron and optical waves contains the factor $\boldsymbol{p} \cdot \boldsymbol{A}$, which is replaced with $\boldsymbol{r} \cdot \boldsymbol{E}$ based on the dipole approximation. This approximation implies that the optical emission can be described by the vibrating motion of the localized electron as explained in Figs. 6.1 and 6.2. Hence, variations in position $\boldsymbol{r}$ and momentum $\boldsymbol{p}$ for the vibrating electron are compatible. Now, we determine the time derivative of the position operator using the Heisenberg equation

$$\dot{\boldsymbol{r}} = \frac{1}{j\hbar}[\boldsymbol{r}, H_o] = \frac{1}{j\hbar}(\boldsymbol{r}\,H_o - H_o\,\boldsymbol{r}), \tag{11.16}$$

where H_o is the principle Hamiltonian for the semiconductor material given by (5.20). The matrix element of the momentum operator is then expressed as

$$\langle b|\boldsymbol{p}|a\rangle = m_o\langle b|\dot{\boldsymbol{r}}|a\rangle = \frac{m_o}{j\hbar}\langle b|\,(\boldsymbol{r}\,H_o - H_o\,\boldsymbol{r})\,|a\rangle = \frac{j\,m_o W_{ba}}{\hbar}\langle b|\boldsymbol{r}|a\rangle. \tag{11.17}$$

Hence, the dipole moment is related to the matrix element for momentum,

$$\langle b|\boldsymbol{R}|a\rangle = \frac{e\,\hbar}{j\,W_{ba}\,m_o}\langle b|\boldsymbol{p}|a\rangle. \tag{11.18}$$

The momentum operator $\boldsymbol{p}$ is given as the spatial operator,

$$\boldsymbol{p} = -j\hbar\,\nabla. \tag{11.19}$$

We substitute this operator into (11.18) with (11.3) and use the property that the function $\psi_n(y)\,e^{j\mathbf{k}_{//}\,\mathbf{r}_{//}}$ is much smoother than $u_a(\boldsymbol{r})$ and orthogonality between $u_b(\boldsymbol{r})$ and $u_a(\boldsymbol{r})$. Then the momentum element becomes

$$\langle b|\boldsymbol{p}|a\rangle = \int \varphi_b^*(\boldsymbol{r})\,\boldsymbol{p}\,\varphi_a(\boldsymbol{r})\,d^3\boldsymbol{r} = \zeta_y \int_{prim} u_b^*(\boldsymbol{r})\,\boldsymbol{p}\,u_a(\boldsymbol{r})\,d^3\boldsymbol{r}\;\delta_{\boldsymbol{k}_{b//},\boldsymbol{k}_{a//}}, \tag{11.20}$$

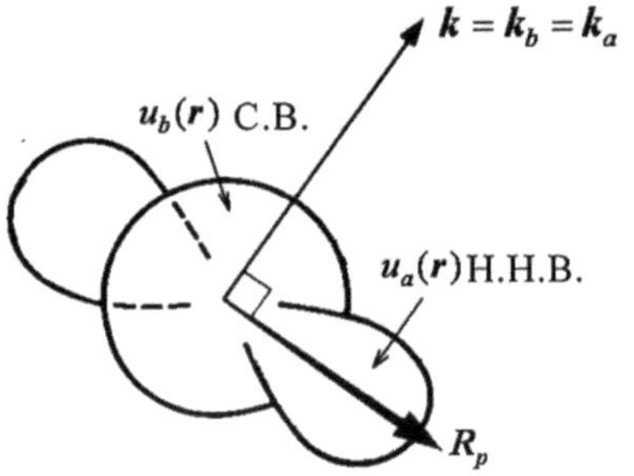

Fig. 11.4 Distributional shapes of the electron wave function and dipole moment; (*C.B.* conduction band; *H.H.B.* in the heavy-hole band)

where ζ_y is the cross-sectional rate of the wave distribution along the y direction,

$$\zeta_y = \int_{-\infty}^{\infty} \psi_{n_C}^*(y)\, \psi_{n_J}(y)\, dy. \tag{11.21}$$

The integral $\int_{prim} d^3\boldsymbol{r}$ is a three-dimensional spatial integration over the primitive cell which is defined as the volume of space surrounding each lattice point. The value of $\int_{prim} u_b^*(\boldsymbol{r})\boldsymbol{p}\; u_a(\boldsymbol{r})\, d^3\boldsymbol{r}$ has been analyzed based on the $\boldsymbol{k} \cdot \boldsymbol{p}$ *perturbation method* and is related to parameters such as the effective mass of the electron and the energy band gap, among others shown in Fig. 5.4. The electron wave functions in a *III–V compound material*, formed under *covalent bonding*, consist of four degenerate electron wave functions. One is the *s-like orbit* having a spherically symmetric shape and the other three are *p-like orbits*, $p_{x'}$, $p_{y'}$ and $p_{z'}$ having dumbbell-shaped distributions in three-dimensional space. The function $u_b(\boldsymbol{r})$ of the *conduction band* mainly consists of *s*-like orbits. The *p*-like orbits are combinations, such as $p_{x'} \pm jp_{y'}$, and have orthogonal eigensolutions that constitute the *heavy-hole band*, the *light-hole band*, and the *split-off band*, as in Fig. 5.4.

Shapes for the functions $u_b(\boldsymbol{r})$ in the conduction band and $u_a(\boldsymbol{r})$ in the heavy-hole band are illustrated in Fig. 11.4. The shape of the function $u_b(\boldsymbol{r})$ is almost spherical, whereas the function $u_a(\boldsymbol{r})$ lies in a direction which is perpendicular to the wave number vector $\boldsymbol{k} = \boldsymbol{k}_b = \boldsymbol{k}_a$. Therefore, the original dipole moment $\boldsymbol{R}_p$ should be in the same direction. The square value of the original dipole moments are calculated using the $\boldsymbol{k} \cdot \boldsymbol{p}$ *perturbation method* [2]:

$$|R_p|^2 = \frac{\hbar^2 e^2}{4 m_o W_{ba}^2}\left(\frac{m_o}{m_c} - 1\right)\frac{W_g\,(W_g + \Delta)}{W_g + (2/3)\,\Delta}. \tag{11.22}$$

The interaction with the optical wave depends on the projection of this dipole moment onto the electric field components. In Fig. 11.5, we have indicated the coordinate system used in assessing the contribution of the dipole moment. The wave number vector $\boldsymbol{k}$ is directed at angles θ and φ from the y and z axes, respectively. The vector $\boldsymbol{R}_p$ lies within the plane of the well at an angle φ from the

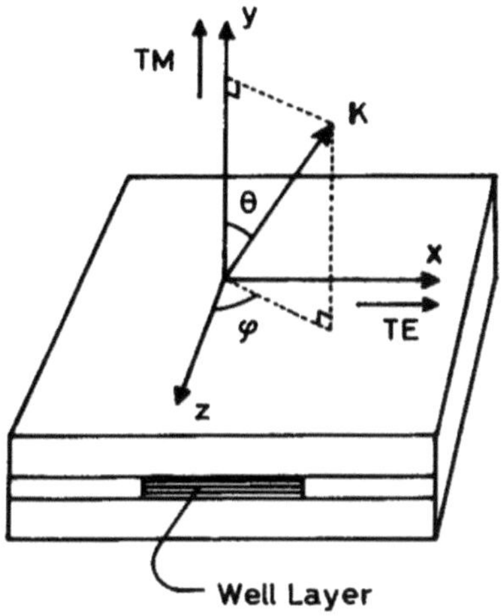

Fig. 11.5 Coordinate system to assess the dipole moment contribution

z axis and perpendicular to the $\boldsymbol{k}$ vector. The projections of the dipole moment onto the three axes are derived by evaluating the possible contributions of $p_{x'}$, $p_{y'}$, and $p_{z'}$:

$$\langle b|R_x|a\rangle = R_p(\cos\theta\,\sin\varphi \pm j\cos\varphi), \tag{11.23}$$

$$\langle b|R_y|a\rangle = -R_p\sin\theta, \tag{11.24}$$

$$\langle b|R_z|a\rangle = -R_p(-\cos\theta\,\cos\varphi \pm j\sin\varphi). \tag{11.25}$$

The square of the magnitude of the dipole moment $|R_{ba}|^2$ in the bulk material, given in (5.37) and (6.49), should be averaged over all possible directions θ and φ of $\boldsymbol{R}_p$, as will be shown next.

11.4.2 Dipole Moment in Bulk Material

The value of $|R_{ba}|^2$ in the *bulk material* for the *TE mode* resulting from the electron transition between the conduction and the *heavy-hole bands* is evaluated from $|\langle b|R_x|a\rangle|^2$ in (11.23),

$$|R_{ba}|^2_{bulk,TE,H} = \frac{|R_p|^2}{4\pi}\int_0^{\pi}\int_0^{2\pi}\left\{\cos^2\theta\,\sin^2\varphi + \cos^2\varphi\right\}\sin\theta\,d\varphi\,d\theta = \frac{2}{3}\,|R_p|^2. \tag{11.26}$$

Similarly, $|R_{ba}|^2$ for *TM modes* resulting from the same electron transition is evaluated from $|\langle b|R_y|a\rangle|^2$ in (11.24) as

$$|R_{ba}|^2_{bulk,TM,H} = \frac{|R_p|^2}{4\pi} \int_0^{\pi} \int_0^{2\pi} \sin^2\theta \sin\theta \, d\varphi \, d\theta = \frac{2}{3}|R_p|^2. \tag{11.27}$$

Note that the *averaged square values of the dipole moment in the bulk material are identical for both polarizations*. We write, as in (5.37),

$$|R_{ba}|^2_{bulk,H} = \frac{2}{3}|R_p|^2. \tag{11.28}$$

Given an electron transition between conduction and *light-hole bands*, the averaged dipole moment in the bulk material becomes

$$|R_{ba}|^2_{bulk,L} = \frac{4}{3}|R_p|^2 - |R_{ba}|^2_{bulk,CH} = \frac{2}{3}|R_p|^2. \tag{11.29}$$

Thus, the *dipole moments for heavy and light-hole bands are identical in the bulk material*.

Because many more holes exist in the heavy-hole band than the light-hole band, lasing gain is *mostly caused* by electron transitions to the *heavy-hole band* in the bulk material. However, this situation differs in quantum well lasers from bulk lasers.

11.4.3 Dipole Moment in the Quantum Well Laser

In the quantum well structure, the wave number component k_y along the y direction depends on discrete values, whereas the wave numbers along the x and z directions take continuous values. Thus, the averaging of the dipole moment is performed only for angle φ by fixing angle θ as θ_n which depends on energy level n. The value of $|R_{ba}|^2$ for a quantum well structure propagating *TE modes* resulting from electron transitions between the conduction and the *heavy-hole bands* is evaluated from $|\langle b|R_x|a\rangle|^2$ in (11.23) as,

$$|R_{ba}|^2_{QW,TE,H} = \frac{|R_p|^2}{2\pi} \int_0^{2\pi} \{\cos^2\theta_n \sin^2\varphi + \cos^2\varphi\} \, d\varphi = \frac{|R_p|^2}{2}(1+\cos^2\theta_n), \tag{11.30}$$

where the angle θ_n is given with the wave number,

$$\cos\theta_n = \frac{k_y}{k} = \frac{k_y}{\sqrt{k_y^2 + k_x^2 + k_z^2}}. \tag{11.31}$$

Similarly, for transitions between the conduction and *light-hole bands*, we have

$$|R_{ba}|^2_{QW,TE,L} = \frac{4}{3}|R_p|^2 - |R_{ba}|^2_{QW,TE,H}. \tag{11.32}$$

Analogously, for *TM modes* produced by electron transitions between the conduction and the *heavy-hole bands*, we have from $|\langle b|R_y|a\rangle|^2$ in (11.24)

$$|R_{ba}|^2_{QW,TM,H} = \frac{|R_p|^2}{2\pi}\int_0^{2\pi} \sin^2\theta_n \, d\varphi = |R_p|^2 \sin^2\theta_n, \tag{11.33}$$

and also for transitions from the conduction to the *light-hole bands* we find

$$|R_{ba}|^2_{QW,TM,L} = \frac{4}{3}|R_p|^2 - |R_{ba}|^2_{QW,TM,H}. \tag{11.34}$$

11.4.4 Lasing Gain

Lasing gain in a discrete energy system has been given in (6.22), whereas that in the bulk material is given in (6.49) by integrating over the continuum of energy levels. For a quantum well structure, lasing gain takes the form

$$g_p(\omega) = \xi\, g(\omega), \tag{11.35}$$

with

$$g(\omega) = \sqrt{\frac{\mu_o}{\varepsilon_o}} \frac{\omega}{n_{eff}} \sum_{J,n_J} \int_{W_g+W_n}^{\infty} \frac{|R_{ba}|^2_J\, g_{n_J}\, \{f_c(W_{ba}) - f_J(W_{ba})\}\, \hbar/\tau_{in}}{(\hbar\omega - W_{ba})^2 + (\hbar/\tau_{in})^2}\, dW_{ba} \tag{11.36}$$

where $g_p(\omega)$ is the mode gain defined for a *guided mode p* in the laser and $g(\omega)$ is called the gain coefficient for a *plane wave* sufficiently extended in space, as defined in (6.22) and (6.49). Additionally, ξ is the *field confinement factor* or spatial overlap ratio between the electric field E and the electron wave, J a suffix indicating the *heavy-hole* or *light-hole band*, and n the quantum number associated with the electron-hole pair along the y direction. $|R_{ba}|^2_J$ is the square of the dipole

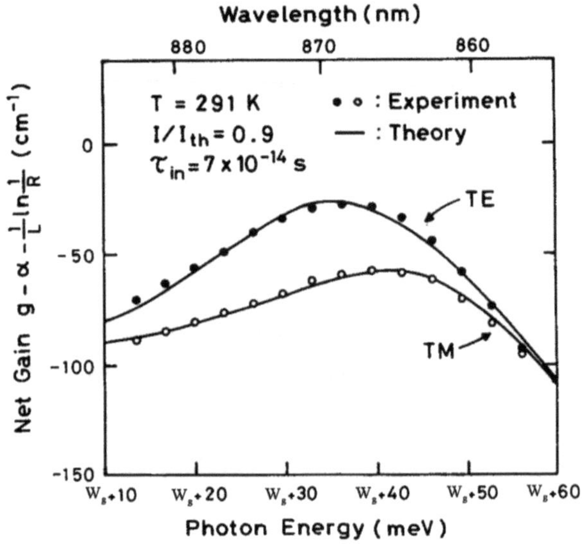

Fig. 11.6 Polarization dependent lasing gain in AlGaAs/GaAs multi-quantum well laser. The gain for TE modes is larger than for TM modes

moment magnitude for the electron transition between the conduction band and the heavy-hole or the light-hole band, where the difference between TE and TM is taken in account in (11.30)–(11.34). Finally, $g_{J,n}$ is the density of states for the electron-hole pairs given by

$$g_{J,n} = \frac{1}{\pi \hbar^2 d} \frac{m_c m_J}{m_c + m_J} u(W_{ba} - W_n - W_g) \tag{11.37}$$

$$\text{with } W_n = W_{nC} + W_{n_J}, \tag{11.38}$$

where m_J is the *effective mass* of the heavy hole or light hole, W_{nC} and W_{n_J} are the energy levels of the nth state in the conduction and valence bands, respectively, and $u(W_{ba} - W_n - W_g)$ is the step function.

Experimentally measured data and theoretically calculated results are shown in Fig. 11.6. The structure is a MQW type with five wells and four internal barrier layers. Materials used are GaAs for the well layers and $Al_{0.2}Ga_{0.8}As$ for the barrier layers. The thickness of each layer is $d_A = 12$ nm for the quantum well and $d_B = 3.5$ nm for the internal barrier layer. The effective masses in the GaAs well layers are assumed to be $m_C = 0.067m_o$, $m_H = 0.45m_o$, and $m_L = 0.11m_o$. From these data, the TE mode exhibits larger gain than the TM mode. Referring to Fig. 11.5, this behavior implies that the electron-hole dipole can move more easily along parallel directions in the x–z plane than the y direction perpendicular to the quantum well walls [3].

The contributing mechanisms for electron transitions are analyzed theoretically as in Fig. 11.7, where the analysis is the same as for Fig. 11.6. We find that gain for the *TE mode* is mostly given by electron transitions from *the conduction band*

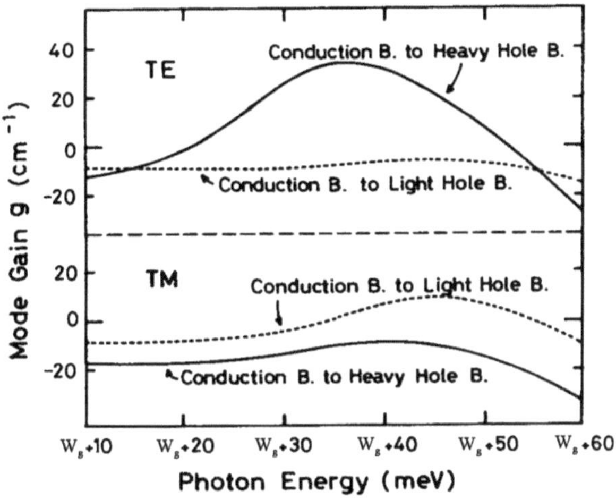

Fig. 11.7 Contribution of each transition mechanism to the gain profile

to the heavy-hole band. However, the gain for the *TM mode* shown in Fig. 11.6 is mostly induced by electron transitions from *the conduction band to the light-hole band*. The difference arising from these electron transitions is understood as follows: The wave function of the light-hole band lies in the *direction perpendicular* to both the wave number $\boldsymbol{k}$ and the main polarization $\boldsymbol{R}_p$ for the heavy-hole band, as in Fig. 11.4. Because the main polarization $\boldsymbol{R}_p$ lies within the x–z plane, the light-hole band is oriented in the y direction which is perpendicular to the walls.

A calculated example of the gain profile in a GaInAs/InP well is shown in Fig. 11.8. Related energy levels are indicated by arrows, where W_g is the band gap of the well layer, 1(e–h) is the transition with $n = 1$ for the conduction to heavy-hole bands, and 2(e–l) is that with $n = 2$ for the conduction to light-hole bands. A calculated example of the gain variation with electron injection density N is shown in Fig. 11.9 for several values of quantum well thickness.

The calculations leading to Figs. 11.8 and 11.9 were performed assuming that the electron wave is completely confined within the well regions. However, the overflow of injected electrons and holes into the barrier layers is an important factor needing to be taken into account [5]. A similar calculated example of the variation of the gain coefficient with injection current with overflow in a GRIN-SCH structure is shown in Fig. 11.10. The material parameters of the model are for a $Al_xGa_{1-x}As/GaAs$ system.

11.5 Strained Quantum Well

Semiconductor lasers have developed in parallel with developments in crystal growth technology of heterojunctions. To make a heterojunction with two different materials, the *crystal structures* and *lattice constants* in principle should match

Fig. 11.8 Calculated example of the gain profile for a single well in GaInAs/InP. Related energy levels are indicated by arrows [4] (Copyright, 1984, IEEE)

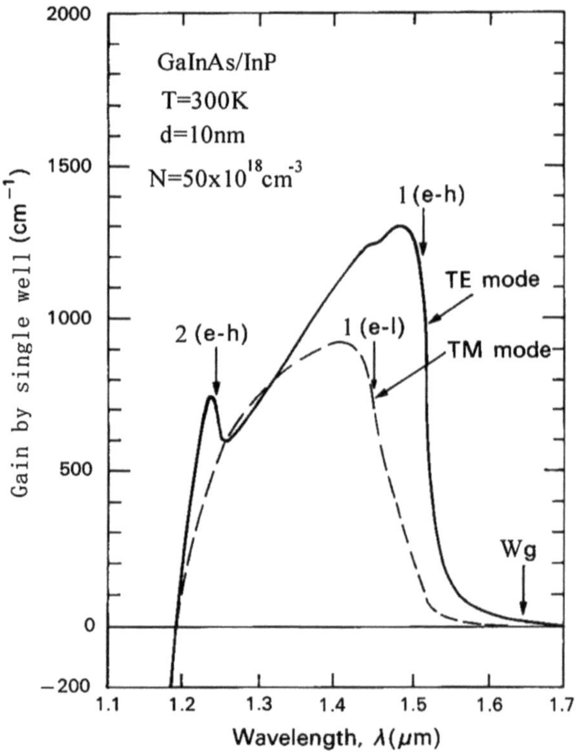

Fig. 11.9 Calculated examples of the gain variation with electron injection density in GaInAs/InP [4] (Copyright, 1984, IEEE)

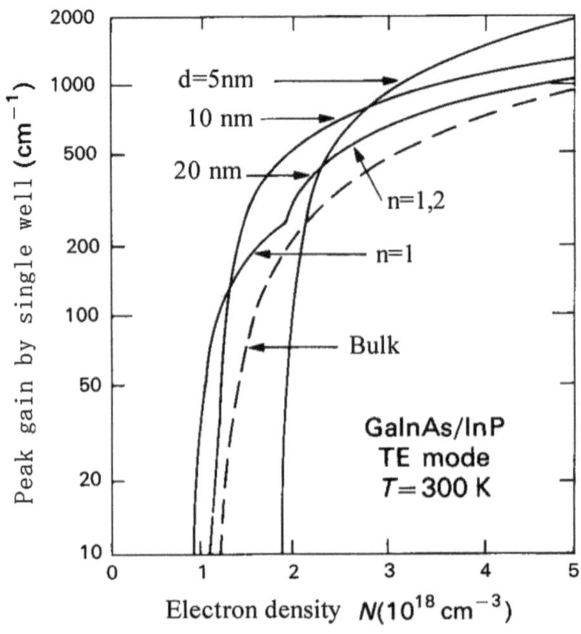

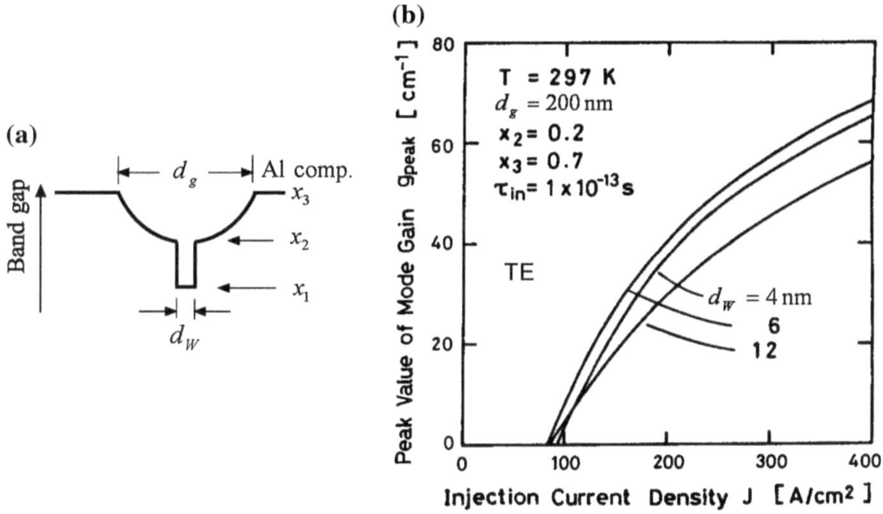

Fig. 11.10 Calculated example of the mode gain in a $Al_xGa_{1-x}As/GaAs$ GRIN-SCH laser. **a** Structure. **b** Calculated gain

Fig. 11.11 Sketch of modified dispersion relationship between electron energy and electron wave number

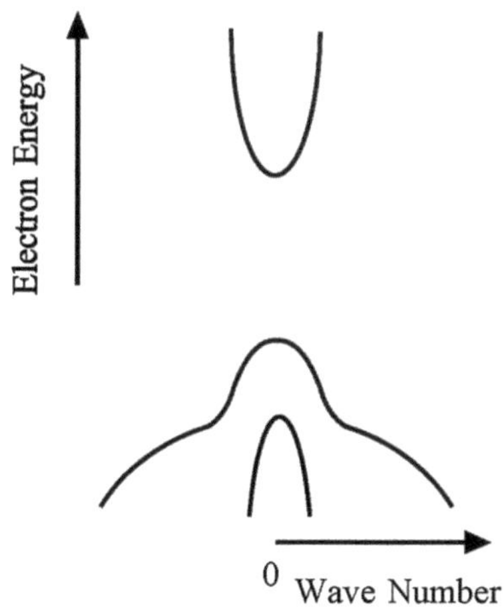

between the two materials; otherwise, the grown material becomes multi-crystalline or has many dislocations. High performance from semiconductor lasers can be achieved using pure single crystals.

However, for thin layers like the quantum well, the materials that are used for epitaxial growth on the substrate material can tolerate slight *differences* in their intrinsic lattice constants, which are then matched with a *strained form* so that dislocations can be avoided. The structures of the thin layers can withstand these stresses. In that situation, the dispersion relation between the electron energy and

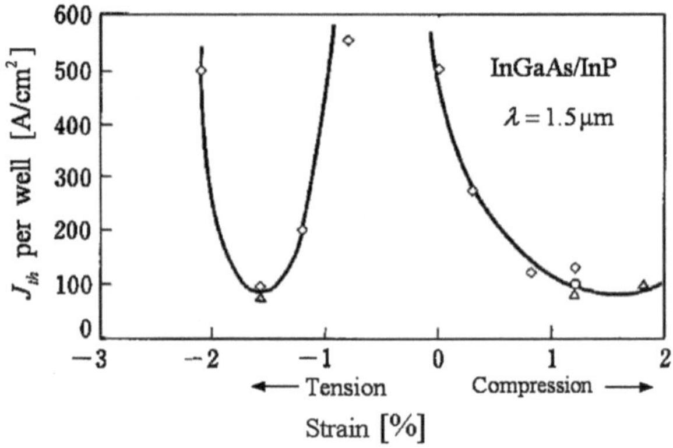

Fig. 11.12 Measured data of the threshold current density [6] (Copyright, 1984, IEEE)

the electron wave number is modified from that in the bulk material, especially for the valence band. Figure 11.11 is an example of a heavy-hole band modified under internal stress. The dispersion curve of the heavy-hole band becomes sharper around the top of the band and close to the dispersion curve of the conduction band. The effective mass and the density-of-states are determined from the dispersion curve. The example of Fig. 11.11 suggests that the *effective mass* and the *density-of-states* of the *heavy-hole band* are close to *those of the conduction band.*

Moreover, the semiconductor lasers operate under electrical neutral conditions with the numbers of injected electrons and holes given in (6.51). For conventional semiconductor materials, the separations in quasi-Fermi-levels from the band edges are much different for the conduction and valence bands, i.e., $\mu_c - W_c \gg W_v - \mu_v$, because the density-of-states in the heavy-hole band is much larger than that in the conduction band. Actually, we have to inject many electrons and holes to achieve the population inversion condition $W_v > \mu_v$. If the density-of-states in the heavy-hole band is reduced, this condition can be achieved with smaller numbers of injected electrons and holes. Thus, the *strained quantum well structure* provides *a reduced threshold current density* by *altering the energy band structure.*

There are two types of strain depending on the direction of stress. One is *compressive* and the another *tensile.* Although the detailed mechanism that changes the band structure is different for the two types, both types act to reduce the threshold current density under suitable criteria. Examples of measured data are shown in Fig. 11.12. The original data were obtained from GRIN-SCH lasers with single, two- or four-well structures. The vertical axis represents the values of the real current density divided by the number of wells. A threshold current density J_{th} below 100 A/cm^2 is possible.

We remark here that a laser with tensile strain tends to oscillate with a TM mode because the light-hole band is raised with respect to the heavy-hole band and becomes dominant.

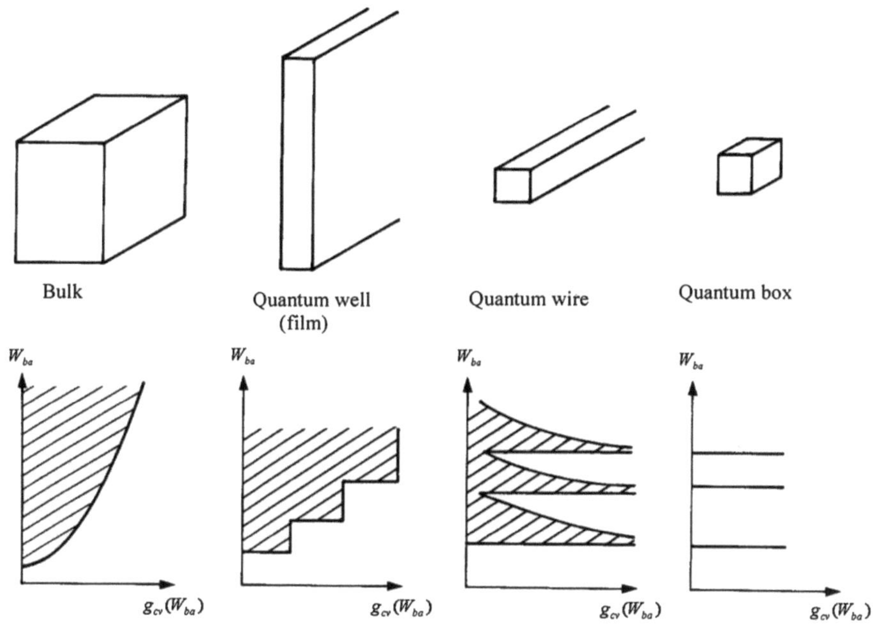

Fig. 11.13 Quantum well structures and corresponding density-of-states showing the degree of quantization

11.6 Quantum Well Structures with Higher Dimension

The quantum well structure treated so far has been set only along the thickness y direction. Fabrication of quantum well structures extended in other directions are possible. The relationship between structure and density-of-states is illustrated in Fig. 11.13. The *bulk material* is defined as one having sufficiently *longer length than the coherence length* of an electron wave. The structure which has a shorter length than the coherence length in *one* of its *dimensional directions* is called in conventional usage a *quantum well structure*, or a *quantum film structure*. The structure in which *two dimensional directions* are restricted is called a *quantum wire*; a structure restricted in *three dimensional directions* is called a *quantum box* or *quantum dots*.

The reduced density-of-states $g_{cv}(W_{ba})$ for an electron-hole pair for the bulk material has been derived in (6.45),

$$g_{cv}(W_{ba}) = \left(\frac{m_c\, m_v}{m_c + m_v}\right)^{3/2} \frac{1}{2\pi^2\hbar^3} \sqrt{W_{ba} - W_g}\,, \quad \text{in } \textit{bulk material}, \qquad (11.39)$$

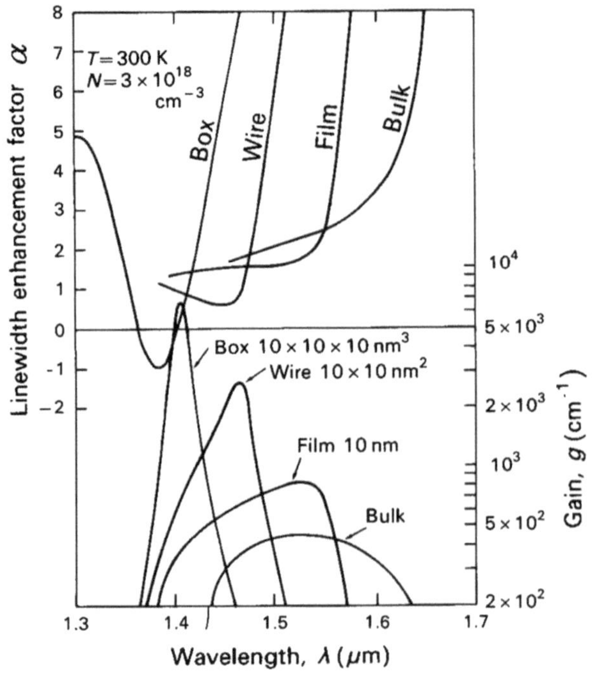

Fig. 11.14 Calculated examples of the gain coefficient for a single well and the linewidth enhancement factor for quantum wells of various dimensional type [7] (Copyright, 1989, The Japan Society of Applied Physics)

and that for the quantum film has been derived in (11.37)

$$g_{cv}(W_{ba}) = \frac{1}{\pi \hbar^2 d_y} \frac{m_c m_v}{m_c + m_v} u(W_{ba} - W_{n_y} - W_g), \quad \text{in } \textit{quantum film}, \qquad (11.40)$$

where d_y is thickness of the quantum film, and n_y is the quantum number.

The reduced density-of-states for the other structures are

$$g_{cv}(W_{ba}) = \left(\frac{m_c m_v}{m_c + m_v}\right)^{1/2} \frac{1}{2\pi \hbar\, d_x d_y} \frac{1}{\sqrt{W_{ba} - W_{n_x,n_y} - W_g}}, \quad \text{in } \textit{quantum wire}, \qquad (11.41)$$

$$g_{cv}(W_{ba}) = \frac{2}{d_x d_y\, d_z} \delta(W_{ba} - W_{n_x,n_y,n_z} - W_g), \quad \text{in } \textit{quantum box}, \qquad (11.42)$$

where $d_x$and d_z are width and length of the quantum well along the x and z directions, respectively. Additionally, W_{n_x,n_y} and W_{n_x,n_y,n_z} are the energy levels determined with two and three quantum numbers, respectively.

Calculated examples of the gain coefficient g and linewidth enhancement factor α are shown in Fig. 11.14. The intraband relaxation time is assumed to be $\tau_{in} = 1 \times 10^{-13}$ s. The gain coefficient from this figure is defined for a single well. The gain by the well itself increases as quantization dimension increases. However, the gain profile broadens owing to intraband relaxation. The linewidth enhancement factor α has been defined in (8.87) as the ratio of the changes in the real part to the imaginary part of the linear susceptibility and *approaches zero* as *the quantization dimension increases*.

Clearly, increasing the quantization dimension seems beneficial in developing better semiconductor lasers. However, there are significant problems to be solved with this prospect. The first is that the *optical confinement factor* ξ defined in (3.14) and (3.16) *is small* for quantum wires and quantum boxes because of the smallness of the wells. The second is that *electron-hole injection* into the wells *is difficult* for both wires and quantum boxes. For quantum film, electrons and holes can be injected through *p-i-n* junctions sandwiching the active region. However, forming the *p-i-n* junctions in wires and boxes is again difficult. Hence, efficient operation using such structures has not yet been reported.

Bibliography

1. M. Yamada, K. Tabata, S. Ogita, M. Yamagishi, Calculation of lasing gain and threshold current in GaAs-AlGaAs multi-quantum-well lasers. Trans. IECE Japan **E 68**(2), 102–108 (1985)
2. M. Asada, Y. Suematsu, Density-matrix theory of semiconductor lasers with relaxation broadening model—Gain and gain-suppression in semiconductor lasers. IEEE J. Quantum Electron. **QE-21**(5), 432–442, (1985)
3. M. Yamada, S. Ogita, M. Yamagishi, K. Tabata, N. Nakaya, M. Asada, Y. Suemtasu, Polarization-dependent gain in GaAs/AlGaAs multi-quantum-well lasers: theory and experiments. Appl. Phys. Lett. **45**(4), 324–325 (1984)
4. M. Asada, K. Kameyama, Y. Suematsu, Gain and tenervalence band absorption in quantum-well lasers. IEEE J. Quantum Electron. **QE-20**(7), 745–753 (1984)
5. M. Yamada, S. Ogita, T. Miyabo, Y. Nishida, A theoretical analysis of lasing gain and threshold current in GaAs-AlGaAs SCH lasers. Trans. IECE Japan **E69**(9), 946–955 (1986)
6. P.J.A. Thijs, Progress in quantum well lasers : application of strain, in *13th IEEE Semiconductor Laser Conference A*-1, Takamatsu/Japan (1992)
7. Y. Miyake, M. Asada, Spectral characteristics of linewidth enhancement factor α of multidimensional quantum well. Japan. J. Appl. Phys. **28**(7), 1280–1281 (1989)

Chapter 12
Distributed Feedback and Mode Selective Lasers

12.1 Distributed Feedback Laser

12.1.1 Model of Analysis

The structure of a *distributed feedback (DFB) laser* is illustrated in Fig. 12.1. In general treatments, the *electric field reflectivities* at the front and back facets, denoted by r_f and r_b respectively, are represented by *complex numbers* to account for phase changes due to reflections. In typical DFB lasers, $r_f = r_b = 0$. The *laser cavity* has four distinct regions, an active region labeled $i = 1$, and cladding labeled $i = 2$, 4 and 5. Between the two layers of the upper cladding, $i = 2$ and $i = 4$, a *corrugated interface* is formed that is of period $\lambda_B/2n_{eq}$, where λ_B is the *Bragg wavelength* and n_{eq} is the equivalent refractive index for optical propagation along the *length of the cavity L*.

In regard to the corrugated interface, its amplitude of variation is sufficient to warrant introducing a new layer, labeled $i = 3$, with thickness equivalent to the amplitude of corrugation and refractive index n_3 that varies according to

$$n_3 = \bar{n}_3 + \Delta n_3 \cos(2\beta_B z + \varphi), \tag{12.1}$$

$$\bar{n}_3 = \frac{n_2 + n_4}{2}, \tag{12.2}$$

$$\Delta n_3 = \frac{n_2 - n_4}{2}, \tag{12.3}$$

$$\beta_B = \frac{2\pi n_{eq}}{\lambda_B}, \tag{12.4}$$

where n_2 and n_4 are the respective refractive indices of layers $i = 2$ and $i = 4$, and φ an initial phase characterizing the variation at $z = 0$. The variation of the dielectric constant for this corrugated layer is

M. Yamada, *Theory of Semiconductor Lasers*,
Springer Series in Optical Sciences 185,
DOI: 10.1007/978-4-431-54889-8_12,

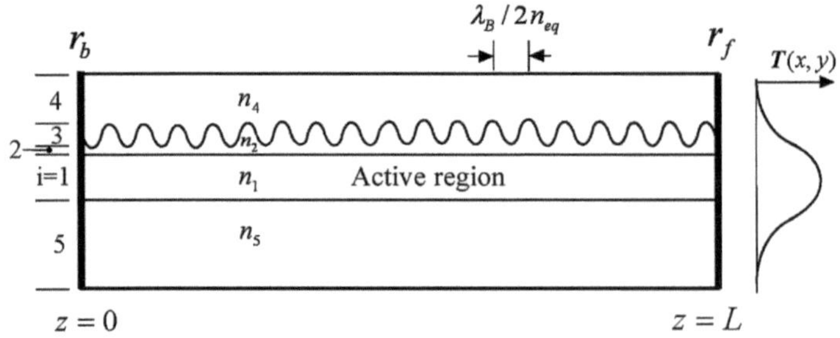

Fig. 12.1 Structure of a distributed feedback (DFB) laser

$$\varepsilon_3 = \varepsilon_o n_3^2 = \bar{\varepsilon}_3 + \Delta\varepsilon_3 \cos(2\beta_B z + \varphi) = \bar{\varepsilon}_3 + \frac{\Delta\varepsilon_3}{2}\left\{e^{j(2\beta_B z + \varphi)} + e^{-j(2\beta_B z + \varphi)}\right\}, \tag{12.5}$$

$$\bar{\varepsilon}_3 = \varepsilon_o \bar{n}_3^2, \tag{12.6}$$

$$\Delta\varepsilon_3 = 2\varepsilon_o \bar{n}_3 \Delta n_3. \tag{12.7}$$

For the laser cavity, Maxwell's wave equation for the main component of the electric field is

$$\nabla^2 E - \mu_o \varepsilon_i(z)\frac{\partial^2 E}{\partial t^2} = \mu_o\left(\sigma_i \frac{\partial E}{\partial t} + \frac{\partial^2 P}{\partial t^2}\right). \tag{12.8}$$

We propose a solution to this equation and laser polarization of the form

$$E = \left\{E^{(+)}(z)e^{j\beta_B z} + E^{(-)}(z)e^{-j\beta_B z}\right\}T(x, y)e^{j\omega t} + c.c., \tag{12.9}$$

$$P = \varepsilon_o \chi_i \left\{E^{(+)}(z)e^{j\beta_B z} + E^{(-)}(z)e^{-j\beta_B z}\right\}T(x, y)e^{j\omega t} + c.c., \tag{12.10}$$

where ω is the angular frequency of the optical wave, $T(x,y)$ the transverse distribution function of the electric field which satisfies the following eigenfunction,

$$\left\{\frac{\partial^2}{\partial x^2} + \frac{\partial^2}{\partial y^2} - \beta^2 + \mu_o \bar{\varepsilon}_i \omega^2\right\}T(x, y)e^{-j\beta z + j\omega t} = 0, \tag{12.11}$$

with

$$\beta = \sqrt{\mu_o \varepsilon_o}\,\omega n_{eq} = \frac{2\pi n_{eq}}{\lambda}, \tag{12.12}$$

and normalization condition

$$\int_{-\infty}^{\infty}\int_{-\infty}^{\infty}|T(x,y)|^2 dx\,dy = 1. \tag{12.13}$$

The $E^{(+)}(z)$ and $E^{(-)}(z)$ in (12.9) are slowly varying amplitudes of forward and backward propagating components, respectively. We note here that the *propagation constant for the optical wave is not β_B but β* introduced in (12.11) and (12.12); β_B is a structural parameter to characterize the corrugation pitch defined in (12.4). The reader might feel a little uneasy with (12.9) in which the electric field E is expressed with β_B and not β. However, using β_B is mathematically easier than starting with β in analyzing the effect of coupling between the forward and backward components.

12.1.2 Coupled Wave Equations

We substitute (12.9) and (12.10) into (12.8), neglect terms with $\partial^2 E^{(\pm)}(z)/\partial z^2$ because of its slowly varying properties and use (12.11). Furthermore, we multiply the equation obtained by $T^*(x,y)e^{-j\omega t}$, perform a spatial integration $\iint_\infty dxdy$ in the transverse cross-section with (12.13), and take a time average between t and $t+\Delta t$ for several periods over the optical vibration. We then obtain

$$\begin{aligned}&\left\{(\beta^2-\beta_B^2)E^{(+)}-2j\beta_B\frac{\partial E^{(+)}}{\partial z}\right\}e^{-j\beta_B z}+\left\{(\beta^2-\beta_B^2)E^{(-)}+2j\beta_B\frac{\partial E^{(-)}}{\partial z}\right\}e^{j\beta_B z}\\&+\frac{\omega^2\mu_o\Delta\varepsilon_{eff}}{2}\left\{E^{(+)}e^{j\beta_B z+j\varphi}+E^{(-)}e^{3j\beta_B z+j\varphi}+E^{(+)}e^{-3j\beta_B z-j\varphi}+E^{(-)}e^{-j\beta_B z-j\varphi}\right\}\\&=\left\{j\omega\mu_o\sigma_{eff}-\omega^2\mu_o\varepsilon_o\chi_{eff}\right\}\left\{E^{(+)}e^{-j\beta_B z}+E^{(-)}e^{j\beta_B z}\right\},\end{aligned} \tag{12.14}$$

where

$$\Delta\varepsilon_{eff} = \int_{-\infty}^{\infty}\int_{-\infty}^{\infty}\Delta\varepsilon_i|T(x,y)|^2 dx\,dy, \tag{12.15}$$

$$\sigma_{eff} = \int_{-\infty}^{\infty}\int_{-\infty}^{\infty}\sigma_i|T(x,y)|^2 dx\,dy, \tag{12.16}$$

$$\chi_{eff} = \int_{-\infty}^{\infty}\int_{-\infty}^{\infty}\chi_i|T(x,y)|^2 dx\,dy. \tag{12.17}$$

We multiply (12.14) by $e^{j\beta_B z}$ and take the spatial average between z and $z + \Delta z$ over several wavelength cycles yielding

$$\frac{\partial E_B^{(+)}}{\partial z} = \frac{\beta^2 - \beta_B^2}{2j\beta_B} E^{(+)} + \frac{\omega^2 \mu_o \Delta\varepsilon_{eff}}{4j\beta_B} e^{-j\varphi} E^{(-)} - \frac{\omega\mu_o}{2\beta_B}\{\sigma_{eff} + j\omega\varepsilon_o\chi_{eff}\}E^{(+)}. \tag{12.18}$$

The last term of this equation is associated with the *mode gain coefficient*, the *loss coefficient*, and the *variation of the propagation constant* defined in (3.11), (3.12), and (3.19):

$$-\frac{\omega\mu_o}{2\beta_B}\{\sigma_{eff} + j\omega\varepsilon_o\chi_{eff}\} = \frac{g - \alpha_{loss}}{2} - j\Delta\beta, \tag{12.19}$$

where

$$g = \frac{\omega^2\mu_o\varepsilon_o}{\beta_B}\int_{-\infty}^{\infty}\int_{-\infty}^{\infty} \mathrm{Im}\,\chi_i(x,y)\,|T(x,y)|^2 dx\,dy, \tag{12.20}$$

$$\alpha_{loss} = \frac{\omega\mu_o}{\beta_B}\int_{-\infty}^{\infty}\int_{-\infty}^{\infty} \sigma_i(x,y)\,|T(x,y)|^2 dx\,dy, \tag{12.21}$$

$$\Delta\beta = \frac{\omega^2\mu_o\varepsilon_o}{2\beta_B}\int_{-\infty}^{\infty}\int_{-\infty}^{\infty} \mathrm{Re}\chi_i(x,y)\,|T(x,y)|^2 dx\,dy. \tag{12.22}$$

For mathematical convenience, we introduce *an effective gain coefficient*

$$\frac{g - \alpha_{loss}}{2} \equiv g'. \tag{12.23}$$

Now the coefficient in the first term on the right-hand side of (12.18) introduces the difference between the propagation constant of the optical wave and the corrugation pitch. We treat this term together with the variation of the propagation constant $\Delta\beta$, and introduce a new parameter $\delta\beta$ to define the *difference* in the *actual propagation constant* of the optical wave and the *corrugation pitch* as follows:

$$\begin{aligned}\frac{\beta^2 - \beta_B^2}{2\beta_B} + \frac{\omega^2\mu_o\varepsilon_o}{2\beta_B}\mathrm{Re}\chi_{eff} &= \frac{(\beta - \beta_B)(\beta + \beta_B)}{2\beta_B} + \Delta\beta \\ &\approx \beta - \beta_B + \Delta\beta \\ &\equiv \delta\beta\end{aligned} \tag{12.24}$$

The second term on the right-hand side of (12.18) couples the forward and backward propagating components. Defining *coupling constant* κ

$$\frac{\omega^2 \mu_o \Delta\varepsilon_{eff}}{4\beta_B} \equiv \kappa, \tag{12.25}$$

and using the above notation, (12.18) is rewritten in simpler mathematical form as

$$\frac{\partial E^{(+)}(z)}{\partial z} = (g' - j\delta\beta)E^{(+)}(z) - j\kappa e^{-j\varphi}E^{(-)}(z). \tag{12.26}$$

By multiplying (12.14) by $e^{-j\beta_B z}$, taking a spatial average between z and $z + \Delta z$ over several wavelength cycles, and similarly rewriting using (12.19)–(12.25), we obtain another coupled equation

$$\frac{\partial E^{(-)}(z)}{\partial z} = (j\delta\beta - g')E^{(-)}(z) + j\kappa e^{j\varphi}E^{(+)}(z). \tag{12.27}$$

Now we have a pair of *coupled wave equations* that describe the propagation of the electric field in the cavity.

12.1.3 Oscillation Condition

By differentiating (12.26) again and combining with (12.27), we derive a second-order equation for $E^{(+)}$ only,

$$\frac{\partial^2 E^{(+)}(z)}{\partial z^2} = \left\{\kappa^2 + (g' - j\delta\beta)^2\right\}E^{(+)}(z). \tag{12.28}$$

We suppose a solution of the form

$$E^{(+)}(z) = a_1 e^{\gamma z} + a_2 e^{-\gamma z}. \tag{12.29}$$

where a_1 and a_2 are unknown constants, and γ is a *propagation constant* found from (12.28) to be

$$\gamma = \sqrt{\kappa^2 + (g' - j\delta\beta)^2}. \tag{12.30}$$

By substituting (12.29) into (12.26), the amplitude of the backward wave is

$$E^{(-)}(z) = \frac{e^{j\varphi}}{j\kappa}\{(-\gamma + g' - j\delta\beta)a_1 e^{\gamma z} + (\gamma + g' - j\delta\beta)a_2 e^{-\gamma z}\}. \tag{12.31}$$

Although the propagation constant $\pm\gamma$ has an imaginary part that is much smaller than β_B, $|\mathrm{Im}\gamma| \ll \beta_B$, the values of $\beta_B \pm \gamma$ never change signs. We can confirm then that $E^{(+)}(z)\exp(-j\beta_B z)$ and $E^{(-)}(z)\exp(j\beta_B z)$ represent the *forward* and *backward propagating waves*, respectively. The boundary conditions at the end facets for the forward and backward propagating waves are given in terms of the reflectivities as

$$E^{(-)}(L)e^{j\beta_B L} = r_f E^{(+)}(L)e^{-j\beta_B L}, \tag{12.32}$$

$$E^{(+)}(0) = r_b E^{(-)}(0). \tag{12.33}$$

By substituting (12.29) and (12.31) into (12.32) and (12.33), we have

$$(\gamma - g' + j\delta\beta + j\kappa r_f e^{-2j\beta_B L - j\varphi})a_1 e^{\gamma L} = (\gamma + g' - j\delta\beta - j\kappa r_f e^{-2j\beta_B L - j\varphi})a_2 e^{-\gamma L}, \tag{12.34}$$

$$\{r_b(\gamma - g' + j\delta\beta) + j\kappa e^{-j\varphi}\}a_1 = \{r_b(\gamma + g' - j\delta\beta) - j\kappa e^{-j\varphi}\}a_2. \tag{12.35}$$

We eliminate the unknown coefficients a_1 and a_2 from (12.34) and (12.35) to arrive at an equation giving the steady-state operation of the DFB laser,

$$\begin{aligned}&\left(1 - r_f r_b e^{-2j\beta_B L}\right)\gamma\cosh(\gamma L)\\ &= \left\{(g' - j\delta\beta)\left(1 + r_f r_b e^{-2j\beta_B L}\right) - j\kappa\left(r_b e^{j\varphi} + r_f e^{-2j\beta_B L - j\varphi}\right)\right\}\sinh(\gamma L).\end{aligned} \tag{12.36}$$

Because (12.36) is complex, this equation yields two equations, one for the amplitude and the other the phase condition for the laser oscillation.

12.1.4 Deriving the Oscillation Condition for a Fabry–Perot Cavity Laser

Before proceeding to determine the general properties of the DFB structure, let us examine in detail (12.36) given certain specific conditions.

To begin we consider a cavity with *no corrugation* obtained by setting $\kappa = 0$. The propagation constant γ in (12.30) becomes

$$\gamma = g' - j\delta\beta. \tag{12.37}$$

and the oscillation condition of (12.36) is rewritten as

$$\left(1 - r_f r_b e^{-2j\beta_B L}\right)\cosh(\gamma L) = \left(1 + r_f r_b e^{-2j\beta_B L}\right)\sinh(\gamma L). \tag{12.38}$$

This equation takes the simpler form

$$r_f r_b e^{-2j\beta_B L + 2\gamma L} = r_f r_b e^{2g'L - 2j\beta L} = 1. \tag{12.39}$$

By referring to the original form in (12.23), (12.39) can be reduced to the oscillation condition for a conventional *Fabry-Perot cavity* laser,

$$g = 2g' + \alpha_{loss} = \alpha_{loss} + \frac{1}{L}\ln\frac{1}{|r_f r_b|}. \tag{12.40}$$

12.1.5 Deriving the Oscillation Condition for a Long Nonreflecting Cavity

The second example corresponds to a nonreflecting corrugated structure with reflectivities $r_f = r_b = 0$. The oscillation condition from (12.36) can be simplified,

$$\gamma \cosh(\gamma L) = (g' - j\delta\beta)\sinh(\gamma L). \tag{12.41}$$

Using either the relation $\cosh^2(\gamma L) - \sinh^2(\gamma L) = 1$ or $1 - \tanh^2(\gamma L) = \cosh^{-2}(\gamma L)$ and (12.30), (12.41) is rewritten as

$$\cosh(\gamma L) = \pm\frac{\delta\beta + jg'}{\kappa}. \tag{12.42}$$

By separating the complex propagation constant into real and imaginary parts $\gamma = \mathrm{Re}\gamma + j\mathrm{Im}\gamma$, and by substituting into (12.42), we obtain two equations,

$$\cosh(\mathrm{Re}\gamma L)\cos(\mathrm{Im}\gamma L) = \pm\frac{\delta\beta}{\kappa}, \tag{12.43}$$

$$\sinh(\mathrm{Re}\gamma L)\sin(\mathrm{Im}\gamma L) = \pm\frac{g'}{\kappa}. \tag{12.44}$$

The left-hand side of both equations becomes infinite for $L \to \infty$, if $\mathrm{Re}\gamma \neq 0$. Thus, a necessary condition for oscillation with nonreflecting facets, and long cavity is $\mathrm{Re}\gamma \approx 0$. From (12.30), this condition is written as the gain condition

$$g' \approx \sqrt{(\delta\beta)^2 - \kappa^2}. \tag{12.45}$$

Although $\mathrm{Re}\gamma \approx 0$, $\mathrm{Re}\gamma L$ is not zero but has finite value for real value L so that (12.43) and (12.44) hold. Another condition corresponding to the phase condition is obtained for $\mathrm{Im}\gamma$. The exact solution of γ is obtained by numerical calculation for the set of simultaneous equations, (12.30), (12.43), and (12.44) together with values of κ, $\delta\beta$, and g'.

A further remark is that there is no solution if $|\delta\beta| < \kappa$ as seen from (12.45). We call this the *cut-off condition*,

$$|\delta\beta| < \kappa, \quad \text{cut-off condition when } r_f = r_b = 0 \text{ and } L \to \infty. \tag{12.46}$$

In a long nonreflecting DFB laser, laser oscillation is not achieved at the Bragg wavelength but at a *slightly shifted wavelength* given by $|\delta\beta| \geq \kappa$ as shown in the next subsection.

12.1.6 Calculated Example of the Threshold Gain

The threshold gain in the DFB laser can be calculated from (12.36). A calculated example is shown in Fig. 12.2, where κL indicates the strength of the coupling between the forward and backward propagating components. The reflectivity of the back facet, which is obtained by cleaving the semiconductor material, is set to $|r_b| = 0.533$. The front facet is set to $r_f = 0$. The initial phase φ characterizing the geometrical shape defined in (12.1) effects the evaluation of the threshold gain as given in (12.36). This value of φ is suitably selected to give a minimum value of the threshold gain for each given value of $\delta\beta L$ in Fig. 12.2. Circles in the figure mark instances for which $\varphi = \pi/2$.

As found from this figure, the threshold gain does not become a minimum at $\delta\beta = 0$ corresponding to the Bragg condition, but shows a minimum value at two points near $|\delta\beta| \geq \kappa$. Thus a DFB laser tends to exhibit two-mode oscillations if the structure is simple.

12.1.7 A DFB Laser with a $\lambda/4$ Phase Shift Structure

When the corrugation has a portion where a structural phase shift by $\lambda/4n_{eq}$ occurs, as illustrated in Fig. 12.3, the threshold gain has a minimum at $\delta\beta = 0$.

The role of this $\lambda/4$-phase shift region is explained as in Fig. 12.4. (a) corresponds to a *simple corrugated structure* in which the laser exhibits a cut-off condition if $|\delta\beta| < \kappa$. We consider a point near the middle of the corrugation ($z \approx L / 2$) and assume an optical wave A is propagating in the forward direction.

The returning light from the corrugation is $A \exp(-j\pi/2)$ with phase delay of $-\pi/2$ and propagates in the backward direction. This backward propagating light is reflected and returned in the forward direction with another phase delay of $-\pi/2$. After making a single round trip, the light wave has form $A \exp(-j\pi) = -A$, which cancels out the initial propagating wave A. Thus, light under the *Bragg condition* is in the *cut-off* mode.

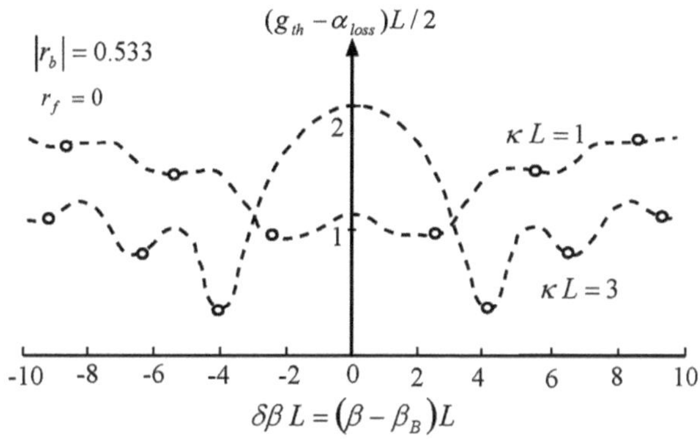

Fig. 12.2 Calculated example of the threshold gain in the DFB laser [1] (Copyright, 1984, IEEE)

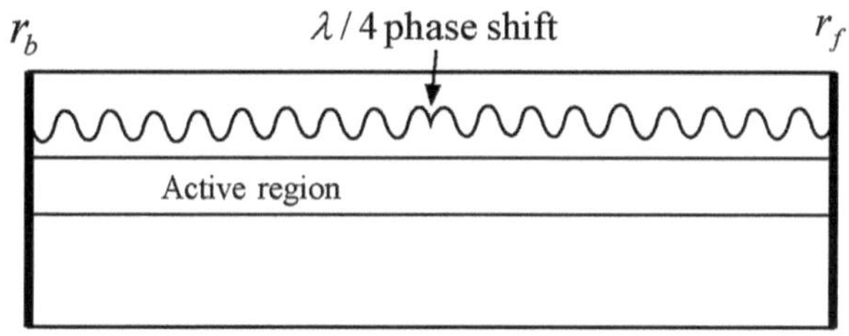

Fig. 12.3 DFB laser with a $\lambda/4$-phase shift region

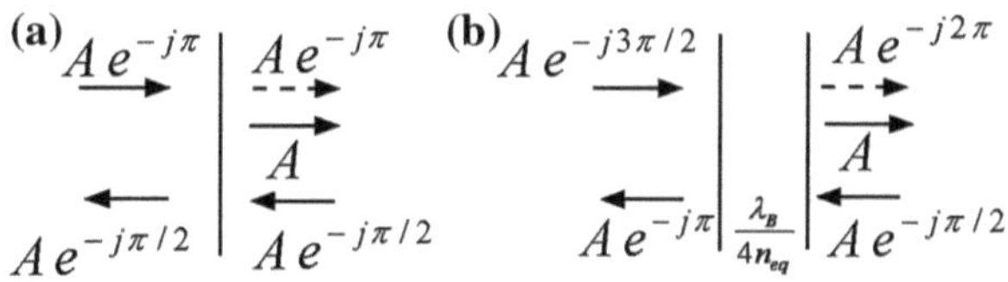

Fig. 12.4 Diagrams explaining the role of the $\lambda/4$-phase shift region. **a** Simple corrugation; **b** a $\lambda/4$ phase shift region

Figure 12.4b corresponds to a region where a $\lambda/4$-*phase shift* exists. We again assume the optical wave A is propagating in the forward direction. The returning light wave from the corrugation becomes $A\exp(-j\pi/2)$. Because we have inserted a region of length $\lambda_B/4n_{eq}$, the phase delay becomes $-\pi/2 - \beta_B\lambda_B/4n_{eq} = -\pi$ for light passing through this region propagating in the backward direction. After reflection, the returning light moving in the forward direction has accumulated a phase delay of $-\pi - \pi/2 = -3\pi/2$ on arriving at the phase shift region. After passing the phase shift region, the phase delay becomes $-3\pi/2 - \beta_B\lambda_B/4n_{eq} = -2\pi$.

Thus, the light wave has the same phase as the initially light, resulting in an effective reduction in the threshold gain.

A calculated example of the threshold gain in the laser having a $\lambda/4$-phase shift region is shown in Fig. 12.5. The threshold gain is *lowest* at the *Bragg condition* $\delta\beta = 0$.

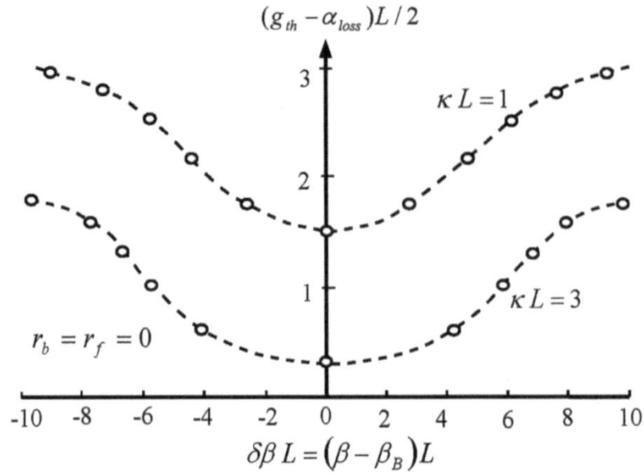

Fig. 12.5 Calculated example of the threshold gain in a laser having a $\lambda/4$-phase shift region [2] (Copyright, 1984, The IET)

12.2 Distributed Bragg Reflector Laser

The corrugation in the DFB laser cavity lies parallel along the active region. The applied voltage for laser operations can induce damage to the material in the corrugated layer. An alternative configuration is to create corrugated wave-guides connected to the ends of the active region, as shown in Fig. 12.6. Lasing gain is obtained only in the active region. The corrugated wave-guides function as passive reflectors. In this setup, they are called *distributed Bragg reflectors (DBRs)* or *Bragg reflectors*.

A model to analyze the Bragg reflector is shown in Fig. 12.7. We suppose the reflector is connected at $z = 0$ to the active region. The length of the reflector is L and the corrugation pitch is $\lambda_B/2n_{eq}$, where λ_B is the *Bragg wavelength*. The reflectivity of the facet is r_L.

We set up an electric field in the waveguide, as in (12.9), to be

$$E = \left\{E^{(+)}(z)e^{j\beta_B z} + E^{(-)}(z)e^{-j\beta_B z}\right\}T(x,y)e^{j\omega t} + c.c.. \tag{12.47}$$

The forward and backward propagating components $E^{(+)}(z)$ and $E^{(-)}(z)$ are related by the coupled equations (12.26) and (12.27). Then, these amplitudes are written as

$$E^{(+)}(z) = a_1 e^{\gamma z} + a_2 e^{-\gamma z}, \tag{12.48}$$

$$E^{(-)}(z) = \frac{e^{j\varphi}}{j\kappa}\{(-\gamma + g' - j\delta\beta)a_1 e^{\gamma z} + (\gamma + g' - j\delta\beta)a_2 e^{-\gamma z}\}, \tag{12.49}$$

where a_1 and a_2 are unknown constants, $\delta\beta$ the deviation of the propagation constant from the Bragg condition, as defined in (12.24), κ the coupling constant

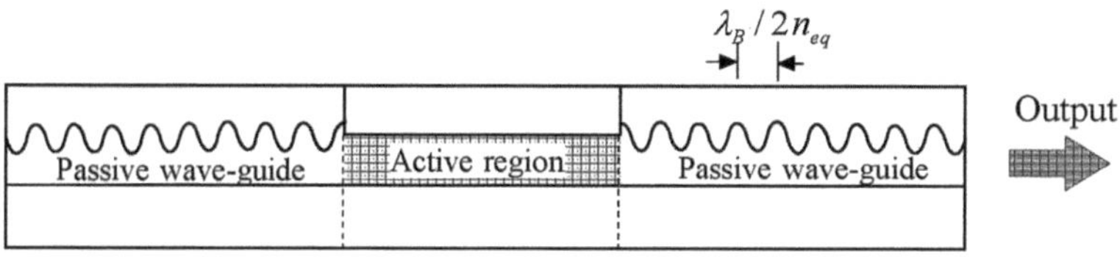

Fig. 12.6 Structure of a distributed Bragg reflector (DBR) laser cavity

Fig. 12.7 Model for the analysis of a Bragg reflector

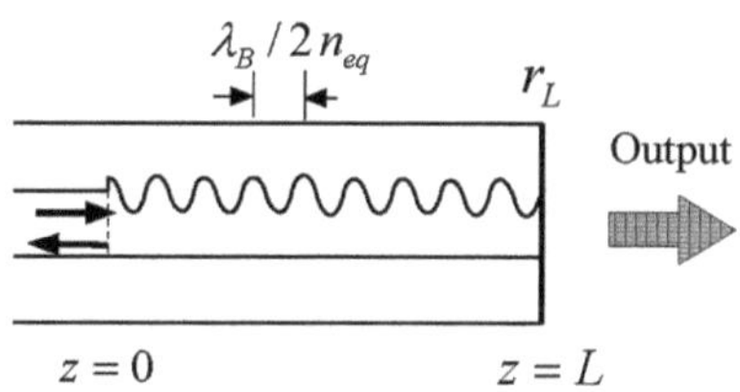

defined in (12.25), and γ the propagation constant. We stipulate a guiding loss α_{loss} but no gain g for the Bragg reflector, which gives a propagation constant

$$\gamma = \sqrt{\kappa^2 + \left(\frac{\alpha_{loss}}{2} + j\delta\beta\right)^2}. \tag{12.50}$$

The forward and the backward waves are related by the boundary condition at the end facet,

$$E^{(-)}(L)e^{j\beta_B L} = r_L E^{(+)}(L)e^{-j\beta_B L}. \tag{12.51}$$

By substituting (12.48) and (12.49) into (12.51), we obtain a relation between the unknown constants

$$a_2 = \frac{j\kappa e^{-j\varphi} r_L e^{-2j\beta_B L} + \gamma + \alpha_{loss}/2 + j\delta\beta}{-j\kappa e^{-j\varphi} r_L e^{-2j\beta_B L} + \gamma - \alpha_{loss}/2 - j\delta\beta} a_1 e^{2\gamma L}. \tag{12.52}$$

Substituting this equation into (12.48) gives the forward wave at $z = 0$

$$\begin{aligned} E^{(+)}(0) &= a_1 + a_2 \\ &= \frac{\gamma\cosh(\gamma L) + (\alpha_{loss}/2 + j\delta\beta + j\kappa e^{-j\varphi} r_L e^{-2j\beta_B L})\sinh(\gamma L)}{\gamma - \alpha_{loss}/2 - j\delta\beta - j\kappa e^{-j\varphi} r_L e^{-2j\beta_B L}} \times 2a_1 e^{\gamma L}. \end{aligned} \tag{12.53}$$

Moreover, substituting (12.52) into (12.49), and with the help of (12.53), gives the backward wave at $z = 0$

$$
\begin{aligned}
E^{(-)}(0) &= \frac{e^{j\varphi}}{j\kappa}\{(-\gamma + g' - j\delta\beta)a_1 + (\gamma + g' - j\delta\beta)a_2\} \\
&= \frac{\gamma r_L e^{-2j\beta_B L}\cosh(\gamma L) - \{(\alpha_{loss}/2 + j\delta\beta)r_L e^{-2j\beta_B L} + j\kappa e^{j\varphi})\sinh(\gamma L)}{\gamma - \alpha_{loss}/2 - j\delta\beta - j\kappa e^{-j\varphi} r_L e^{-2j\beta_B L}} \times 2a_1 e^{\gamma L}.
\end{aligned}
\tag{12.54}
$$

Hence, an equivalent reflectivity r_{eq} at $z = 0$ for the Bragg reflector is obtained:

$$
\begin{aligned}
r_{eq} &= |r_{eq}|e^{-j\phi} \equiv \frac{E^{(-)}(0)}{E^{(+)}(0)} \\
&= \frac{{}_L e^{-2j\beta_B L}\cosh(\gamma L) - \{(\alpha_{loss}/2 + j\delta\beta)r_L e^{-2j\beta_B L} + j\kappa e^{j\varphi})\sinh(\gamma L)}{\gamma\cosh(\gamma L) + (\alpha_{loss}/2 + j\delta\beta + j\kappa r_L e^{-2j\beta_B L - j\varphi})\sinh(\gamma L)},
\end{aligned}
\tag{12.55}
$$

where ϕ indicates the phase delay of the reflected light.

In particular, when the reflectivity of the end facet is zero, $r_L = 0$, the equivalent reflectivity is

$$
r_{eq} = \frac{-j\kappa e^{j\varphi}\tanh(\gamma L)}{\gamma + (\alpha_{loss}/2 + j\delta\beta)\tanh(\gamma L)}, \quad \text{for } r_L = 0. \tag{12.56}
$$

Calculated examples of the equivalent reflectivity are shown in Fig. 12.8. The Bragg reflector reveals *a maximum reflection at the Bragg condition.*

By preparing two sets of Bragg reflectors, we can construct a DBR laser as shown in Fig. 12.6. An oscillation condition is obtained similar to a conventional Fabry-Perot laser by substituting the equivalent reflectivity r_{eq} instead of the reflectivity of the facet mirror.

Several readers might question here whether we need to adjust the length of the active region or the initial phase φ of the corrugated structure within an accuracy of several tenths of the optical wavelength to obtain a minimum threshold gain. The answer is that it is not necessary to make adjustments to such precision for the DBR laser, because the phase variation in the active region can be suitably chosen with very small changes in the injected electron density.

12.3 Wavelength Tunable Laser

The lasing wavelength in the DFB laser or the DBR laser is the Bragg wavelength given as twice the corrugation pitch times the equivalent refractive index n_{eq}, which is defined in (3.13). We thus can manipulate the lasing wavelength λ if we can change the *equivalent refractive index* n_{eq}. This in turn depends on the refractive indices of the materials and thicknesses in the waveguide.

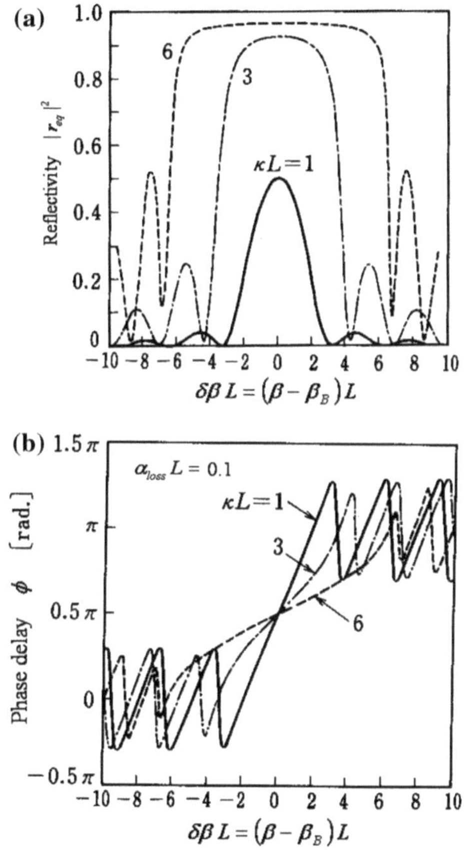

Fig. 12.8 Calculated example of the equivalent reflectivity by a Bragg reflector [3] ($r_L = 0$, $\varphi = 0$ and $\alpha_{loss} L = 0.1$ are assumed). **a** Power reflectivity. **b** Phase delay

One way to construct a wavelength tunable laser is to use a DBR structure as sketched in Fig. 12.9. The active region and the DBR region have isolated electrodes. The driving current giving the lasing gain is injected into the active region. Another electrode on the DBR region is used to change the equivalent refractive index n_{eq} of the waveguide. Possible mechanisms to effect the change are as follows:

(1) Construct *p-i-n* junctions of higher band-gap materials than the active region and inject the electrons and the holes *to reduce the refractive index*. The drawback of this method is that absorption losses increase with the number of electrons and the holes.
(2) Construct a quantum-well structure with a suitable band gap, and apply a voltage to induce a *quantum-Stark effect*, an electro-optic effect that changes the refractive index. The drawback of this method is that the change in refractive index is not large and the range over which the wavelength varies is very narrow.
(3) Attach a thin-film heater to the DBR region to change its *temperature*. The refractive index becomes smaller as temperature increases. However, the power consumption required for this method is rather large.

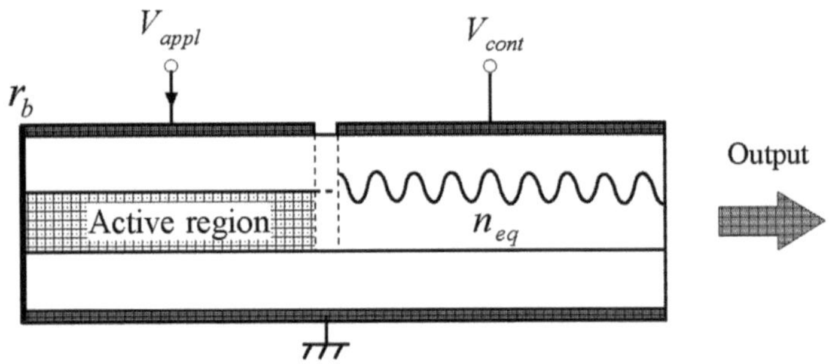

Fig. 12.9 Schematic illustration of a wavelength tunable laser

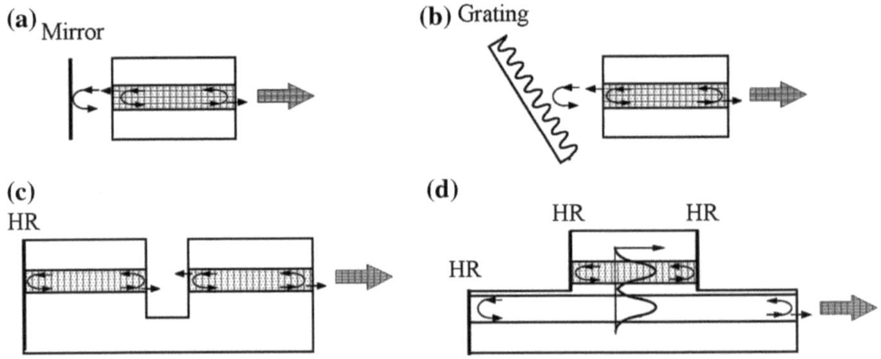

Fig. 12.10 Several structures designed to produce mode selectivity. **a** External mirror. **b** External grating. **c** Multi section. **d** Twin guide

12.4 Various Structures for Mode Selective Laser

If the laser cavity has more than two multi-reflecting regions, the threshold gain level becomes non-uniform because the reflected light creates *interference*. Several structures that try to establish mode selectivity in semiconductor lasers are illustrated in Fig. 12.10. The hatched regions are the active regions; HR indicates high-reflection mirrors.

Diagram (a) presents the simplest case where an *external mirror* is placed behind the laser. Output light from the back facet returns back into the laser by reflecting off the external mirror. The threshold gain is reduced when the phase of the returned light coincides with that of the reflected light by the back facet. However, this method is unable to stabilize the frequency fluctuation of the laser light as discussed in Chap. 10 for *optical feedback noise*.

Instead of the simple mirror, diagram (b) has a *grating* for an *external reflector*. Stable single-mode operations can be obtained from this configuration Furthermore, the lasing wavelength can be tuned by changing the angle between the grating and the optical light. The DBR could be a structure for which the grating is integrated with the laser itself.

In diagram (c), two laser chips are placed in the longitudinal direction. This structure consists of three cavities. Two of them are lasers, whereas the third is an air gap between the two lasers. Therefore, this structure can provide mode selectivity. The drawback with this structure is the difficulty in controlling the lasing wavelength precisely, because the length of the gap must be fabricated within an error of 100 nm.

Finally, diagram (d) employs a passive waveguide attached under the active region. Laser light is generated in the active region and is coupled to the passive wave guide from where the output light is taken out. This structure was invented originally to construct an optical integrated circuit to connect other optical devices through the passive waveguide. However, if we use this structure alone, it consists of three cavities. One contains a twin guide. The second and third are arm-like extensions containing the passive waveguide. This structure also provides mode selectivity [4]. However, the problem of this structure is the difficulty in fabricating active regions and passive waveguides of lengths accurate to within 100 nm.

For DFB and the DBR structures, the corrugations can be fabricated with very precise accuracy by *photo-lithography*. However, the bonding of crystals or the cutting to well-prescribed lengths is very difficult. Hence accurate settings of the lasing wavelength in structures (a), (c), and (d) are difficult.

Bibliography

1. K. Utaka, S. Akiba, K. Sakai, Y. Matsushima, Effect of mirror facets on lasing characteristics of distributed feedback InGaAsP/InP laser diode at 1.5 μm range. IEEE J. Quantum Electron. QE-**20**(3), 236–245 (1984)
2. K. Utaka, S. Akiba, K. Sakai, Y. Matsushima, Analysis of quarter-wave-shifted DFB laser. Electron. Lett. **20**(8), 326–327 (1984)
3. S. Arai, in *Semiconductor Lasers*, p. 285 Chap. 13, ed. by K. Iga (Ohmsha, Ltd., Tokyo, 1994). ISBN: 4-274-1298-X (in Japanese)
4. M. Yamada, Y. Suematsu, Analysis of an integrated twin-guide laser with coupled-wave theory. IEEE J. Quantum Electron. QE-**13**(4), 201–206 (1977)

Chapter 13
Surface Emitting Lasers

13.1 Fundamental Configuration and Oscillation Condition

The semiconductor lasers treated so far in this book have laser cavities along the longitudinal z direction, as depicted in Fig. 1.8. Nevertheless, structures with cavities along the thickness direction y of the active region have been developed. These lasers are called *surface-emitting lasers* or *vertical-cavity surface-emitting lasers (VCSEL)*. The fundamental structure, illustrated in Fig. 13.1, has an active region sandwiched between a pair of cladding mirror regions. We suppose that the length between the two mirrors is L, the complex reflectivities at the mirrors for the electric field are r_f and r_b, the thickness of the active region d, the gain coefficient for the plane wave in the active region g, and the coefficients of loss for the active and cladding regions α_a and α_c, respectively.

As the optical wave now propagates along the y direction, the electric field component is written as

$$E(t,\boldsymbol{r}) = E^{(+)}(y)\,T(x,z)\,e^{-j\beta y+j\omega t} + E^{(-)}(y)\,T(x,z)\,e^{-j\beta(L-y)y+j\omega t} + c.c., \quad (13.1)$$

where $T(x,z)$ is the normalized field distribution function in the x-z plane,

$$\int_{-\infty}^{\infty}\int_{-\infty}^{\infty} |T(x,z)|^2\,dx\,dz = 1, \quad (13.2)$$

$E^{(+)}(y)$ and $E^{(-)}(y)$ are the respective amplitudes of the forward and backward propagating components. The oscillation condition is then given by evaluating the phase accumulated in a single round trip for the optical field in the cavity,

$$e^{(g-\alpha_a)d-\alpha_c(L-d)-2j\beta L}\, r_f\, r_b = 1. \quad (13.3)$$

M. Yamada, *Theory of Semiconductor Lasers*,
Springer Series in Optical Sciences 185,
DOI: 10.1007/978-4-431-54889-8_13,

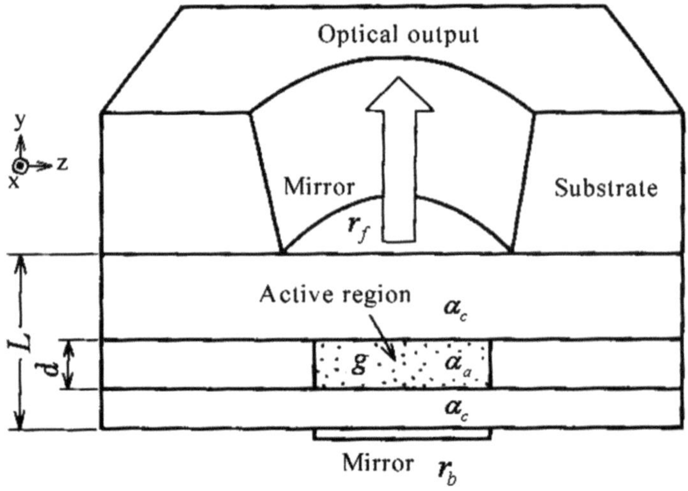

Fig. 13.1 Fundamental structure of VCSEL

Here, we represent the reflectivities in polar form with magnitudes and phase delays as

$$r_f = |r_f| e^{-j\phi_f}, \tag{13.4}$$

$$r_b = |r_b| e^{-j\phi_b}, \tag{13.5}$$

By taking the logarithm, this oscillation condition becomes two equations for the gain and phase conditions:

$$g_{th} = \alpha_a + \left(\frac{L}{d} - 1\right)\alpha_c + \frac{1}{d}\ln\frac{1}{|r_f r_b|}, \tag{13.6}$$

$$2\beta L + \phi_f + \phi_b = 2p\pi. \tag{13.7}$$

The lasing operation is analyzed not only for the *traveling wave model* but also for the *standing wave model* to form the rate equations. The standing-wave expression for the electric field is given by

$$E(t, \boldsymbol{r}) = \widehat{E}_p(t)\Phi_p(x, y, z)e^{j\omega_p t} + c.c., \tag{13.8}$$

where $\Phi_p(x, y, z)$ is the normalized field distribution function. Let us examine the field distribution function in VCSEL. The field amplitudes $|E^{(+)}(y)|$and$|E^{(-)}(y)|$ are assumed to be uniform in the cavity. The relationship between the forward and backward propagating components are given in terms of the reflectivities at the mirrors,

$$E^{(-)}(L) = r_f\, E^{(+)}(L) e^{-j\beta L}, \tag{13.9}$$

$$E^{(+)}(0) = r_b\, E^{(-)}(0)\, e^{-j\beta L}. \tag{13.10}$$

Assuming the approximation $|r_b| \approx 1$ and almost-constant amplitudes $|E^{(+)}(y)|$and$|E^{(-)}(y)|$, the function describing the field along the y direction is found to be proportional to $\cos(\beta y + \phi_b/2)$. The normalization of the distribution is evaluated by calculating the spatial integration

$$\int_0^L \cos^2\left(\beta y + \frac{\phi_b}{2}\right) dy = \frac{2\beta L - \sin\phi_b - \sin\phi_f}{4\beta}, \tag{13.11}$$

where (13.7) is applied to simplify the notation. We then get

$$\Phi_p(x, y, z) = \sqrt{\frac{4\beta}{2\beta L - \sin\phi_f - \sin\phi_b}} \cos\left(\beta y + \frac{\phi_b}{2}\right) T(x, z), \tag{13.12}$$

with normalization condition

$$\int_{-\infty}^{\infty} \int_0^L \int_{-\infty}^{\infty} |\Phi_p(x, y, z)|^2\, dx\, dy\, dz = 1. \tag{13.13}$$

The photon number S_p is obtained by determining the optical energy in the cavity, as was similarly done in (3.56). This results in similar rate equations as for conventional lasers,

$$\frac{dS_p}{dt} = (G_p - G_{th}) S_p + \xi a N, \tag{13.14}$$

$$\frac{dN}{dt} = -G_p \frac{S_p}{V_a} - \frac{N}{\tau_s} + \frac{I}{eV_a}, \tag{13.15}$$

where

$$G_p = \xi a \left(N - N_g\right), \tag{13.16}$$

$$G_{th} = v g_{th} = v\left\{\alpha_a + \left(\frac{L}{d} - 1\right)\alpha_c + \frac{1}{d}\ln\frac{1}{|r_f\, r_b|}\right\}, \tag{13.17}$$

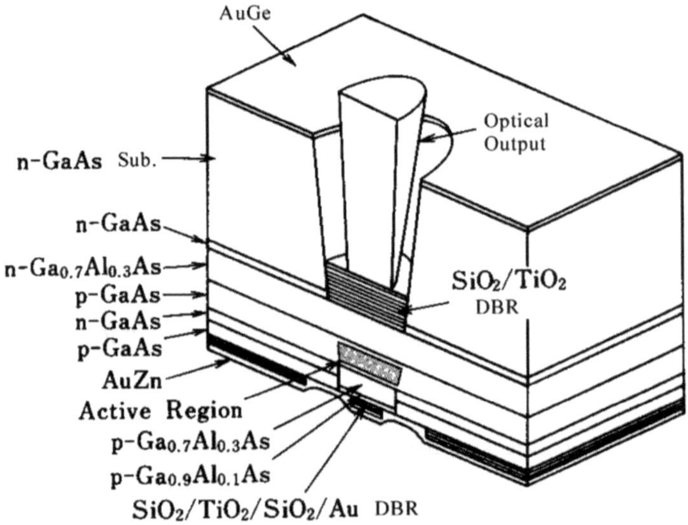

Fig. 13.2 Structure of a constructed VCSEL [1] (Copyright, 1989, AIP Publication LLC)

and

$$\xi = \iiint_{act} \left|\Phi_p(x, y, z)\right|^2 dx\, dy\, dz. \tag{13.18}$$

The differences between VCSELs and conventional lasers are as follows. The first is that, from (13.17), the radiation loss, needed in the determination of threshold gain G_{th}, is divided by the thickness d of the active region, whereas that in conventional lasers is divided by the cavity length L. The second is that the field confinement factor along the y direction depends not only on the thickness d, but also on the position of the active region relative to the standing wave $\cos(\beta y + \phi_b/2)$, because βd is much smaller than unity.

13.2 Examples of Fabricated Structures

Because of the low thickness d, the threshold gain G_{th} is very high as long as the reflectivities $\left|r_f\right|^2$ and $\left|r_b\right|^2$ are not large, as can be seen in (13.17). The power reflectivity by a metal surface is less than 0.95, which is not sufficient to achieve lasing in a VCSEL. For this reason, DBR mirrors are used as VCSELs like that sketched in Fig. 13.2. A reflectivity above 0.995 can be achieved from such mirrors.

Fig. 13.3 Two-dimensional laser array using the VCSELs [2] (Copyright, 1989, The IET)

The most advantageous feature of the VCSEL is its capacity to form *two-dimensional laser arrays*. An example of a fabricated two-dimensional laser array is shown in Fig. 13.3. Each laser column has a distinct radius yielding a specific lasing wavelength. Applications for two-dimensional laser arrays will bring further developments in optoelectronics.

Bibliography

1. F. Koyama, S. Kinoshita, K. Iga, Room temperature continuous wave lasing characteristics of a GaAs vertical cavity surface-emitting laser. Appl. Phys. Lett. **55**(3), 221–222 (1989)
2. J.L. Jewell, A. Scherer, S.L. McCall, Y.H. Lee, S. Walker, J.J.P. Harbison, L.T. Florez, Low-threshold electrically pumped vertical-cavity surface-emitting microlasers. Electron. Lett. **25**(17), 1123–1124 (1989)

Appendix 1
Derivations of the Annihilation and Creation Operators

To begin, we introduce an operator S defined by the relation

$$S = a^{+}a, \tag{A.1.1}$$

where a^{+} and a are operators given by (2.36) and (2.37). We assume here that for the operator S an eigenvalue s corresponding to eigenstate $|s\rangle$ exists satisfying

$$S|s\rangle = s|s\rangle. \tag{A.1.2}$$

Here, we do not suppose that s is an integer but derive the fact in due course. We now consider a state $|as\rangle = a|s\rangle$ and operator S acting on this state. We obtain the following relation from the commutation relation for a^{+} and a in (2.40),

$$\begin{aligned} S|as\rangle &= Sa|s\rangle = (aa^{+} - 1)a|s\rangle = a(a^{+}a - 1)|s\rangle \\ &= a(S-1)|s\rangle = a(s-1)|s\rangle = (s-1)a|s\rangle = (s-1)|as\rangle \end{aligned} \tag{A.1.3}$$

This equation implies that $|as\rangle$ is an eigenstate of S corresponding to eigenvalue $s - 1$. Similarly, we establish the following:

$$\begin{aligned} S|a^{2}s\rangle &= Sa^{2}|s\rangle = (aa^{+} - 1)a^{2}|s\rangle = a(a^{+}a - 1)|as\rangle \\ &= a(S-1)|as\rangle = a(s-2)|as\rangle = (s-2)|a^{2}s\rangle, \end{aligned} \tag{A.1.4}$$

$$\begin{aligned} S|a^{3}s\rangle &= Sa^{3}|s\rangle = (aa^{+} - 1)a^{3}|s\rangle = a(a^{+}a - 1)|a^{2}s\rangle \\ &= a(S-1)|a^{2}s\rangle = a(s-3)|a^{2}s\rangle = (s-3)|a^{3}s\rangle. \end{aligned} \tag{A.1.5}$$

By iteration, we get the general relation

$$S|a^{k}s\rangle = (s-k)|a^{k}s\rangle, \text{where } k \text{ is an integer.} \tag{A.1.6}$$

That is, $|a^{k}s\rangle$ is an eigenstate of S corresponding to eigenvalue $s - k$. Therefore, assuming $|a^{k}s\rangle$ is proportional to $|s-k\rangle$, we can write

$$S|a^{k}s\rangle = (s-k)B_{k}|s-k\rangle, \tag{A.1.7}$$

or

$$|a^{k}s\rangle = B_{k}|s-k\rangle, \tag{A.1.8}$$

M. Yamada, *Theory of Semiconductor Lasers*,
Springer Series in Optical Sciences 185,
DOI: 10.1007/978-4-431-54889-8, © Springer Japan 2014

where B_k is a constant of proportionality. For $k = 1$, setting $B = B_1$, we have

$$|as\rangle = a|s\rangle = B|s-1\rangle. \tag{A.1.9}$$

Specifically, a is an operator that reduces the eigenvalue s by one.

Here, we determine the value of B. Requiring that the eigenstate should be normalized to unity

$$\langle s|s\rangle = 1, \tag{A.1.10}$$

we find that

$$\begin{aligned}\langle as|as\rangle &= \langle s|a^+a|s\rangle = s\\ &= |B|^2\langle s-1|s-1\rangle = |B|^2.\end{aligned} \tag{A.1.11}$$

This yields, after a simple choice of phase,

$$B = \sqrt{s}. \tag{A.1.12}$$

For the general case, we obtain similarly

$$\begin{aligned}\langle a^k s|a^k s\rangle &= \langle a^{k-1}s|a^+a|a^{k-1}s\rangle = \langle a^{k-1}s|S|a^{k-1}s\rangle\\ &= (s-k+1)\langle a^{k-1}s|a^{k-1}s\rangle\\ &= (s-k+1)(s-k+2)\langle a^{k-2}s|a^{k-2}s\rangle\\ &\ \ \vdots\\ &= s(s-1)(s-2)\cdot\cdots\cdots(s-k+1)\\ &= |B_k|^2.\end{aligned} \tag{A.1.13}$$

which gives

$$B_k = \sqrt{s(s-1)(s-2)\cdot\cdots\cdots(s-k+1)}. \tag{A.1.14}$$

Note that the eigenvalue $(s - k)$ in (A.1.7) must be a nonnegative integer, which follows from the result

$$\begin{aligned}s-k &= \langle a^k s|(s-k)|a^k s\rangle/|B_k|^2 = \langle a^k s|S|a^k s\rangle/|B_k|^2\\ &= \langle a^k s|a^+a|a^k s\rangle/|B_k|^2 = \langle a^{k+1}s|a^{k+1}s\rangle/|B_k|^2 = |B_{k+1}|^2/|B_k|^2\\ &\geq 0\end{aligned} \tag{A.1.15}$$

Therefore, for integers k larger than s, the state $|a^k s\rangle$ should not exist. If s is a nonnegative integer, we find for $s+1 \leq k$,

$$|a^k s\rangle = \sqrt{s(s-1)\cdots(s-s)\cdots(s-k+1)}|s-k\rangle = 0, \tag{A.1.16}$$

that is, the state $|a^k s\rangle$ vanishes. However, if s is a number of another type, then for $s+1<k$,

$$|a^k s\rangle = \sqrt{s(s-1)\cdots(s-k+1)}|s-k\rangle \neq 0. \tag{A.1.17}$$

This relation cannot satisfy (A.1.15), hence the number s must be a nonnegative integer.

Let us now perform a similar analysis for a^+. We first find that

$$\begin{aligned} Sa^+|s\rangle &= a^+aa^+|s\rangle = a^+(a^+a+1)|s\rangle \\ &= a^+(S+1)|s\rangle = a^+(s+1)|s\rangle = (s+1)|a^+s\rangle \end{aligned} \tag{A.1.18}$$

and

$$Sa^{+\ell}|s\rangle = (s+\ell)|a^{+\ell}s\rangle, \text{ where } \ell \text{ is an integer.} \tag{A.1.19}$$

Hence, a^+ is an operator that increments the eigenvalues by one. Therefore, we can write

$$a^{+\ell}|s\rangle = C_\ell|s+\ell\rangle, \tag{A.1.20}$$

and

$$a^+|s\rangle = C|s+1\rangle. \tag{A.1.21}$$

From the normalization of the eigenstate, we obtain

$$\begin{aligned} \langle a^+s|a^+s\rangle &= \langle s|aa^+|s\rangle = \langle s|1+a^+a|s\rangle = 1+s \\ &= |C|^2\langle s+1|s+1\rangle = |C|^2. \end{aligned} \tag{A.1.22}$$

Then the constant of proportionality C becomes

$$C = \sqrt{s+1}. \tag{A.1.23}$$

Thus we have derived the action of the *annihilation operator a* and the *creation operator* a^+ on the eigenstates of S

$$a|s\rangle = \sqrt{s}|s-1\rangle, \tag{A.1.24}$$

$$a^+|s\rangle = \sqrt{s+1}|s+1\rangle, \tag{A.1.25}$$

which clearly satisfy (A.1.2) given (A.1.1), that is

$$S|s\rangle = a^+a|s\rangle = s|s\rangle. \tag{A.1.26}$$

The operator $S = a^+a$ has nonnegative integer eigenvalues; hence S is called the *number operator*.

Appendix 2
Proof of the Uncertainty Principle

To begin, we recall the *Cauchy–Schwarz inequality* for vectors of an inner product space

$$\langle X|X\rangle\langle Y|Y\rangle \geq |\langle X|Y\rangle|^2. \tag{A.2.1}$$

This inequality is better understood by analogy with spatial vectors, as illustrated in Fig. A.2.1.

With the help of the Hermiticity of the operators and the Cauchy–Schwarz inequality, the product of the square of the standard deviations ΔA and ΔB of two operators A and B, defined in (2.47)–(2.50), can be rewritten as follows

$$\begin{aligned}(\Delta A)^2(\Delta B)^2 &= \left\langle \Psi|(A-\bar{A})^2|\Psi\right\rangle\left\langle \Psi|(B-\bar{B})^2|\Psi\right\rangle \\ &= \left\langle \Psi|(A-\bar{A})^+(A-\bar{A})|\Psi\right\rangle\left\langle \Psi|(B-\bar{B})^+(B-\bar{B})|\Psi\right\rangle \\ &\geq |\langle \Psi|(A-\bar{A})(B-\bar{B})|\Psi\rangle|^2\end{aligned} \tag{A.2.2}$$

Here the Cauchy–Schwarz inequality is applied by identifying

$$(A-\bar{A})|\Psi\rangle = |X\rangle, \tag{A.2.3}$$

$$(B-\bar{B})|\Psi\rangle = |Y\rangle. \tag{A.2.4}$$

We proceed to further expand the operator product

$$\begin{aligned}&(A-\bar{A})(B-\bar{B}) \\ &= \frac{1}{2}\{(A-\bar{A})(B-\bar{B})+(B-\bar{B})(A-\bar{A})\}+\frac{1}{2}\{(A-\bar{A})(B-\bar{B})-(B-\bar{B})(A-\bar{A})\} \\ &= \frac{1}{2}\{(A-\bar{A})(B-\bar{B})+(B-\bar{B})(A-\bar{A})\}+\frac{j}{2}C\end{aligned} \tag{A.2.5}$$

M. Yamada, *Theory of Semiconductor Lasers*,
Springer Series in Optical Sciences 185,
DOI: 10.1007/978-4-431-54889-8, © Springer Japan 2014

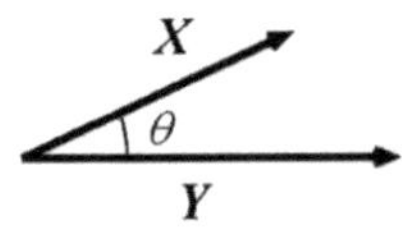

$$|\boldsymbol{X}||\boldsymbol{Y}| \geq |\boldsymbol{X}\cdot\boldsymbol{Y}|$$

$$|\boldsymbol{X}|^2|\boldsymbol{Y}|^2 \geq |\boldsymbol{X}\cdot\boldsymbol{Y}|^2$$

Fig. A.2.1 Illustration of the Cauchy–Schwarz inequality with spatial vectors

and hence taking the expectation value

$$\begin{aligned} &|\langle\Psi|(A-\bar{A})(B-\bar{B})|\Psi\rangle|^2 \\ &\quad = \frac{1}{4}|\langle\Psi|(A-\bar{A})(B-\bar{B})+(B-\bar{B})(A-\bar{A})|\Psi\rangle + j\langle\Psi|C|\Psi\rangle|^2 \\ &\quad = \frac{1}{4}|\langle\Psi|(A-\bar{A})(B-\bar{B})+(B-\bar{B})(A-\bar{A})|\Psi\rangle|^2 + \frac{1}{4}|\langle\Psi|C|\Psi\rangle|^2 \end{aligned} \tag{A.2.6}$$

From (A.2.2) and (A.2.6), we obtain

$$(\Delta A)^2(\Delta B)^2 \geq \frac{1}{4}|\langle\Psi|C|\Psi\rangle|^2. \tag{A.2.7}$$

That is, we arrive at the uncertainty relation for a pair of Hermitian operators

$$(\Delta A)(\Delta B) \geq \frac{1}{2}|\langle\Psi|C|\Psi\rangle|. \tag{A.2.8}$$

This inequality relation is sometimes called the *Robertson-Schrödinger uncertainty relation*.

Appendix 3
Classical Hamiltonian Dynamics for a Charged Particle in an Electromagnetic Field

The Hamiltonian is a generating function by which the dynamic motion of a physical system can be analyzed. However, there is no explicit way to derive the Hamiltonian. We suppose a hypothetical function, then examine its validity by comparing it with other theories.

We consider the Hamiltonian of a charged particle under the influence of an external electromagnetic field,

$$\begin{aligned} H &= \frac{1}{2m}(\boldsymbol{p} - e\boldsymbol{A})^2 + eU \\ &= \frac{1}{2m}\left[(p_x - eA_x)^2 + (p_y - eA_y)^2 + (p_z - eA_z)^2\right] + eU \end{aligned}, \tag{A.3.1}$$

where m and $\boldsymbol{p}$ are mass and momentum of the charged particle, respectively, $\boldsymbol{A}$ and U the vector and scalar potentials for the electromagnetic field, and e is the charge unit. The electric field $\boldsymbol{E}$ and the magnetic field $\boldsymbol{H}$ are obtained from these potentials using

$$\boldsymbol{E} = -\nabla U - \frac{\partial \boldsymbol{A}}{\partial t}, \tag{A.3.2}$$

$$\boldsymbol{H} = \frac{1}{\mu}\nabla \times \boldsymbol{A}, \tag{A.3.3}$$

where μ is the magnetic permeability.

Now, from the given Hamiltonian, we derive the equation for the *Lorentz force* represented in the form as found in *Newtonian mechanics*. We derive the time derivative of the position x and the momentum p_x based on the classical Hamilton principle given in (2.7) and (2.8). These equations are

$$\dot{x} = \frac{\partial H}{\partial p_x} = \frac{1}{m}(p_x - eA_x). \tag{A.3.4}$$

$$\begin{aligned} \dot{p}_x &= -\frac{\partial H}{\partial x} \\ &= \frac{e}{m}\left[(p_x - eA_x)\frac{\partial A_x}{\partial x} + (p_y - eA_y)\frac{\partial A_y}{\partial x} + (p_z - eA_z)\frac{\partial A_z}{\partial x}\right] - e\frac{\partial U}{\partial x}. \end{aligned} \tag{A.3.5}$$

M. Yamada, *Theory of Semiconductor Lasers*,
Springer Series in Optical Sciences 185,
DOI: 10.1007/978-4-431-54889-8, © Springer Japan 2014

Differentiating (A.3.4) a second time yields

$$m\ddot{x} = \dot{p}_x - e\dot{A}_x. \tag{A.3.6}$$

Substituting (A.3.5) to (A.3.6) gives

$$m\ddot{x} = \frac{e}{m}\left[(p_x - eA_x)\frac{\partial A_x}{\partial x} + (p_y - eA_y)\frac{\partial A_y}{\partial x} + (p_z - eA_z)\frac{\partial A_z}{\partial x}\right] - e\frac{\partial U}{\partial x} - e\dot{A}_x. \tag{A.3.7}$$

We also substitute (A.3.4) into (A.3.7), resulting in

$$m\ddot{x} = e\left[\dot{x}\frac{\partial A_x}{\partial x} + \dot{y}\frac{\partial A_y}{\partial x} + \dot{z}\frac{\partial A_z}{\partial x}\right] - e\frac{\partial U}{\partial x} - e\dot{A}_x. \tag{A.3.8}$$

Moreover, the total time derivative is given as

$$\begin{aligned}\dot{A}_x &= \frac{dA_x}{dt} \\ &= \frac{\partial A_x}{\partial t} + \frac{\partial A_x}{\partial x}\frac{\partial x}{\partial t} + \frac{\partial A_x}{\partial y}\frac{\partial y}{\partial t} + \frac{\partial A_x}{\partial z}\frac{\partial z}{\partial t} \\ &= \frac{\partial A_x}{\partial t} + \frac{\partial A_x}{\partial x}\dot{x} + \frac{\partial A_x}{\partial y}\dot{y} + \frac{\partial A_x}{\partial z}\dot{z}\end{aligned} \tag{A.3.9}$$

Substitution of this relation into (A.3.8) gives

$$\begin{aligned}m\ddot{x} &= e\left[\dot{y}\left(\frac{\partial A_y}{\partial x} - \frac{\partial A_x}{\partial y}\right) - \dot{z}\left(\frac{\partial A_x}{\partial z} - \frac{\partial A_z}{\partial x}\right)\right] - e\frac{\partial A_x}{\partial t} - e\frac{\partial U}{\partial x} \\ &= e\left[E_x + \mu(\dot{\boldsymbol{r}} \times \boldsymbol{H})_x\right]\end{aligned} \tag{A.3.10}$$

We can develop similar relations for the y and z directions. From these, we arrive at the *Newtonian dynamic equation* for a charged particle subjected to a *Lorentz force*,

$$m\ddot{\boldsymbol{r}} = e[\boldsymbol{E} + \mu(\dot{\boldsymbol{r}} \times \boldsymbol{H})]. \tag{A.3.11}$$

Appendix 4
Derivation of the Classical Lorentz Force from the Quantum Mechanical Schrödinger Equation

The Schrödinger equation for a charged particle in an applied electromagnetic field is given by

$$j\hbar\frac{\partial|\Psi(\boldsymbol{r},t)\rangle}{\partial t} = H|\Psi(\boldsymbol{r},t)\rangle, \tag{A.4.1}$$

$$H = \frac{1}{2m}(\boldsymbol{p} - e\boldsymbol{A})^2 + eU, \tag{A.4.2}$$

where m is the mass of the charged particle, e the electric charge unit, and $\boldsymbol{p}$ the momentum operator given by

$$\boldsymbol{p} = -j\hbar\nabla. \tag{A.4.3}$$

$\boldsymbol{A}$ and U are the vector and the scalar potential of the electromagnetic field. Here, we assume that these potentials are represented by a classical function, and the electric field $\boldsymbol{E}$ and the magnetic field $\boldsymbol{H}$ are given by these potentials as

$$\boldsymbol{E} = -\nabla U - \frac{\partial \boldsymbol{A}}{\partial t}, \tag{A.4.4}$$

$$\boldsymbol{H} = \frac{1}{\mu}\nabla \times \boldsymbol{A}, \tag{A.4.5}$$

where μ is the magnetic permeability.

In (A.4.1), $|\Psi(\mathbf{r},t)\rangle$ is a state vector given as a linear combination of a time-varying coefficient and spatial functions,

$$|\Psi(\boldsymbol{r},t)\rangle = \sum_n c_n(t)|\phi_n(\boldsymbol{r})\rangle, \tag{A.4.6}$$

where $|\phi_n(\mathbf{r})\rangle$ is a state vector corresponding to the spatial distribution function (plane-wave solution)

$$\phi_n(\boldsymbol{r}) = \frac{1}{\sqrt{V_e}}e^{j\boldsymbol{k}_n\cdot\boldsymbol{r}}. \tag{A.4.7}$$

M. Yamada, *Theory of Semiconductor Lasers*,
Springer Series in Optical Sciences 185,
DOI: 10.1007/978-4-431-54889-8,

Here, V_e is the occupation volume of the charged particle, and $\boldsymbol{k}_n$ is the wave number of the *n–th* state, written in component form as

$$\boldsymbol{k}_n = \boldsymbol{e}_x k_{nx} + \boldsymbol{e}_y k_{ny} + \boldsymbol{e}_z k_{nz}, \tag{A.4.8}$$

where $\boldsymbol{e}_i$ is a spatial unit vector indicating the i-th direction. The orthonormalization condition is defined with respect to an inner product of the state vector space as

$$\langle \phi_n(\boldsymbol{r}) | \phi_\ell(\boldsymbol{r}) \rangle = \frac{1}{V_e} \int_{V_e} e^{j(\boldsymbol{k}_\ell - \boldsymbol{k}_n)\cdot \boldsymbol{r}} d^3\boldsymbol{r} = \delta_{n,\ell}. \tag{A.4.9}$$

The expectation value for the position of the charged particle is $\langle \boldsymbol{r} \rangle$ and for the velocity of the moving particle is $d\langle \boldsymbol{r} \rangle / dt$. Then, the expectation value of the one-dimensional velocity must be given as

$$\begin{aligned} \frac{d}{dt}\langle x \rangle &= \frac{\partial \langle \Psi|}{\partial t} x |\Psi\rangle + \langle \Psi | x \frac{\partial |\Psi\rangle}{\partial t} \\ &= \frac{1}{j\hbar} \langle \Psi | (xH - Hx) | \Psi \rangle \end{aligned} \tag{A.4.10}$$

Here, the Hamiltonian in (A.4.2) is given in more explicit coordinate form,

$$\begin{aligned} H = &-\frac{\hbar^2}{2m}\left(\frac{\partial^2}{\partial x^2} + \frac{\partial^2}{\partial y^2} + \frac{\partial^2}{\partial z^2}\right) + \frac{j\hbar e}{2m}\left(\frac{\partial A_x}{\partial x} + \frac{\partial A_y}{\partial y} + \frac{\partial A_z}{\partial z}\right) \\ &+ \frac{j\hbar e}{m}\left(A_x \frac{\partial}{\partial x} + A_y \frac{\partial}{\partial y} + A_z \frac{\partial}{\partial z}\right) + \frac{e^2}{2m}\left(A_x^2 + A_y^2 + A_z^2\right) + eU. \end{aligned} \tag{A.4.11}$$

We then obtain the relation

$$\langle \Psi | (xH - Hx) | \Psi \rangle = \frac{\hbar^2}{m} \left\langle \Psi | \frac{\partial}{\partial x} | \Psi \right\rangle - j \frac{\hbar e}{m} \langle \Psi | A_x | \Psi \rangle. \tag{A.4.12}$$

By substituting (A.4.12) into (A.4.10), the velocity is given as

$$\frac{d}{dt}\langle x \rangle = \frac{1}{m}\left[\langle p_x \rangle - e\langle A_x \rangle\right], \tag{A.4.13}$$

where

$$\langle p_x \rangle = -j\hbar \langle \Psi | \frac{\partial}{\partial x} | \Psi \rangle. \tag{A.4.14}$$

We evaluate the second time derivative of the position expectation value as

$$\frac{d^2}{dt^2}\langle x \rangle = \frac{1}{m}\frac{d}{dt}\left[\langle p_x \rangle - e\langle A_x \rangle\right]. \tag{A.4.15}$$

The temporal derivative of the momentum is given as

$$\begin{aligned}\frac{d}{dt}\langle p_x\rangle &= -j\hbar\left[\frac{\partial\langle\Psi|}{\partial t}\frac{\partial}{\partial x}|\Psi\rangle + \langle\Psi|\frac{\partial}{\partial x}\frac{\partial|\Psi\rangle}{\partial t}\right]\\ &= \langle\Psi|\left(H\frac{\partial}{\partial x} - \frac{\partial}{\partial x}H\right)|\Psi\rangle \\ &= -\langle\Psi|\left(\frac{\partial H}{\partial x}\right)|\Psi\rangle\end{aligned} \tag{A.4.16}$$

We substitute (A.4.11) and (A.4.6) into (A.4.16) and perform the calculation. Here we use the relation

$$\frac{\partial|\Psi\rangle}{\partial x} = j\sum_n c_n(t)k_{nx}|\phi_n\rangle. \tag{A.4.17}$$

Then, (A.4.16) becomes

$$\begin{aligned}\frac{d}{dt}\langle p_x\rangle = &-j\frac{\hbar^2}{2m}\sum_n |c_n|^2 k_{nx}\left(k_{nx}^2 + k_{ny}^2 + k_{nz}^2\right)\\ &-j\frac{\hbar e}{2m}\left\langle\left(\frac{\partial^2 A_x}{\partial x^2} + \frac{\partial^2 A_y}{\partial x\partial y} + \frac{\partial^2 A_z}{\partial x\partial z}\right)\right\rangle\\ &+\frac{\hbar e}{m}\sum_n |c_n|^2\langle\varphi_n|\left(k_{nx}\frac{\partial A_x}{\partial x} + k_{ny}\frac{\partial A_y}{\partial y} + k_{nz}\frac{\partial A_z}{\partial z}\right)|\varphi_n\rangle\\ &+j\frac{\hbar e}{m}\sum_n |c_n|^2 k_{nx}\langle\varphi_n|\left(k_{nx}A_x + k_{ny}A_y + k_{nz}A_z\right)|\varphi_n\rangle\\ &-\frac{e^2}{m}\left\langle\left(A_x\frac{\partial A_x}{\partial x} + A_y\frac{\partial A_y}{\partial x} + A_z\frac{\partial A_z}{\partial x}\right)\right\rangle\\ &-e\left\langle\frac{\partial U}{\partial x}\right\rangle\end{aligned} \tag{A.4.18}$$

The temporal variation of the vector potential is

$$\begin{aligned}\frac{d}{dt}\langle A_x\rangle &= \frac{d}{dt}\langle\Psi|A_x|\Psi\rangle\\ &= \left[\frac{\partial\langle\Psi|}{\partial t}A_x|\Psi\rangle + \langle\Psi|\frac{\partial A_x}{\partial t}|\Psi\rangle + \langle\Psi|A_x\frac{\partial|\Psi\rangle}{\partial t}\right].\\ &= \frac{1}{j\hbar}\langle\Psi|(A_xH - HA_x)|\Psi\rangle + \langle\Psi|\frac{\partial A_x}{\partial t}|\Psi\rangle\end{aligned} \tag{A.4.19}$$

Substitution of (A.4.6) and (A.4.11) into (A.4.19) gives

$$
\begin{aligned}
\frac{d}{dt}\langle A_x\rangle = & -j\frac{\hbar}{2m}\left\langle\left(\frac{\partial^2 A_x}{\partial x^2}+\frac{\partial^2 A_y}{\partial y^2}+\frac{\partial^2 A_z}{\partial z^2}\right)\right\rangle \\
& +\frac{\hbar}{m}\sum_n |c_n|^2\langle\varphi_n|\left(k_{nx}\frac{\partial A_x}{\partial x}+k_{ny}\frac{\partial A_x}{\partial y}+k_{nz}\frac{\partial A_x}{\partial z}\right)|\varphi_n\rangle \\
& -\frac{e}{m}\left\langle\left(A_x\frac{\partial A_x}{\partial x}+A_y\frac{\partial A_y}{\partial x}+A_z\frac{\partial A_z}{\partial x}\right)\right\rangle \\
& +\left\langle\frac{\partial A_x}{\partial t}\right\rangle
\end{aligned} \qquad (A.4.20)
$$

We combine (A.4.18) and (A.4.20), resulting in

$$
\begin{aligned}
& \frac{d}{dt}(\langle p_x\rangle - e\langle A_x\rangle) \\
& = \frac{\hbar e}{m}\sum_n |c_n|^2\langle\phi_n|\left[k_{ny}(\frac{\partial A_y}{\partial x}-\frac{\partial A_x}{\partial y})+k_{nz}(\frac{\partial A_z}{\partial x}-\frac{\partial A_x}{\partial z})\right]|\phi_n\rangle \\
& +\frac{e^2}{m}\left\langle\left[A_y(\frac{\partial A_y}{\partial x}-\frac{\partial A_x}{\partial y})+A_z(\frac{\partial A_z}{\partial x}-\frac{\partial A_x}{\partial z})\right]\right\rangle \\
& -e\left\langle\left(\frac{\partial U}{\partial x}+\frac{\partial A_x}{\partial t}\right)\right\rangle \\
& -j\frac{\hbar}{2m}\sum_n |c_n|^2 k_{nx}\left[k_{nx}(k_{nx}-2e\langle A_x\rangle)+k_{ny}(k_{ny}-2e\langle A_y\rangle)+k_{nz}(k_{nz}-2e\langle A_z\rangle)\right] \\
& -j\frac{\hbar e}{2m}\left\langle\left[\frac{\partial}{\partial y}(\frac{\partial A_y}{\partial x}-\frac{\partial A_x}{\partial y})+\frac{\partial}{\partial z}(\frac{\partial A_z}{\partial x}-\frac{\partial A_x}{\partial z})\right]\right\rangle
\end{aligned}
$$

(A.4.21)

Here, *we suppose that the spatial distributions of the potentials and its spatial gradients are almost uniform over the charged particle.* We also use the three-dimensional form of (A.4.14),

$$
\langle \boldsymbol{p}\rangle = -j\hbar\langle\Psi|\nabla|\Psi\rangle = \hbar\sum_n |c_n|^2\boldsymbol{k} = m\langle\boldsymbol{v}\rangle + e\langle\boldsymbol{A}\rangle, \qquad (A.4.22)
$$

as well as expressions for the electric field $\boldsymbol{E}$ and magnetic field $\boldsymbol{H}$ as defined in (A.4.4) and (A.4.5). Therefore, (A.4.21) is rewritten as

$$
\begin{aligned}
& \frac{d}{dt}(\langle p_x\rangle - e\langle A_x\rangle) \\
& = \frac{e\mu}{m}\langle[(p_y-eA_y)H_z-(p_x-eA_x)H_y]\rangle + e\langle E_x\rangle \\
& = e\langle[E_x+\mu(v_yH_z-v_xH_y)]\rangle
\end{aligned} \qquad (A.4.23)
$$

In this calculation, *we have dropped the imaginary terms because the expectation value of any physical quantity should give only a real number.*

A similar analysis can be given for the y and z directions. This then leads to the classical Newtonian dynamic equation for a charged particle subjected to an applied Lorentz force,

$$m\frac{d^2\langle \boldsymbol{r}\rangle}{dt^2} = e[\langle \boldsymbol{E}\rangle + \mu(\langle \boldsymbol{v}\rangle \times \langle \boldsymbol{H}\rangle)]. \qquad \text{(A.4.24)}$$

Appendix 5
Analysis of Electron Transition Probability Based on the Fully Quantized Treatment

A.5.1 Hamiltonian and Eigenstates

The Hamiltonian of an optical wave has been given in (2.41), while that of a charged particle or electron is given in (2.69). The Hamiltonian H for the combined system is given by,

$$H = \hbar\omega\left(a^{+}a + \frac{1}{2}\right) + \frac{1}{2m}(\boldsymbol{p} - e\boldsymbol{A})^2 + eU(\boldsymbol{r}), \tag{A.5.1}$$

where a and a^{+} are the respective annihilation and creation operators for the optical wave of angular frequency ω, e is the unit of electric charge, that is, the charge of an electron is $-e$, $\boldsymbol{p}$ the momentum operator, and m the mass of the electron. $U(\boldsymbol{r})$ is the scalar potential for the electron in the surrounding material, and $\boldsymbol{A}$ the vector potential of the optical wave and is written in quantum mechanical form as

$$\boldsymbol{A} = \sqrt{\frac{\hbar}{2\varepsilon\omega}}(a + a^{+})\boldsymbol{\Phi}(\boldsymbol{r}), \tag{A.5.2}$$

where $\boldsymbol{\Phi}(\boldsymbol{r})$ is the normalized spatial distribution function of the optical wave, as has been defined in (2.24)–(2.27). Here, we need to be aware that the amplitude of the vector potential $\boldsymbol{A}$ is given by the annihilation and creation operators and is not an explicit function of time t in quantum mechanics.

The Hamiltonian consists of terms for the optical wave, the electron, and cross terms between the optical wave and the electron. We separate the Hamiltonian into the *principal Hamiltonian* H_o for terms just the optical wave and the electron themselves, and an *interaction Hamiltonian* H_{int} for the cross term:

$$H_o = \hbar\omega\left(a^{+}a + \frac{1}{2}\right) + \frac{\boldsymbol{p}^2}{2m} + eU(\boldsymbol{r}), \tag{A.5.3}$$

$$H_{\text{int}} = -\frac{e}{2m}(\boldsymbol{p}\cdot\boldsymbol{A} + \boldsymbol{A}\cdot\boldsymbol{p}) + \frac{e^2}{2m}\boldsymbol{A}^2. \tag{A.5.4}$$

M. Yamada, *Theory of Semiconductor Lasers*,
Springer Series in Optical Sciences 185,
DOI: 10.1007/978-4-431-54889-8, © Springer Japan 2014

As the principal Hamiltonian is a sum of Hamiltonians for the optical wave and non-interacting electron, its eigenstates $|E_\alpha\rangle$ are given as a product of eigenstates for the photon and for the electron, that is

$$H_o|E_\alpha\rangle = E_\alpha|E_\alpha\rangle, \tag{A.5.5}$$

$$|E_\alpha\rangle = |s\rangle|\varphi_n(\boldsymbol{r})\rangle, \tag{A.5.6}$$

$$E_\alpha = \hbar\omega\left(s + \frac{1}{2}\right) + W_n. \tag{A.5.7}$$

Here, s is the photon number and $|s\rangle$ its eigenstate given by

$$\hbar\omega\left(a^+a + \frac{1}{2}\right)|s\rangle = \hbar\omega\left(s + \frac{1}{2}\right)|s\rangle, \tag{A.5.8}$$

and W_n is the n-th energy of the electron wave state $|\phi_n(\boldsymbol{r})\rangle$ obtained as solutions of

$$\left(\frac{\boldsymbol{p}^2}{2m} + U(\boldsymbol{r})\right)|\varphi_n(\boldsymbol{r})\rangle = W_n|\varphi_n(\boldsymbol{r})\rangle. \tag{A.5.9}$$

Note that the energies and the eigenstates of the optical wave and the electron are independently defined in the principal Hamiltonian.

A.5.2 Interaction

The interaction between optical wave and electron is analyzed with the Schrödinger equation. Any quantum mechanical state $|\Psi\rangle$ can be represented as a linear combination of energy eigenstates,

$$|\Psi(t,\boldsymbol{r})\rangle = \sum_\alpha C_\alpha(t)|E_\alpha\rangle. \tag{A.5.10}$$

This arbitrary state must satisfy the Schrödinger equation.

$$j\hbar\frac{d|\Psi(t,\boldsymbol{r})\rangle}{dt} = H|\Psi(t,\boldsymbol{r})\rangle = (H_o + H_{\text{int}})|\Psi(t,\boldsymbol{r})\rangle. \tag{A.5.11}$$

We transform this state into another state $|\Xi(t,\boldsymbol{r})\rangle$ specified by the relation

$$|\Psi(t,\boldsymbol{r})\rangle = e^{-j(H_o/\hbar)t}|\Xi(t,\boldsymbol{r})\rangle. \tag{A.5.12}$$

Here, $\exp[At]$ is a function of operator A having a series expansion similar to its classical function,

$$e^{At} = 1 + At + \frac{1}{2!}(At)^2 + \cdots + \frac{1}{n!}(At)^n + \cdots. \tag{A.5.13}$$

From this we obtain

$$\frac{d}{dt}e^{At} = Ae^{At}. \tag{A.5.14}$$

We substitute (A.5.12) into (A.5.11)

$$\begin{aligned} j\hbar\frac{d|\Psi\rangle}{dt} &= H_o e^{-j(H_o/\hbar)t}|\Xi\rangle + j\hbar e^{-j(H_o/\hbar)t}\frac{d|\Xi\rangle}{dt} \\ &= H_o e^{-j(H_o/\hbar)t}|\Xi\rangle + H_{\text{int}}e^{-j(H_o/\hbar)t}|\Xi\rangle \end{aligned} \tag{A.5.15}$$

By multiplying from the left with $\exp[j(H_o/\hbar)t]$, we get a dynamic equation for the new state $|\Xi(t,\boldsymbol{r})\rangle$,

$$j\hbar\frac{d|\Xi(t,\boldsymbol{r})\rangle}{dt} = H'_{\text{int}}|\Xi(t,\boldsymbol{r})\rangle, \tag{A.5.16}$$

with

$$H'_{\text{int}} \equiv e^{j(H_o/\hbar)t}H_{\text{int}}e^{-j(H_o/\hbar)t}. \tag{A.5.17}$$

Whereas the right-hand side of (A.5.11) consists of two terms, the right-hand side of (A.5.16) has only one term. We can then perform time integration more easily with (A.5.16) than the other. The state $|\Psi(t,\boldsymbol{r})\rangle$defined from the original equation (A.5.11) determines the so-called *Schrödinger picture*, whereas the state $|\Xi(t,\boldsymbol{r})\rangle$defined with respect to (A.5.16) corresponds to the *interaction picture*.

Because the term H'_{int} includes time t, finding an exact analytical solution of $|\Xi(t)\rangle$ in the (A.5.16) is difficult in general. We perform this analysis by following a *perturbation method*.

By performing a time integration of (A.5.16), $|\Xi(t)\rangle$ is expressible as

$$|\Xi(t)\rangle = |\Xi(0)\rangle + \frac{1}{j\hbar}\int_0^t H'_{\text{int}}(\tau)|\Xi(\tau)\Big\rangle d\tau. \tag{A.5.18}$$

Because $|\Xi(\tau)\rangle$ is as yet unknown, we cannot perform the integration. As a first approximation, we insert the initial value of $|\Xi(0)\rangle$, yielding

$$\begin{aligned} |\Xi(t)\rangle &\approx |\Xi(0)\rangle + \frac{1}{j\hbar}\int_0^t H'_{\text{int}}(\tau)|\Xi(0)\rangle d\tau \\ &= \left\{1 + \frac{1}{j\hbar}\int_0^t H'_{\text{int}}(\tau)d\tau\right\}|\Xi(0)\rangle. \end{aligned} \tag{A.5.19}$$

We again substitute this $|\Xi(t)\rangle$ for $|\Xi(\tau)\rangle$ in (A.5.18) to obtain

$$|\Xi(t)\rangle \approx \left\{1 + \frac{1}{j\hbar}\int_0^t H'_{\text{int}}(\tau)d\tau + \left(\frac{1}{j\hbar}\right)^2 \int_0^t \int_0^{\tau_1} H'_{\text{int}}(\tau_1)H'_{\text{int}}(\tau_2)d\tau_2 d\tau_1 \right\}|\Xi(0)\rangle. \tag{A.5.20}$$

By repeating the procedure, we get a series of integrated equations from which the state $|\Xi(t)\rangle$ is represented in the form of a transformation from the initial state $|\Xi(0)\rangle$,

$$|\Xi(t)\rangle \equiv S|\Xi(0)\rangle, \tag{A.5.21}$$

where S, called scattering matrix, is written as

$$\begin{aligned} S = 1 &+ \frac{1}{j\hbar}\int_0^t H'_{\text{int}}(\tau)d\tau \\ &+ \left(\frac{1}{j\hbar}\right)^2 \int_0^t \int_0^{\tau_1} H'_{\text{int}}(\tau_1)H'_{\text{int}}(\tau_2)d\tau_2 d\tau_1 \\ &+ \left(\frac{1}{j\hbar}\right)^3 \int_0^t \int_0^{\tau_1} \int_0^{\tau_2} H'_{\text{int}}(\tau_1)H'_{\text{int}}(\tau_2)H'_{\text{int}}(\tau_3)d\tau_3 d\tau_2 d\tau_1 \\ &+ \cdots\cdots \end{aligned} \tag{A.5.22}$$

In accordance with the order of H'_{int}, we can write S as a series expansion

$$S = S^{(0)} + S^{(1)} + S^{(2)} + \cdots + S^{(i)} + \cdots, \tag{A.5.23}$$

where i indexes the number of interactions between optical wave and electron.

We introduce element $S_{\beta\alpha}$ of the scattering matrix S describing the transition from an eigenstate α to another eigenstate β after time interval t:

$$S_{\beta\alpha} = \langle E_\beta|S|E_\alpha\rangle. \tag{A.5.24}$$

As expected from (A.5.10), (A.5.12), and (A.5.21), the transition probability $P_{\beta\alpha}(t)$ from an eigenstate α to another eigenstate β is given by

$$P_{\beta\alpha}(t) = |S_{\beta\alpha}|^2. \tag{A.5.25}$$

A.5.3 First-Order Transition Probability

As found from (A.5.22) and (A.5.23), the zeroth-order scattering matrix has only diagonal elements and does not contribute to the interaction between optical wave and electron,

$$S_{\beta\alpha}^{(0)} = \delta_{\beta,\alpha}, \tag{A.5.26}$$

where $\delta_{\beta,\alpha}$ is the *Kronecker delta*. The first-order term gives the one-time interaction, and is the principal interaction. Let us determine the first-order term:

$$\begin{aligned} S_{\beta\alpha}^{(1)} &= \left\langle E_\beta | S^{(1)} | E_\alpha \right\rangle \\ &= \frac{1}{j\hbar} \langle E_\beta | \int_0^t e^{j(H_o/\hbar)\,\tau} H_{\text{int}} e^{-j(H_o/\hbar)\,\tau} d\tau | E_\alpha \rangle \end{aligned} \tag{A.5.27}$$

Here, we recall that the vector potential $\boldsymbol{A}$ in the interaction Hamiltonian H_{int} is not a function of t in the quantum mechanical treatment of the optical field given in (A.5.2). Hence, H_{int} factorizes from the temporal integration

$$S_{\beta\alpha}^{(1)} = \frac{\langle E_\beta | H_{\text{int}} | E_\alpha \rangle}{j\hbar} \int_0^t e^{j(E_\beta - E_\alpha)\,\tau/\hbar} d\tau. \tag{A.5.28}$$

The temporal integration now becomes

$$\int_0^t e^{j(E_\beta - E_\alpha)\,\tau/\hbar} d\tau = 2\hbar e^{j(E_\beta - E_\alpha)t/2\hbar} \frac{\sin\left[(E_\beta - E_\alpha)t/2\hbar\right]}{E_\beta - E_\alpha}. \tag{A.5.29}$$

Transition probability from state α to state β is given by $\left|S_{\beta\alpha}^{(1)}\right|^2$, namely

$$\left|S_{\beta\alpha}^{(1)}\right|^2 = 4\left|\langle E_\beta | H_{\text{int}} | E_\alpha \rangle\right|^2 \left\{\frac{\sin\left[(E_\beta - E_\alpha)t/2\hbar\right]}{E_\beta - E_\alpha}\right\}^2. \tag{A.5.30}$$

The transition rate $P_{\beta\alpha}^{(1)}$ per unit time is defined as the time derivative of $\left|S_{\beta\alpha}^{(1)}\right|^2$, resulting in

$$\begin{aligned} w_{\beta\alpha}^{(1)} &= \frac{\partial \left|S_{\beta\alpha}^{(1)}\right|^2}{\partial t} = 2\left|\langle E_\beta | H_{\text{int}} | E_\alpha \rangle\right|^2 \frac{\sin\left[(E_\beta - E_\alpha)t/\hbar\right]}{(E_\beta - E_\alpha)\hbar} \\ &= \frac{2\pi}{\hbar} \left|\langle E_\beta | H_{\text{int}} | E_\alpha \rangle\right|^2 \delta(E_\beta - E_\alpha) \end{aligned} \tag{A.5.31}$$

Here, we use the representation

$$\delta(x) = \frac{1}{\pi}\lim_{K\to\infty}\frac{\sin(Kx)}{x}, \tag{A.5.32}$$

for the *Dirac delta-function*. The delta function $\delta(E_\beta - E_\alpha)$ implies that the total energy, a sum of the optical energy and the electron energy, should be conserved before and after the interaction.

For the initial state $|E_\alpha\rangle$ and corresponding energy E_α, we use (A.5.6) and (A.5.7). We represent the final state $\langle E_\beta|$ corresponding to energy E_β as

$$\langle E_\beta| = \langle\varphi_m(\boldsymbol{r})|\langle s'|, \tag{A.5.33}$$

$$E_\beta = \hbar\omega\left(s' + \frac{1}{2}\right) + W_m \tag{A.5.34}$$

The annihilation and creation operators, a and a^+, act only on the optical wave, whereas the momentum operator $\boldsymbol{P}$ acts as a spatial divergence operator,

$$\boldsymbol{p} = -j\hbar\nabla = -j\hbar\left(\boldsymbol{e}_x\frac{\partial}{\partial x} + \boldsymbol{e}_y\frac{\partial}{\partial y} + \boldsymbol{e}_z\frac{\partial}{\partial z}\right). \tag{A.5.35}$$

Then, one of the terms in the interaction Hamiltonian becomes

$$\langle E_\beta|\boldsymbol{A}\cdot\boldsymbol{p}|E_\alpha\rangle = \sqrt{\frac{\hbar}{2\varepsilon\omega}}\langle s'|(a+a^+)|s\rangle\boldsymbol{\Phi}(\boldsymbol{r})\cdot\langle\varphi_m(\boldsymbol{r})|\boldsymbol{p}|\varphi_n(\boldsymbol{r})\rangle. \tag{A.5.36}$$

Evaluation of the other term, $\boldsymbol{p}\cdot\boldsymbol{A}|\varphi_n(\boldsymbol{r})\rangle$, is performed assuming the *Coulomb gauge condition*. That is,

$$\nabla\cdot\boldsymbol{A} = \sqrt{\frac{\hbar}{2\varepsilon\omega}}\langle s'|(a+a^+)|s\rangle\nabla\cdot\boldsymbol{\Phi}(\boldsymbol{r}) = 0. \tag{A.5.37}$$

This condition means that the spatial variation of the optical wave is very small over the volume in which an electron is localized. We then have

$$\langle E_\beta|\boldsymbol{p}\cdot\boldsymbol{A}|E_\alpha\rangle = \langle E_\beta|\boldsymbol{A}\cdot\boldsymbol{p}|E_\alpha\rangle. \tag{A.5.38}$$

The remaining term involving $\boldsymbol{A}^2$ does not contribute to the first-order interaction because this term must change the photon number by two. As a result, the transition rate per unit time becomes

$$w_{\beta\alpha}^{(1)} = \frac{\pi}{\varepsilon\omega}\left(\frac{e}{m_o}\right)^2\left|\left\langle s'|(a+a^+)|s\right\rangle\right|^2|\boldsymbol{\Phi}(\boldsymbol{r})\cdot\langle\varphi_m(\boldsymbol{r})|\boldsymbol{p}|\varphi_n(\boldsymbol{r})\rangle|^2\delta(E_\beta - E_\alpha). \tag{A.5.39}$$

A.5.4 Photon Emission and Absorption

The operator part of the photon number is given by (2.42) and (2.43) as

$$\langle s'|(a+a^{+})|s\rangle = \sqrt{s}\langle s'|s-1\rangle + \sqrt{s+1}\langle s'|s+1\rangle. \tag{A.5.40}$$

This equation means that a single photon is absorbed when $s' = s - 1$ or emitted when $s' = s + 1$. For photon absorption, the total energy of the final state is

$$E_\beta = \hbar\omega\left(s - 1 + \frac{1}{2}\right) + W_m, \tag{A.5.41}$$

which should be the same as the total energy E_α of the initial state given in (A.5.7). The energy difference for the electron is

$$W_m - W_n = \hbar\omega, \text{ for photon absorption.} \tag{A.5.42}$$

For photon emission, the total energy of the final state is

$$E_\beta = \hbar\omega\left(s + 1 + \frac{1}{2}\right) + W_m. \tag{A.5.43}$$

The energy difference for the electron is

$$W_n - W_m = \hbar\omega, \text{ for photon emission.} \tag{A.5.44}$$

We denote the two energy levels of the electron as b for the upper level and a for the lower level.

As transition probability is proportional to the squared value of (A.5.40), the transition rate per unit time becomes

$$w_{\beta\alpha}^{(1)} = \frac{\pi}{\varepsilon\omega}\left(\frac{e}{m_o}\right)^2 |\boldsymbol{\Phi}(\boldsymbol{r})\cdot\langle\varphi_b(\boldsymbol{r})|\boldsymbol{p}|\varphi_a(\boldsymbol{r})\rangle|^2\delta(E_\beta - E_\alpha) \times \begin{cases} s+1 & \text{for photon emission} \\ s & \text{for photon absorption} \end{cases} \tag{A.5.45}$$

The above discussions have been carried out by supposing that the occupation probability of the initial state is one and that of the final state is zero at $t = 0$, that is, $|C_\alpha(0)|^2 = 1$ and $|C_\beta(0)|^2 = 0$ in (A.5.10). We suppose also that the occupation probability of the electron in energy levels b and a are f_b and f_a, respectively. The electron transition must be proportional to the occupation probability of the initial state and vacant probability of the final state. The transition rate from the upper to the lower level must then be proportional to $f_b(1 - f_a)(s + 1)$, and that from the

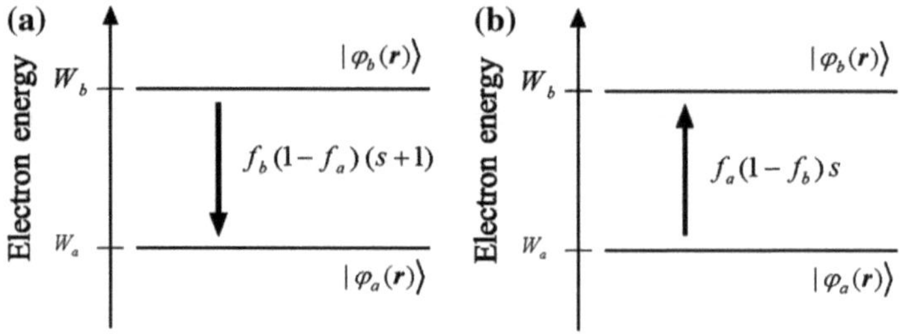

Fig. A.5.1 Electron transition between two energy levels, **a** from upper to lower levels, **b** from lower to upper levels

lower to the upper level must be proportional to $f_a(1\text{-}f_b)s$, as illustrated in Fig. A.5.1. The net transition probability must be given by the difference in these probabilities as

$$\{f_b(1-f_a)(s+1)\} - \{f_a(1-f_b)\}s = (f_b - f_a)s + f_b(1-f_a)$$

Therefore, the electron transition rate from the upper level b to the lower level a becomes

$$w_{ab}^{(1)} = \frac{\pi}{\varepsilon\omega}\left(\frac{e}{m_o}\right)^2 |\boldsymbol{\Phi}(\boldsymbol{r}) \cdot \langle\varphi_b(\boldsymbol{r})|\boldsymbol{p}|\phi_a(\boldsymbol{r})\rangle|^2\{(f_b - f_a)s + f_b(1-f_a)\} \times \delta(W_b - W_a - \hbar\omega) \tag{A.5.46}$$

For a deeper study, we need to determine the electron wave function and the distribution of energy levels in the semiconductor material. However, such studies are outside the introductory remit of the Appendix and are given in the main text, for example, in Chap. 5.

Appendix 6
Identity Operator

We suppose that any quantum mechanical state $|\psi\rangle$ can be represented as a linear combination of orthonormalized energy states as

$$|\psi\rangle = \sum_n C_n |n\rangle, \tag{A.6.1}$$

where $|n\rangle$ is an *orthonormalized energy state* satisfying

$$\langle m \mid n \rangle = \delta_{n,m}. \tag{A.6.2}$$

The coefficient C_n in (A.6.1) is then given using the inner product,

$$C_n = \langle n \mid \psi \rangle. \tag{A.6.3}$$

By substituting this relation into (A.6.1), we obtain

$$|\psi\rangle = \sum_n \langle n \mid \psi \rangle |n\rangle. \tag{A.6.4}$$

As the positions of $\langle n \mid \psi \rangle$ and $|n\rangle$ can be interchanged, (A.6.4) is represented as

$$|\psi\rangle = \sum_n |n\rangle \langle n \mid \psi \rangle. \tag{A.6.5}$$

Equation (A.6.5) implies that any state is unchanged by the operator

$$I = \sum_n |n\rangle \langle n|, \tag{A.6.6}$$

that is

$$|\psi\rangle = I|\psi\rangle. \tag{A.6.7}$$

The operator I in (A.6.6) is called *identity operator*. We can insert the identity operator between other operators, or remove an operator if the operator is identical to the identity operator.

M. Yamada, *Theory of Semiconductor Lasers*,
Springer Series in Optical Sciences 185,
DOI: 10.1007/978-4-431-54889-8,

Appendix 7
Dynamic Motion of a Classical Dipole

We consider a *dipole* consisting of a *positive charge* with *mass* m_1 and a *negative charge* with *mass* m_2. Their distances to the center of gravity are r_1 and r_2, respectively (Fig. A.7.1). The angle between the incident optical wave and the dipole is θ, which is assumed not large. Projections of r_1 and r_2 onto the incident electric field are x_1 and x_2, respectively,

$$x_1 = r_1 \sin\theta \approx r_1\theta \tag{A.7.1}$$

$$x_2 = r_2 \sin\theta \approx r_2\theta \tag{A.7.2}$$

As the dipole can oscillate around its center of gravity, the following relation holds:

$$m_1 x_1 = m_2 x_2 \tag{A.7.3}$$

When the electric field is not applied, the dipole returns to $\theta = 0$ with *restoring force* whose coefficient is κ and *damping time* τ. The dynamic motion of the dipole under the electric field E_i is then given as

$$\frac{d^2}{dt^2}(m_1 x_1 + m_2 x_2) = 2eE_i - \kappa(x_1 + x_2) - \frac{1}{\tau}\frac{d}{dt}(m_1 x_1 + m_2 x_2). \tag{A.7.4}$$

We rearrange this equation using the *total distance* x,

$$x_1 + x_2 \equiv x, \tag{A.7.5}$$

and *reduced mass* m_R defined by

$$\frac{1}{m_R} \equiv \frac{1}{m_1} + \frac{1}{m_2}, \tag{A.7.6}$$

to obtain

$$m_1 x_1 + m_2 x_2 = \frac{2m_1 m_2}{m_1 + m_2} x = 2m_R x. \tag{A.7.7}$$

M. Yamada, *Theory of Semiconductor Lasers*,
Springer Series in Optical Sciences 185,
DOI: 10.1007/978-4-431-54889-8, © Springer Japan 2014

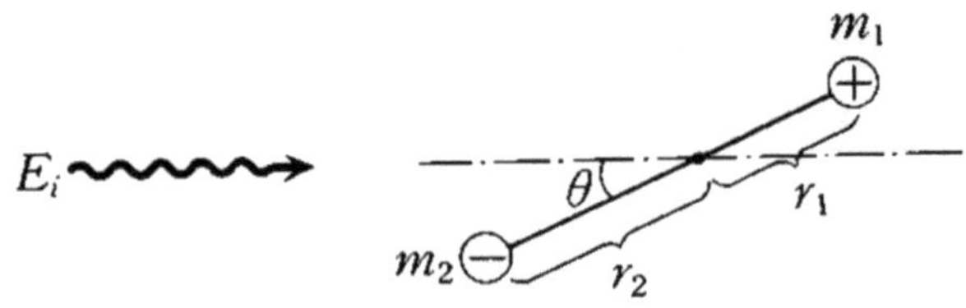

Fig. A.7.1 Interaction between electric field and dipole

The *dynamic equation* of (A.7.4) is rewritten as

$$\frac{d^2x}{dt^2} + \frac{1}{\tau}\frac{dx}{dt} + \frac{1}{2m_R}\kappa x = \frac{e}{m_R}E_i. \tag{A.7.8}$$

We assume that the restoring force is much stronger than the damping force:

$$\frac{\kappa}{2m_R} \gg \frac{1}{\tau^2}. \tag{A.7.9}$$

We set the *incident electric field* in the form

$$E_i = E_\omega e^{j\omega t}, \tag{A.7.10}$$

and the motion of the dipole in the steady state to be

$$x(t) = x_o e^{j\omega t}. \tag{A.7.11}$$

By substituting these equations into (A.7.8), the amplitude x_o of the oscillation becomes

$$\begin{aligned} x_o &= \frac{eE_\omega}{m_R\{\omega_o^2 - \omega^2 + j\omega/\tau\}} \\ &\approx \frac{eE_\omega}{2j\omega_o m_R\{j(\omega - \omega_o) + 1/\tau\}}, \end{aligned} \tag{A.7.12}$$

where ω_o is the resonance frequency given by

$$\omega_o = \sqrt{\frac{\kappa}{2m_R}}. \tag{A.7.13}$$

By supposing the density of dipoles to be N_t, the *polarization* P for these dipoles is given by

$$\begin{aligned} P &= eN_t x(t) \\ &= \frac{e^2N_t}{2j\omega_o m_R}\frac{E_\omega e^{j\omega t}}{j(\omega - \omega_o) + 1/\tau}. \end{aligned} \tag{A.7.14}$$

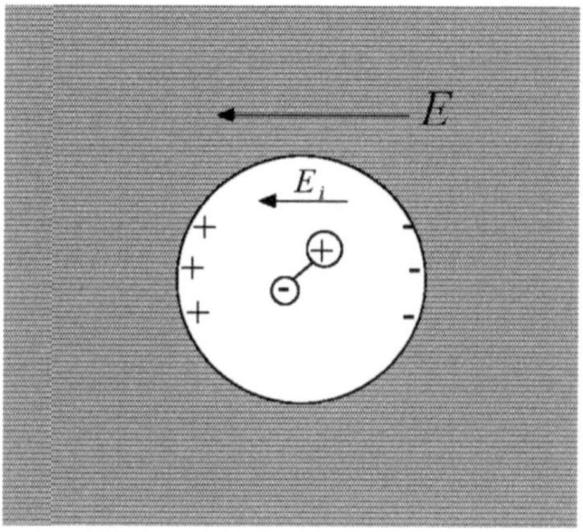

Fig. A.7.2 Applied and inner electric fields in a cavity of a dielectric

Furthermore, in classical dielectric material theory, the electric field E_i inside of a dielectric material is reduced to a third of field E,

$$E_i = \frac{1}{3}E, \tag{A.7.15}$$

as illustrated in Fig. A.7.2. This so-called Clausius-Mossotti relation is explained in the context of classical dielectric material theory as follows: We imagine a spherical hole around the dipole in the dielectric material and calculate the distribution of the induced electric charges on the wall of sphere. Because of cancelations from these induced charges, the electric field in the hole is reduced to 1/3 of the applied field.

The *susceptibility* $\chi(\omega)$ defined by the relation

$$P = \varepsilon_o \chi(\omega) E = 3\varepsilon_o \chi(\omega) E_i, \tag{A.7.16}$$

is given by

$$\chi(\omega) = \frac{e^2 N_t}{6j\omega_o \varepsilon_o}\left(\frac{1}{m_1} + \frac{1}{m_2}\right)\frac{1}{j(\omega - \omega_o) + 1/\tau}. \tag{A.7.17}$$

In this classical treatment, the inner electric field E_i is given by (A.7.15).

In contrast, in quantum mechanics, this reduction is already taken into account by the *dipole moment* $|R_{ba}|^2$, as given in (5.37) via the three-dimensional spreading of the electron wave. Therefore, in a quantum mechanical treatment, differentiating between the applied field and the inner electric field is unnecessary. We can directly formulate the polarization as

$$P = \varepsilon_o \chi(\omega) E = \varepsilon_o \chi(\omega) E_i.$$

The three-dimensional spreading of the electron wave in the dipole moment $|R_{ba}|^2$ is explained in more detail in Chap. 11.

Appendix 8
Kramers–Kronig Relation

In the definition of the electric field, (6.1), the angular frequency ω is supposed to take only positive values; the terms with negative angular frequency $-\omega$ are considered as complex conjugates of the positive frequency terms. The terms with $\omega + \omega_{ba}$ in the denominator of the density matrix element are dropped in determining the effective interaction.

However, the dropping of these $\omega + \omega_{ba}$ terms is incorrect if exact mathematical relations are required. Therefore, we go back to (6.16) to start with the *original equation* and neglect no term,

$$\rho_{ab}(t) = j\frac{R_{ab}}{\hbar}(\rho_{bb} - \rho_{aa})\left[\frac{E_{\omega}e^{j\omega t}}{j(\omega - \omega_{ba}) + 1/\tau_{in}} + \frac{E_{\omega}^{*}e^{-j\omega t}}{-j(\omega + \omega_{ba}) + 1/\tau_{in}}\right]. \tag{A.8.1}$$

Reversing states, we have

$$\begin{aligned}\rho_{ba}(t) &= \rho_{ab}^{*}(t)\\ &= -j\frac{R_{ba}}{\hbar}(\rho_{bb} - \rho_{aa})\left[\frac{E_{\omega}^{*}e^{-j\omega t}}{-j(\omega - \omega_{ba}) + 1/\tau_{in}} + \frac{E_{\omega}e^{j\omega t}}{j(\omega + \omega_{ba}) + 1/\tau_{in}}\right].\end{aligned} \tag{A.8.2}$$

By substituting these equations into (6.4), the laser susceptibility is rewritten to cover both cases, $\omega \geq 0$ and $\omega \leq 0$,

$$\chi(\omega) = \frac{jN_t}{\varepsilon_o\hbar}\sum_{bora}|R_{ba}|^2(\rho_{bb} - \rho_{aa})\left[\frac{1}{j(\omega - \omega_{ba}) + 1/\tau_{in}} - \frac{1}{j(\omega + \omega_{ba}) + 1/\tau_{in}}\right]. \tag{A.8.3}$$

Now, we perform the following transformation in the complex plane

$$Y(\omega) = \oint \frac{\chi(\omega')}{\omega' - \omega}d\omega', \tag{A.8.4}$$

to investigate in more detail the characteristics of the susceptibility. Here, $\oint d\omega'$denotes integration along a closed line given as $D \to 0 \to A \to B \to C \to D$ in

M. Yamada, *Theory of Semiconductor Lasers*,
Springer Series in Optical Sciences 185,
DOI: 10.1007/978-4-431-54889-8, © Springer Japan 2014

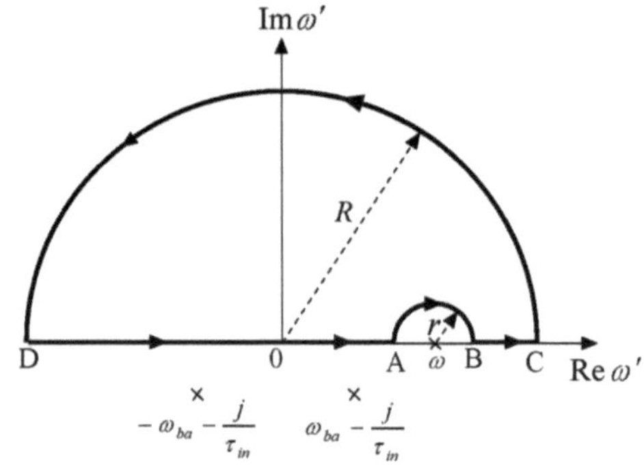

Fig. A.8.1 Path of complex integration to derive the Kramers–Kronig relation

Fig. A.8.1. As there is no singular point within the enclosed line, $Y(\omega) = 0$ results. The integration $C \to D$ along the semi-circle of radius R tends to zero as $R \to \infty$. We then obtain

$$\left\{ \int_D^A + \int_B^C \right\} \frac{\chi(\omega')}{\omega' - \omega} d\omega' = - \int_A^B \frac{\chi(\omega')}{\omega' - \omega} d\omega'. \tag{A.8.5}$$

We now put

$$\omega' = \omega + re^{j\theta}, \tag{A.8.6}$$

to perform the integration along $A \to B$. With $d\omega' = jre^{j\theta} d\theta$, the integration $A \to B$ is rewritten as $re^{j\pi} \to re^{j0}$. Then,

$$\int_A^B \frac{\chi(\omega')}{\omega' - \omega} d\omega' = \lim_{r \to 0} j \int_\pi^0 \chi(\omega + re^{j\theta}) d\theta = -j\pi\chi(\omega). \tag{A.8.7}$$

Therefore,

$$\int_{-\infty}^{\infty} \frac{\chi(\omega')}{\omega' - \omega} d\omega' = \lim_{R \to \infty, r \to 0} \left\{ \int_D^A + \int_B^C \right\} \frac{\chi(\omega')}{\omega' - \omega} d\omega' = j\pi\chi(\omega). \tag{A.8.8}$$

Separating real and imaginary parts gives

$$\mathrm{Re}\ \chi(\omega) = \frac{1}{\pi} \int_{-\infty}^{\infty} \frac{\mathrm{Im}\ \chi(\omega')}{\omega' - \omega} d\omega', \tag{A.8.9}$$

$$\mathrm{Im}\ \chi(\omega) = -\frac{1}{\pi} \int_{-\infty}^{\infty} \frac{\mathrm{Re}\ \chi(\omega')}{\omega' - \omega} d\omega'. \tag{A.8.10}$$

In the next step, we shall rewrite these equations for only positive frequencies. From (A.8.3), we get

$$\mathrm{Re}\ \chi(\omega) = \frac{N_t}{\varepsilon_o \hbar} \sum_{b\,or\,a} |R_{ba}|^2 (\rho_{bb} - \rho_{aa}) \left[\frac{\omega - \omega_{ba}}{(\omega - \omega_{ba})^2 + 1/\tau_{in}^2} - \frac{\omega + \omega_{ba}}{(\omega + \omega_{ba})^2 + 1/\tau_{in}^2} \right], \tag{A.8.11}$$

$$\mathrm{Im}\ \chi(\omega) = \frac{N_t}{\varepsilon_o \hbar} \sum_{b\,or\,a} \frac{|R_{ba}|^2 (\rho_{bb} - \rho_{aa})}{\tau_{in}} \left[\frac{1}{(\omega - \omega_{ba})^2 + 1/\tau_{in}^2} - \frac{1}{(\omega + \omega_{ba})^2 + 1/\tau_{in}^2} \right], \tag{A.8.12}$$

With these explicit expressions, we find

$$\mathrm{Re}\ \chi(-\omega) = \mathrm{Re}\ \chi(\omega), \tag{A.8.13}$$

$$\mathrm{Im}\ \chi(-\omega) = -\mathrm{Im}\ \chi(\omega). \tag{A.8.14}$$

With these properties,

$$\begin{aligned} \int_{-\infty}^{\infty} \frac{\mathrm{Im}\ \chi(\omega')}{\omega' - \omega} d\omega' &= \int_{-\infty}^{0} \frac{\mathrm{Im}\ \chi(\omega')}{\omega' - \omega} d\omega' + \int_{0}^{\infty} \frac{\mathrm{Im}\ \chi(\omega')}{\omega' - \omega} d\omega' \\ &= \int_{0}^{\infty} \mathrm{Im}\ \chi(\omega') \left(\frac{1}{\omega' + \omega} + \frac{1}{\omega' - \omega} \right) d\omega'. \\ &= \int_{0}^{\infty} \frac{2\omega' \mathrm{Im}\ \chi(\omega')}{(\omega'^2 - \omega^2)} d\omega' \end{aligned} \tag{A.8.15}$$

$$\begin{aligned} \int_{-\infty}^{\infty} \frac{\mathrm{Re}\ \chi(\omega')}{\omega' - \omega} d\omega' &= \int_{-\infty}^{0} \frac{\mathrm{Re}\ \chi(\omega')}{\omega' - \omega} d\omega' + \int_{0}^{\infty} \frac{\mathrm{Re}\ \chi(\omega')}{\omega' - \omega} d\omega' \\ &= \int_{0}^{\infty} \mathrm{Re}\ \chi(\omega') \left(\frac{1}{\omega' - \omega} - \frac{1}{\omega' + \omega} \right) d\omega'. \\ &= \int_{0}^{\infty} \frac{2\omega \mathrm{Re}\ \chi(\omega')}{(\omega'^2 - \omega^2)} d\omega' \end{aligned} \tag{A.8.16}$$

Therefore, (A.8.9) and (A.8.10) are written as integrals over the positive frequency range as

$$\mathrm{Re}\ \chi(\omega) = \frac{2}{\pi}\int_0^\infty \frac{\omega' \mathrm{Im}\chi(\omega')}{\omega'^2 - \omega^2} d\omega', \tag{A.8.17}$$

$$\mathrm{Im}\ \chi(\omega) = \frac{2\omega}{\pi}\int_0^\infty \frac{\mathrm{Re}\chi(\omega')}{\omega'^2 - \omega^2} d\omega'. \tag{A.8.18}$$

These equations are called the *Kramers–Kronig relation*.

Appendix 9
Classical Analysis of Plasma Effect

The electrons in the conduction band and the holes in the valence band can, under the influence of the electric field of the optical wave, move inside of the bands as free particles. This kind of motion falls under the plasma effect. Because this dynamic motion is not caused by electron transitions between energy levels in the semiconductor, a direct quantum mechanical analysis is very difficult. Therefore, we analyze the plasma effect based on classical mechanics.

Here, we suppose the direction of the electric polarization of the optical wave is along the x direction. Then, the dynamic equation associated with the electron or hole is

$$m_i \frac{d^2x}{dt^2} = eE - \frac{m_i}{\tau}\frac{dx}{dt}, \tag{A.9.1}$$

where x is the position of the electron or hole, m_i its effective mass, and τ the damping time. In the plasma effect, the restoring force is supposed not to exist.

The electric field is given by

$$E = E_{\omega} e^{j\omega t}. \tag{A.9.2}$$

The solution for the position x is assumed to be

$$x(t) = x_o e^{j\omega t}. \tag{A.9.3}$$

By substituting (A.9.2) and (A.9.3) into (A.9.1), we obtain the vibration amplitude for the position

$$x_o = \frac{-eE_{\omega}}{m_c\omega(\omega - j/\tau)}. \tag{A.9.4}$$

This vibration motion contributes to the *polarization* P_{pli}, which defines *susceptibility* χ_{pli} in the relation

$$P_{pli} = eNx(t) \equiv \varepsilon_o \chi_{pli}(\omega)E, \tag{A.9.5}$$

where N is the electron/hole density. The susceptibility then becomes

M. Yamada, *Theory of Semiconductor Lasers*,
Springer Series in Optical Sciences 185,
DOI: 10.1007/978-4-431-54889-8, © Springer Japan 2014

$$\chi_{pli}(\omega) = -\frac{e^2 N}{\varepsilon_o m_i \omega(\omega - j/\tau)}. \quad (A.9.6)$$

The above analysis is applied both to electrons in the conduction band and holes in the valence band. Therefore, the total susceptibility is

$$\chi_{pl}(\omega) = \chi_{plc}(\omega) + \chi_{plv}(\omega) = -\frac{e^2 N}{\varepsilon_o m_R \omega(\omega - j/\tau)}, \quad (A.9.7)$$

where m_R is the *reduced mass* defined by

$$\frac{1}{m_R} \equiv \frac{1}{m_c} + \frac{1}{m_v}. \quad (A.9.8)$$

As the angular frequency of the optical wave is much larger than the relaxation effect, $\omega >> 1/\tau$, the real and imaginary parts of the susceptibility are approximately given as

$$\mathrm{Re}\chi_{pl}(\omega) = -\frac{e^2 N}{\varepsilon_o m_R(\omega^2 + 1/\tau^2)} \approx -\left(\frac{\omega_{pl}}{\omega}\right)^2, \quad (A.9.9)$$

$$\mathrm{Im}\chi_{pl}(\omega) = -\frac{e^2 N}{\varepsilon_o m_R \omega(\omega^2 + 1/\tau^2)\tau} \approx -\frac{\omega_{pl}^2}{\omega^3 \tau}. \quad (A.9.10)$$

Here, ω_{pl} is called the *plasma angular frequency* defined by

$$\omega_{pl} = e\sqrt{\frac{N}{\varepsilon_o m_R}}. \quad (A.9.11)$$

The real part gives a reduction in the dielectric constant or the refractive index. The electric displacement $\boldsymbol{D}$ taking into account the plasma effect is written as

$$\boldsymbol{D} = \varepsilon_o n_r^2 \boldsymbol{E} + \boldsymbol{P} = \varepsilon_o\left(n_r^2 + \mathrm{Re}\chi_{pl}\right)\boldsymbol{E} = \varepsilon_o\left(n_r + \Delta n_{pl}\right)^2 \boldsymbol{E}. \quad (A.9.12)$$

Then, the *variation* Δn_{pl} of the *refractive index* by the plasma effect is written as

$$\Delta n_{pl} = \frac{\mathrm{Re}\chi_{pl}}{2n_r} = -\frac{1}{2n_r}\left(\frac{\omega_{pl}}{\omega}\right)^2. \quad (A.9.13)$$

The imaginary part gives the *absorption loss*. The *conductivity* σ_{pl} from the plasma effect is

$$\sigma_{pl} = -\varepsilon_o \omega \mathrm{Im}\chi_{pl} = \frac{e^2 N}{m_R \omega^2 \tau}, \quad (A.9.14)$$

or the *coefficient of loss* α_{pl} is

$$\alpha_{pl} = -\frac{\omega}{c n_r}\mathrm{Im}\ \chi_{pl} = \frac{1}{c n_r \tau}\left(\frac{\omega_{pl}}{\omega}\right)^2. \quad (A.9.15)$$

Let us introduce numerical values to provide estimates of the plasma effect. We suppose a semiconductor material of GaAs and an electron density of $N = 4 \times 10^{18}\,\text{cm}^{-3}$. Other parameters are set to $m_R = 0.058\ m_o = 5.28 \times 10^{-32}\,\text{kg}$, $\omega = 2\pi c/\lambda = 2.24 \times 10^{15}\,\text{s}^{-1}$ with $\lambda = 0.84\,\mu\text{m}$, $n_r = 3.5$ and $\tau = 7 \times 10^{-14}\,\text{s}$. With a plasma frequency of $\omega_{pl} = 4.69 \times 10^{14}\,\text{s}^{-1}$, we obtain Re $\chi_{pl} = -0.044$, $\Delta n_{pl} = -6.26 \times 10^{-3}$, and $\alpha_{pl} = 5.99\,\text{cm}^{-1}$. As discussed in Sect. 6.5, a reduction of the refractive index Δn_{pl} by the plasma effect is almost 1/3 that by laser polarizationΔn_{la}. However, the plasma effect causes *absorption losses* of the optical wave in the semiconductor laser.

Appendix 10
Amplified Spontaneous Emission

Analysis with Cavity Modes

In the following analysis of the laser cavity model shown in Fig. 3.1, we represent the electric field of the optical wave in form of the traveling wave,

$$E(t,\boldsymbol{r}) = \sum_{p} \left\{ E_p^{(+)}(z)T(x,y)e^{-j\beta_p z + j\omega_p t} + E_p^{(-)}(z)T(x,y)e^{-j\beta_p (L-z) + j\omega_p t} \right\} + c.c., \tag{A.10.1}$$

where $E_p^{(+)}(z)$ and $E_p^{(-)}(z)$ are amplitudes of the forward and backward propagating components of the mode p, respectively. A direct introduction of the spontaneous emission in the electric field is difficult because the optical phase of the spontaneous emission is random. We introduce instead the spontaneous emission in the optical power.

The carrying optical power $P_p^{(\pm)}(z)$ of the forward and backward components are given as

$$\begin{aligned} P_p^{(\pm)}(z) &= \int_{-\infty}^{\infty}\int_{-\infty}^{\infty} \left\{ E_p^{(\pm)}(z) \times H_p^{(\pm)*}(z) + c.c. \right\} |T(x,y)|^2 dxdy \\ &= 2\sqrt{\frac{\varepsilon_{eff}}{\mu_o}} \left| E_p^{(\pm)}(z) \right|^2 \end{aligned} \quad . \tag{A.10.2}$$

Variations of this carrying power are given by (3.10)–(3.12) and by including the spontaneous emission with C'_p,

$$\frac{dP_p^{(+)}(z)}{dz} = \left(g_p - \alpha_{loss}\right)P_p^{(+)}(z) + C'_p, \tag{A.10.3}$$

$$-\frac{dP_p^{(-)}(z)}{dz} = \left(g_p - \alpha_{loss}\right)P_p^{(-)}(z) + C'_p. \tag{A.10.4}$$

M. Yamada, *Theory of Semiconductor Lasers*,
Springer Series in Optical Sciences 185,
DOI: 10.1007/978-4-431-54889-8, © Springer Japan 2014

We suppose a solution of (A.10.3) to have the form

$$P_p^{(+)}(z) = u^{(+)}(z)e^{(g_p-\alpha_{loss})\,z}, \tag{A.10.5}$$

which when substituted into (A.10.3) yields a dynamic equation for $u^{(+)}(z)$,

$$\frac{du^{(+)}(z)}{dz} = C'_p e^{-\,(g_p-\alpha_{loss})\,z}. \tag{A.10.6}$$

The solution of this equation is

$$u^{(+)}(z) = -\frac{C'_p e^{-(g_p-\alpha_{loss})z}}{g_p-\alpha_{loss}} + C_1, \tag{A.10.7}$$

where C_1 is an unknown integration constant. By substituting this equation into (A.10.5), we obtain

$$P_p^{(+)}(z) = -\frac{C'_p}{g_p-\alpha_{loss}} + C_1 e^{(g_p-\alpha_{loss})z}. \tag{A.10.8}$$

Similarly, the backward propagating power is written with integration constant C_2 as

$$P_p^{(-)}(z) = -\frac{C'_p}{g_p-\alpha_{loss}} + C_2 e^{(g_p-\alpha_{loss})(L-z)}. \tag{A.10.9}$$

The optical power at the two facet mirrors are then

$$P_p^{(+)}(0) = -\frac{C'_p}{g_p-\alpha_{loss}} + C_1, \tag{A.10.10}$$

$$P_p^{(+)}(L) = -\frac{C'_p}{g_p-\alpha_{loss}} + C_1 e^{(g_p-\alpha_{loss})L}, \tag{A.10.11}$$

$$P_p^{(-)}(L) = -\frac{C'_p}{g_p-\alpha_{loss}} + C_2, \tag{A.10.12}$$

$$P_p^{(-)}(0) = -\frac{C'_p}{g_p-\alpha_{loss}} + C_2 e^{(g_p-\alpha_{loss})L}. \tag{A.10.13}$$

The forward and backward components are related via the coefficients of reflection for the mirrors,

$$P_p^{(-)}(L) = R_f P_p^{(+)}(L), \tag{A.10.14}$$

$$P_p^{(+)}(0) = R_b P_p^{(-)}(0) \tag{A.10.15}$$

From (A.10.12)–(A.10.14), we have

$$P_p^{(-)}(0) = \frac{C_p'}{g_p - \alpha_{loss}} \left\{ e^{(g_p - \alpha_{loss})L} - 1 \right\} + R_f P_p^{(+)}(L) e^{(g_p - \alpha_{loss})L}, \tag{A.10.16}$$

and similarly (A.10.10), (A.10.11), and (A.10.15) give

$$P_p^{(+)}(L) = \frac{C_p'}{g_p - \alpha_{loss}} \left\{ e^{(g_p - \alpha_{loss})L} - 1 \right\} + R_b P_p^{(-)}(0) e^{(g_p - \alpha_{loss})L}. \tag{A.10.17}$$

By substituting (A.10.16) into (A.10.17), we obtain

$$P_p^{(+)}(L) = \frac{C_p' \{ e^{(g_p - \alpha_{loss})L} - 1 \} \{ 1 + R_b e^{(g_p - \alpha_{loss})L} \}}{(g_p - \alpha_{loss}) \{ 1 - R_b R_f e^{2(g_p - \alpha_{loss})L} \}}. \tag{A.10.18}$$

The output power P_{out} from the front facet is

$$\begin{aligned} P_{out} &= (1 - R_f) \sum_p P_p^{(+)}(L) \\ &= (1 - R_f) \sum_p \frac{C_p' \{ e^{(g_p - \alpha_{loss})L} - 1 \} \{ 1 + R_b e^{(g_p - \alpha_{loss})L} \}}{(g_p - \alpha_{loss}) \{ 1 - R_b R_f e^{2(g_p - \alpha_{loss})L} \}} \end{aligned} \tag{A.10.19}$$

The contribution of the spontaneous emission from C_p' is proportional to the emission probability $T_{sp}(\omega)$ given in (6.73). The profile in Fig. 6.11 is thus obtained.

A.10.2 Analysis of the Frequency Spectrum

The longitudinal mode in the laser cavity is determined by the resonance condition of the optical field for the facet mirrors, as given in (3.30), and is characterized by discrete values of the angular frequency, as illustrated in Fig. A.10.1a. However, the measured angular frequencies of the emitted optical waves fall within a continuous range because the spectrum of each cavity mode broadens due to spontaneous emission. In particular, when the injection current I is lower than the threshold current I_{th}, the spectrum distribution of the cavity mode overlaps with other modes, as illustrated in Fig. A.10.1b. In this subsection, we analyze this emission profile based on a simple treatment.

We suppose that spontaneous emission is generated at position z and continues over a time interval τ_s much longer than the time required for the emitted light to make multiple reflections within the laser cavity. If the amplitude of electric field of the spontaneous emission in the forward direction is $w^{(+)}(z)$, as illustrated in Fig. A.10.2a, the amplitude $E_1^{(+)}(L, z)$ when the light reaches the front facet at $z = L$ becomes

$$E_1^{(+)}(L, z) = w^{(+)}(z) e^{\{(g_\omega - \alpha_{loss})/2 - j\beta\}(L - z)} \tag{A.10.20}$$

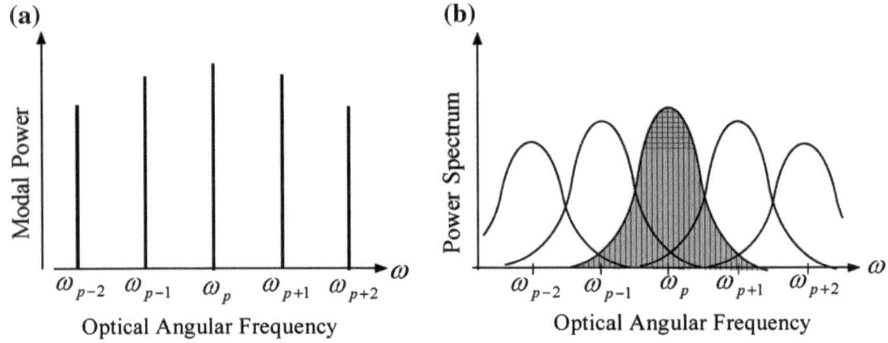

Fig. A.10.1 Optical spectrum, **a** Conceptual spectrum of modal power, **b** Power spectrum

where g_ω is the gain coefficient at ω defined similarly as for the modal gain coefficient in (3.11). The light is reflected by the front facet, propagates in the backward direction, is reflected by the back facet, and returns to the front facet. The amplitude after one complete cycle is denoted $E_2^{(+)}(L,z)$,

$$E_2^{(+)}(L,z) = \sqrt{R_f R_b} e^{(g_\omega - \alpha_{loss} - 2j\beta)\,L} E_1^{(+)}(L,z). \tag{A.10.21}$$

After multiple cycles, the amplitude for the m-th cycle is

$$E_m^{(+)}(L,z) = (R_f R_b)^{(m-1)/2} e^{(g_\omega - \alpha_{loss} - 2j\beta)\,(m-1)L} E_1^{(+)}(L,z). \tag{A.10.22}$$

Therefore, the summed value of the electric field originally emitted in the forward direction is

$$\begin{aligned} E_{total}^{(+)}(L,z) &= \sum_{m=1}^{\infty} E_m^{(+)}(L,z) \\ &= \frac{E_1^{(+)}(L,z)}{1 - \sqrt{R_f R_b} e^{(g_\omega - \alpha_{loss} - 2j\beta)L}} \\ &= \frac{w^{(+)}(z) e^{\{(g_\omega - \alpha_{loss})/2 - j\beta)\,(L-z)}}{1 - \sqrt{R_f R_b} e^{(g_\omega - \alpha_{loss} - 2j\beta)L}} \end{aligned} \tag{A.10.23}$$

Spontaneous emission is also emitted in the backward direction as illustrated in Fig. A.10.2b, and is reflected by the back facet and propagates toward the front facet. After one cycle, the electric field is

$$E_1^{(-)}(L,z) = w^{(-)}(z) \sqrt{R_b} e^{\{(g_\omega - \alpha_{loss})/2 - j\beta\}(L+z)}, \tag{A.10.24}$$

which, after the m-th cycle becomes

$$E_m^{(-)}(L,z) = (R_f R_b)^{(m-1)/2} e^{(g_\omega - \alpha_{loss} - 2j\beta)\,(m-1)L} E_1^{(-)}(L,z). \tag{A.10.25}$$

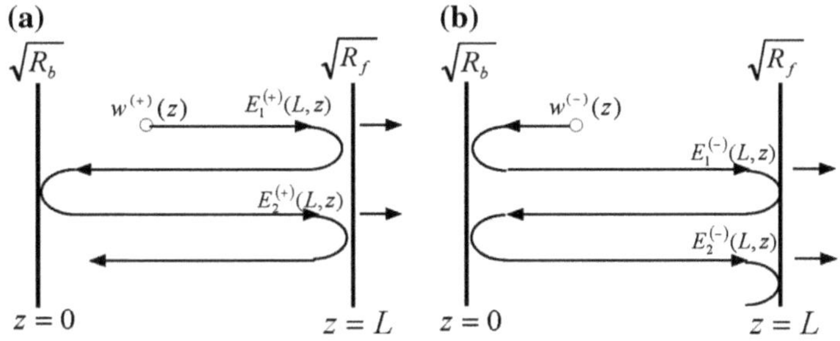

Fig. A.10.2 Traces of the spontaneously emitted light, **a** Trace of emitted light in the forward direction, **b** Trace of emitted light in the backward direction

Then, the sum value of the electric field originally emitted in the backward direction is

$$E_{total}^{(-)}(L,z) = \sum_{m=1}^{\infty} E_m^{(-)}(L,z) = \frac{w^{(-)}(z)\sqrt{R_b}e^{\{(g_\omega - \alpha_{loss})/2 - j\beta\}(L+z)}}{1 - \sqrt{R_f R_b}e^{(g_\omega - \alpha_{loss} - 2j\beta)L}} \tag{A.10.26}$$

Thus the total strength of the electric field at the front facet emitted at z is

$$E_{total}(L,z) = E_{total}^{(+)}(L,z) + E_{total}^{(-)}(L,z). \tag{A.10.27}$$

The optical power for this field is given by the square of the absolute value. However, we need to check that for this statistical treatment no correlations exist between spontaneous emission in the forward and backward directions along z direction, and that the absolute value for both directions must be identical and almost constant along the z direction, that is,

$$\left\langle w^{(+)}(z) w^{(-)*}(z) \right\rangle = 0, \tag{A.10.28}$$

$$\left|w^{(+)}(z)\right|^2 = \left|w^{(-)}(z)\right|^2 \equiv |w|^2. \tag{A.10.29}$$

With that, we obtain

$$|E_{total}(L,z)|^2 = \left|E_{total}^{(+)}(L,z)\right|^2 + \left|E_{total}^{(-)}(L,z)\right|^2 = \frac{|w|^2\{e^{(g_\omega - \alpha_{loss})(L-z)} + R_b e^{(g_\omega - \alpha_{loss})(L+z)}\}}{1 + R_f R_b e^{2(g_\omega - \alpha_{loss})L} - 2\sqrt{R_f R_b}e^{(g_\omega - \alpha_{loss})L}\cos(2\beta L)} \tag{A.10.30}$$

As this optical power is the result of the spontaneous emission at z, we need to perform an integration over all positions, $0 \le z \le L$, along the cavity length. Finally, the optical output P_ω from the front facet for angular frequency ω is obtained as

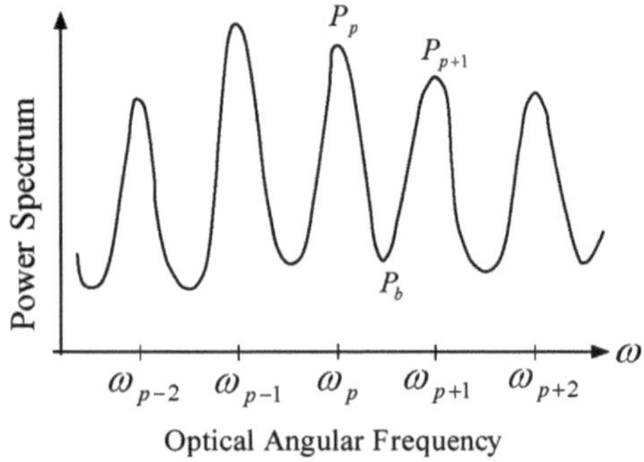

Fig. A.10.3 Frequency of wavelength profile of the emitted light

$$\begin{aligned} P_\omega &= (1 - R_f)\sqrt{\frac{\varepsilon_{eff}}{\mu_o}} \int_0^L |E_{total}(L, z)|^2 dz \\ &= \sqrt{\frac{\varepsilon_{eff}}{\mu_0}} \frac{(1 - R_f)|w|^2 \{e^{(g_\omega - \alpha_{loss})L} - 1\}\{1 + R_b e^{(g_\omega - \alpha_{loss})L}\}}{(g_\omega - \alpha_{loss})\{1 + R_f R_b e^{2(g_\omega - \alpha_{loss})L} - 2\sqrt{R_f R_b} e^{(g_\omega - \alpha_{loss})L} \cos(2\beta L)\}} \end{aligned} \tag{A.10.31}$$

A.10.3 Measuring Method of the Optical Gain (Hakki–Paoli method)

If the injection current I is lower than the threshold current I_{th}, the semiconductor lasers exhibit amplified spontaneous emission with a periodically varying wavelength or frequency spectrum. Using this spectral shape, we can estimate the value of the optical gain employing the *Hakki–Paoli method*. We describe this method in this subsection.

The emitted spectrum should show a periodic variation $\cos(2\beta L)$, as given in (A.10.31). As the real spectrum profile includes other dispersion characteristics from spontaneous emission $|w|^2$ and gain g_ω, as shown in Fig. A.10.3, we define the ratio r for the emission peak and bottom by taking the averaged value of two successive peaks to detect resonance phenomena,

$$r = \frac{P_p + P_{p+1}}{2P_b}. \tag{A.10.32}$$

From (A.10.31), the ratio must be

$$r = \frac{1 + R_f R_b e^{2(g_\omega - \alpha_{loss})L} + 2\sqrt{R_f R_b} e^{(g_\omega - \alpha_{loss})L}}{1 + R_f R_b e^{2(g_\omega - \alpha_{loss})L} - 2\sqrt{R_f R_b} e^{(g_\omega - \alpha_{loss})L}} = \left\{ \frac{1 + \sqrt{R_f R_b} e^{(g_\omega - \alpha_{loss})L}}{1 - \sqrt{R_f R_b} e^{(g_\omega - \alpha_{loss})L}} \right\}^2. \tag{A.10.33}$$

This equation is rearranged to give

$$\sqrt{R_f R_b} e^{(g_\omega - \alpha_{loss})L} = \frac{\sqrt{r} - 1}{\sqrt{r} + 1}, \tag{A.10.34}$$

and hence gain

$$\begin{aligned} g_\omega &= \alpha_{loss} + \frac{1}{2L} \ln\left(\frac{1}{R_f R_b}\right) + \frac{1}{L} \ln\left(\frac{\sqrt{r} - 1}{\sqrt{r} + 1}\right) \\ &= g_{th} - \frac{1}{L} \ln\left(\frac{\sqrt{r} + 1}{\sqrt{r} - 1}\right) \end{aligned}. \tag{A.10.35}$$

Appendix 11
Spectrum of Correlated Value

In the determination of the Fourier transformation, we used a certain mathematical representation of the *Dirac delta function*,

$$\delta(x) = \frac{1}{\pi} \lim_{K\to\infty} \frac{\sin(Kx)}{x}, \tag{A.11.1}$$

where x is a real variable and K a real number.

We consider now temporally varying functions expanded in terms of its frequency components:

$$X(t) = \int_{-\infty}^{\infty} X_{\Omega} e^{j\Omega t} d\Omega, \tag{A.11.2}$$

$$Y(t) = \int_{-\infty}^{\infty} Y_{\Omega} e^{j\Omega t} d\Omega. \tag{A.11.3}$$

We substitute these equations in the following convolution equation:

$$\begin{aligned} &\frac{1}{T}\int_0^T \int_{-\infty}^{\infty} X(t) Y(t+\tau) e^{j\Omega\tau} d\tau dt \\ &= \frac{1}{T}\int_0^T \int_{-\infty}^{\infty}\int_{-\infty}^{\infty}\int_{-\infty}^{\infty} X_{\Omega'} Y_{\Omega''} e^{j(\Omega'+\Omega'')t + j(\Omega''+\Omega)\tau} d\Omega'' d\Omega' d\tau dt. \end{aligned} \tag{A.11.4}$$

We first perform the integration over τ,

$$\begin{aligned} \int_{-\infty}^{\infty} e^{j(\Omega''+\Omega)\tau} d\tau &= \lim_{K\to\infty} \frac{e^{j(\Omega''+\Omega)K} - e^{-j(\Omega''+\Omega)K}}{j(\Omega''+\Omega)} \\ &= 2\lim_{K\to\infty} \frac{\sin\{(\Omega''+\Omega)K\}}{\Omega''+\Omega} \\ &= 2\pi\delta(\Omega''+\Omega). \end{aligned} \tag{A.11.5}$$

M. Yamada, *Theory of Semiconductor Lasers*,
Springer Series in Optical Sciences 185,
DOI: 10.1007/978-4-431-54889-8, © Springer Japan 2014

Hence, $\Omega'' = -\Omega$ is obtained. Next we integrate over t,

$$\int_0^T e^{j(\Omega'-\Omega)t}dt = \frac{e^{j(\Omega'-\Omega)T}-1}{j(\Omega'-\Omega)} = e^{j(\Omega'-\Omega)T/2}\frac{e^{j(\Omega'-\Omega)T/2}-e^{j(\Omega'-\Omega)T/2}}{j(\Omega'-\Omega)}$$
$$= 2e^{j(\Omega'-\Omega)T/2}\frac{\sin\{(\Omega'-\Omega)T\}}{\Omega'-\Omega}. \tag{A.11.6}$$

Because the measuring time T is here assumed long compared with the variable $1/(\Omega'-\Omega)$, this equation is written as

$$\int_0^T e^{j(\Omega'-\Omega)t}dt = 2\pi\delta(\Omega'-\Omega). \tag{A.11.7}$$

Then (A.11.4) expressed as frequency components is

$$\frac{1}{T}\int_0^T\int_{-\infty}^{\infty} X(t)Y(t+\tau)e^{j\Omega\tau}d\tau dt = \frac{4\pi^2}{T}X_\Omega Y_\Omega^*, \tag{A.11.8}$$

where we have used $Y_{-\Omega} = Y_\Omega^*$.

As (A.11.8) involves the measuring time T, its value appears not to be unique, but actually is. The inverse value of T is the resolution Δf of the measured spectrum from the spectrum analyzer,

$$\Delta f = \frac{1}{T}. \tag{A.11.9}$$

Then the value of (A.11.8) can be cast in the form

$$\frac{4\pi^2}{T}X_\Omega Y_\Omega^* = T\cdot X_\Omega\frac{2\pi}{T}\cdot Y_\Omega^*\frac{2\pi}{T} = \frac{X_\Omega\Delta\Omega Y_\Omega^*\Delta\Omega}{\Delta f}.$$
$$\equiv \langle X_\Omega Y_\Omega^*\rangle \tag{A.11.10}$$

Next, we derive a relation between the correlated values in the frequency domain and the time domain. The time correlation of the two functions is given as

$$\langle X(t)Y(t')\rangle = \lim_{T\to\infty}\frac{1}{T}\int_0^T X(t)Y(t+\tau)dt. \tag{A.11.11}$$

We substitute (A.11.2) and (A.11.3) into (A.11.11), resulting in

$$\langle X(t)Y(t')\rangle = \lim_{T\to\infty}\frac{1}{T}\int_0^T\int_{-\infty}^{\infty}\int_{-\infty}^{\infty} X_\Omega Y_{\Omega'}e^{j(\Omega+\Omega')t+j\Omega'\tau}d\Omega' d\Omega dt \tag{A.11.12}$$

The time integration gives a delta function

$$\int_0^T e^{j(\Omega+\Omega')t}dt = 2\pi\delta(\Omega' + \Omega). \tag{A.11.13}$$

so that (A.11.12) becomes

$$\langle X(t)Y(t')\rangle = \lim_{T\to\infty}\frac{2\pi}{T}\int_{-\infty}^{\infty} X_\Omega Y_\Omega^* e^{-j\Omega\tau}d\Omega. \tag{A.11.14}$$

Here, we assume that the term $X_\Omega Y_\Omega^*$ is almost constant in Ω for then the integral becomes

$$\int_{-\infty}^{\infty} e^{-j\Omega\tau}d\Omega = 2\pi\delta(\tau). \tag{A.11.15}$$

Hence, we arrive at the following relation

$$\langle X(t)Y(t')\rangle = \frac{4\pi^2}{T}X_\Omega Y_\Omega^*\delta(t' - t) = \langle X_\Omega Y_\Omega^*\rangle\delta(t' - t). \tag{A.11.16}$$

Appendix 12
Calculating Procedure of the Mode Gain for MQW Structures

The profile of the band gap in a AlGaAs/GaAs MQW laser is shown in Fig. A.12.1. The core region consists of *M* wells made of GaAs and *M-1* barrier layers made of $Al_xGa_{1-x}As$. This core region is sandwiched between cladding made of *p* and *n* types $Al_{0.45}Ga_{0.55}As$. d_A and d_B are the thicknesses of the well and barrier layers, respectively; U_C and U_V are the barrier heights in the conduction and valence bands. In this analysis, we suppose that the effective masses m_c, m_H, and m_L in the conduction, heavy-hole, and light-hole bands, respectively, are the same in all layers.

The Schrödinger equation for an envelope function $\psi_c(y)$ associated with the conduction electron with energy eigenvalue is given for the *y* direction as

$$\left\{-\frac{\hbar^2}{2m_c}\frac{\partial^2}{\partial y^2}+U_c(y)\right\}\psi_c(y)=W_{yc}\psi_c(y). \tag{A.12.1}$$

We assume its solution in the range $y_{i-1}\le y\le y_i$ to be

$$\psi_c(y)=F_ie^{j\gamma_i(y-y_{i-1})}+B_ie^{-j\gamma_i(y-y_{i-1})}, \tag{A.12.2}$$

where F_i and B_i are unknown constants and γ_i a complex wave number for the *i*-th region formally given as

$$\gamma_i=\begin{cases}k_A & \text{in wells}\\ jk_B & \text{in barriers}\\ jk_o & \text{in cladding regions}\end{cases}. \tag{A.12.3}$$

The *energy value* associated with these *wave numbers* is

$$W_{yc}=\frac{\hbar^2k_A^2}{2m_c}=U_C-\frac{\hbar^2k_B^2}{2m_c}=U_0-\frac{\hbar^2k_o^2}{2m_c}, \tag{A.12.4}$$

where U_0 is the barrier height of the cladding. We analyze (A.12.1) using *continuity conditions* for $\psi_c(y)$ and $\partial\psi_c(y)/\partial y$(or $\{\partial\psi_c(y)/\partial y\}/m_i$ if the effective mass is different) on the boundaries of each section as follows:

The derivative of (A.12.1) with respect to *y* is

M. Yamada, *Theory of Semiconductor Lasers*,
Springer Series in Optical Sciences 185,
DOI: 10.1007/978-4-431-54889-8, © Springer Japan 2014

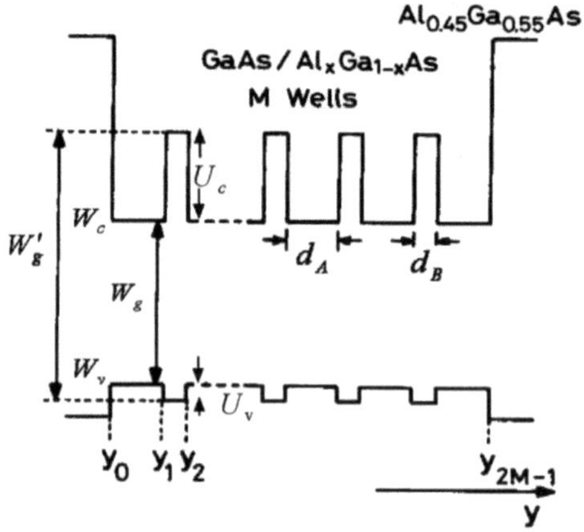

Fig. A.12.1 Band gap profile in a MQW laser

$$\frac{\partial \psi_c(y)}{\partial y} = j\gamma_i \left\{ F_i e^{j\gamma_i(y-y_{i-1})} - B_i e^{-j\gamma_i(y-y_{i-1})} \right\}. \tag{A.12.5}$$

Substitutions of $y = y_{i-1}$ into (A.12.2) and (A.12.5) give relations

$$\psi_c(y_{i-1}) = F_i + B_i, \tag{A.12.6}$$

$$\frac{\partial \psi_c(y_{i-1})}{\partial y} = j\gamma_i(F_i - B_i), \tag{A.12.7}$$

from which the unknown constants are obtained,

$$F_i = \frac{1}{2}\left\{ \psi_c(y_{i-1}) + \frac{1}{j\gamma_i}\frac{\partial \psi_c(y_{i-1})}{\partial y} \right\}, \tag{A.12.8}$$

$$B_i = \frac{1}{2}\left\{ \psi_c(y_{i-1}) - \frac{1}{j\gamma_i}\frac{\partial \psi_c(y_{i-1})}{\partial y} \right\}. \tag{A.12.9}$$

Substituting these solutions into (A.12.2) and (A.12.5), we derive the *transfer equation in matrix form*,

$$\begin{bmatrix} \psi_c(y) \\ \dfrac{\partial \psi_c(y)}{\partial y} \end{bmatrix} = \begin{bmatrix} \cos\{\gamma_i(y - y_{i-1})\} & \frac{1}{\gamma_i}\sin\{\gamma_i(y - y_{i-1})\} \\ -\gamma_i \sin\{\gamma_i(y - y_{i-1})\} & \cos\{\gamma_i(y - y_{i-1})\} \end{bmatrix} \begin{bmatrix} \psi_c(y_{i-1}) \\ \dfrac{\partial \psi_c(y_{i-1})}{\partial y} \end{bmatrix}. \tag{A.12.10}$$

As the wave decays smoothly in the cladding regions, the function in region $i = 0$ (left-hand side cladding) should be

$$\psi_c(y) = B_0 e^{k_o(y-y_0)}, \tag{A.12.11}$$

$$\frac{\partial \psi_c(y)}{\partial y} = k_o B_0 e^{k_o(y-y_0)}, \tag{A.12.12}$$

which give initial conditions at $y = y_0$

$$\psi_c(y_0) = B_0, \tag{A.12.13}$$

$$\frac{\partial \psi_c(y_0)}{\partial y} = k_o B_0. \tag{A.12.14}$$

Another function in the region $i = 2M$ (right-hand side cladding) should be

$$\psi_c(y) = F_{2M} e^{-k_o(y-y_{2M})}, \tag{A.12.15}$$

$$\frac{\partial \psi_c(y)}{\partial y} = -k_o F_{2M} e^{-k_o(y-y_{2M})}. \tag{A.12.16}$$

This condition implies that the backward propagating component in the $i = 2M$ region must be 0,

$$B_{2M} = \frac{1}{2}\left\{\psi_c(y_{2M-1}) + \frac{1}{k_o}\frac{\partial \psi_c(y_{2M-1})}{\partial y}\right\} = 0. \tag{A.12.17}$$

By tracing $\psi_c(y)$ and $\partial\psi_c(y)/\partial y$ from $y = y_o$ to $y = y_{2M-1}$, then with help from (A.12.10) and substitution into (A.12.17), we obtain the *characteristic equation*

$$[k_o \quad 1]\{[F_A][F_B]\}^{M-1}[F_A]\begin{bmatrix} 1 \\ k_o \end{bmatrix} = 0, \tag{A.12.18}$$

where $[F_A]$ and $[F_B]$ are *transfer matrices* for the well and barrier layers, respectively,

$$[F_A] = \begin{bmatrix} \cos(k_A d_A) & \frac{1}{k_A}\sin(k_A d_A) \\ -k_A \sin(k_A d_A) & \cos(k_A d_A) \end{bmatrix}, \tag{A.12.19}$$

$$[F_B] = \begin{bmatrix} \cosh(k_B d_B) & \frac{1}{k_B}\sinh(k_B d_B) \\ k_B \sinh(k_B d_B) & \cosh(k_B d_B) \end{bmatrix}. \tag{A.12.20}$$

The numerical calculation is done by changing the energy value W_{yc} in (A.12.4) and finding values to satisfy the characteristic equation of (A.12.18). The wave function is drawn using (A.12.10).

The wave functions $\psi_J(y)$ and energy value W_{yJ} in the valence band are similarly derived; here J indexes the heavy ($J = H$) and light ($J = L$) holes. The

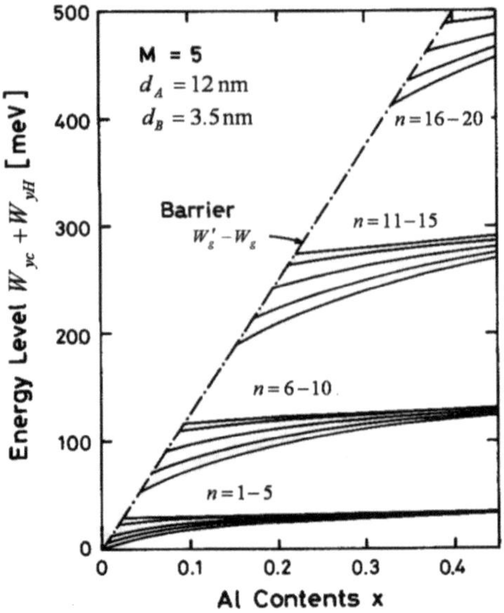

Fig. A.12.2 Calculated example of energy levels. The energy levels are split to M = 5 sublevels

value of the potential barrier is the difference between that in the conduction and valence bands,

$$W_g' - W_g = U_C + U_V. \tag{A.12.21}$$

The potential barrier U_H for the heavy hole is the same as U_V, but that of the light hole must differ from U_V.

Calculated examples based on the $Al_xGa_{1-x}As/GaAs$ MQW structure are shown in Figs. A.12.2, A.12.3 and A.12.4. The energy separation is assumed to be $U_C : U_V = 0.85 : 0.15$. Fortunately for the $Al_xGa_{1-x}As/GaAs$ system, this separation coincides with the inverse ratio of the effective masses, $U_C/U_H = m_H/m_c$. Then, we can put $m_cU_C = m_HU_H = m_LU_L$, resulting in the same shape for the electron wave functions in the conduction, heavy-hole, and light-hole bands, $\psi_C(y) = \psi_H(y) = \psi_L(y)$. The cross-sectional rate ζ_y of the electron waves defined in (11.21) is $\zeta_y = 1$.

The energy value W_y is given as

$$W_y = W_{yc} + W_{yJ}. \tag{A.12.22}$$

As found from Figs. A.12.2 and A.12.3, each energy level of a single quantum well is split into M sublevels in the MQW structure because of the coupling of electron wave functions among the wells. We relabel integer n as the energy level W_y from the lowest energy. When the energy W_y exceeds $W_g' - W_g$, W_y takes continuous values.

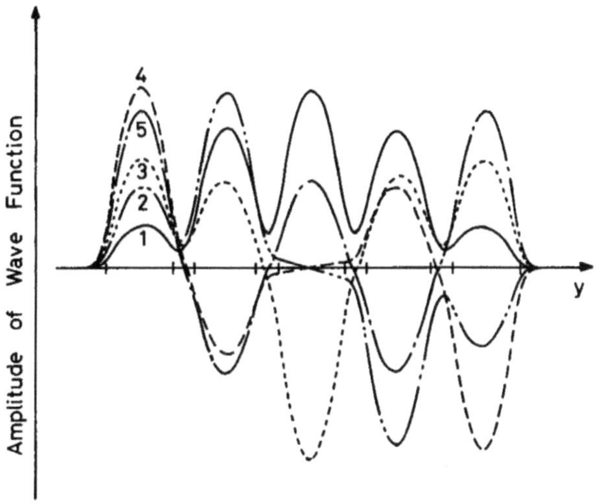

Fig. A.12.3 Calculated example of an electron wave. Each wave is numbered by lowest energy level

The total energy of an electron-hole pair is

$$W_{ba} = W_g + W_y + \frac{\hbar^2(k_x^2 + k_z^2)}{2}\left(\frac{1}{m_c} + \frac{1}{m_J}\right). \tag{A.12.23}$$

The density of states at energy W_{ba} for the energy level n is

$$g_{J,n}(W_{ba})dW_{ba} = \frac{1}{\pi\hbar^2 M d_A}\frac{m_c m_J}{m_c + m_J}dW_{ba}, \quad \text{for } W_g + W_y(n) < W_{ba} < W_g + U_C + U_J \tag{A.12.24}$$

As found from Fig. A.12.4, the energy levels take continuous values similar to the bulk crystal when W_{ba} is higher than the barrier height; then

$$\sum_n g_{J,n}(W_{ba})\, dW_{ba} = \frac{1}{2\pi\hbar^3}\left(\frac{m_c m_J}{m_c + m_J}\right)^{3/2}\sqrt{W_{ba} - W_g}dW_{ba}, \quad \text{for } W_{ba} > W_g + U_C + U_J \tag{A.12.25}$$

The electron injection density N is evaluated with

$$N = \sum_n \int_{W_g + W_y(n)}^{\infty} g_{C,H}(n, W_{ba})\{f_c(W_{ba}) - \tilde{f}_c(W_{ba})\}dW_{ba} \tag{A.12.26}$$

A calculated example of the gain profile for plane waves is shown in Fig. A.12.5. The profile shows effects depending on the intra-band relaxation time τ_{in}. If the relaxation time is long enough, $\tau_{in} > 1 \times 10^{-11}$ s, the gain profile must directly correspond to the profile for the density of states. However, the gain profile is smoothed out by the relaxation effect resulting in a diminishing of peak values. For bulk GaAs, the gain profile is marked by a broken line in Fig. A.12.5.

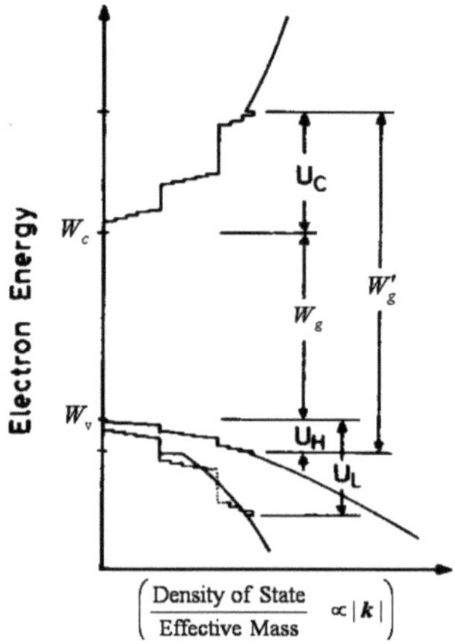

Fig. A.12.4 Calculated example of density of states

Fig. A.12.5 Calculated example of the gain profile

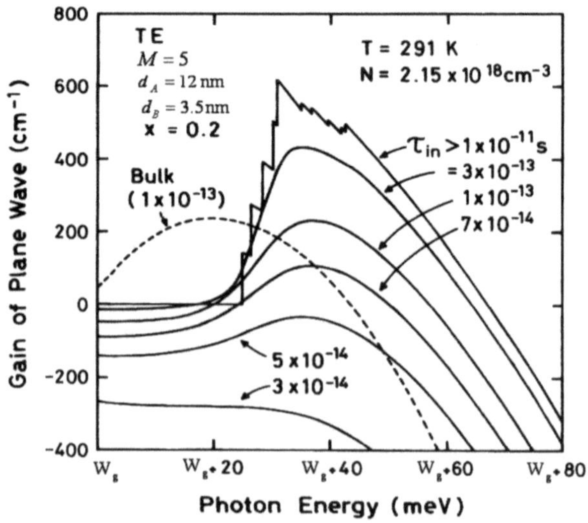

Unfortunately, in this example, the gain in MQW is not higher than that of the bulk crystal when the injected electron densities N are identical.

Therefore, we need to examine gain in terms of the injection current density rather than the injected electron density. Injection current density J for the laser is

$$J = \frac{eMd_A N}{\tau_s}, \tag{A.12.27}$$

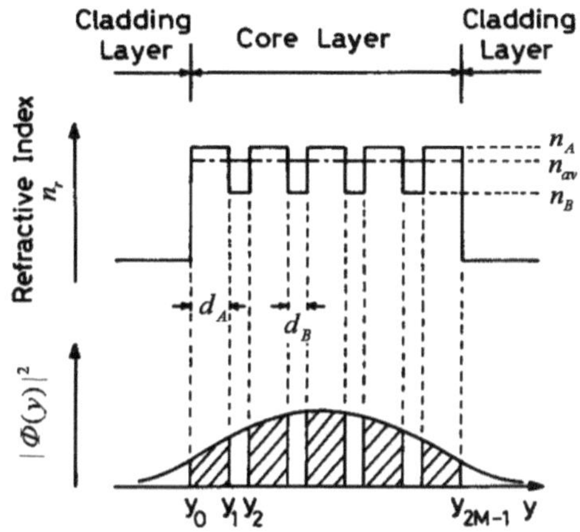

Fig. A.12.6 An approximate model to analyze the field confinement factor ξ

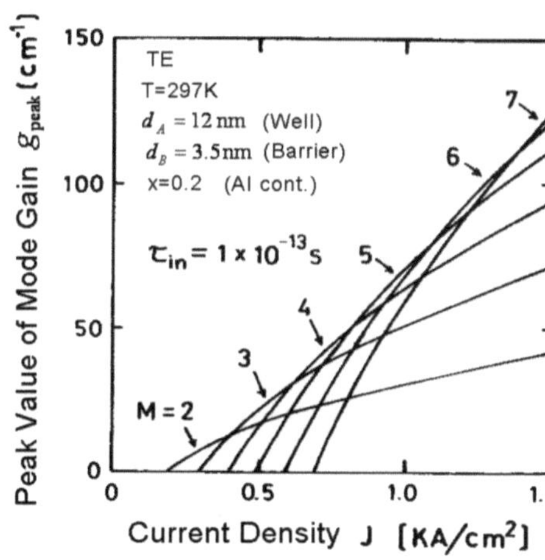

Fig. A.12.7 Variation of the peak value for mode gain with injection current density

where τ_s is the electron lifetime and can be theoretically evaluated by calculating the rate of spontaneous emission and the density of radiation modes as in (6.82). As a calculated value of τ_s for the $Al_xGa_{1-x}As/GaAs$ system, we use the approximation

$$\tau_s \approx 2.6 \times 10^6 \times N^{-0.81} \tag{A.12.28}$$

when N is given in units of cm^{-3}.

To obtain the mode gain $g_p(\omega)$, as introduced in (11.35), we need to evaluate the field confinement factor ξ. This factor is the spatial overlap rate between the electric field E and the electron wave $\phi_b(\mathbf{r})$. An accurate analysis of the rate calculation is very complicated because the electron distribution differs for each energy level, as shown in Fig. A.12.3. Therefore, we evaluate ξ approximately by the following procedure. We estimate the average refractive index n_{av} of the core layer which here consists of M wells and M-1 barrier layers. Next, we analyze the distribution function $\Phi(y)$ of the electric field as a slab wave guide consisting of one core layer and two cladding layers, as illustrated in Fig. A.12.6. Finally, we evaluate the normalized distribution over all wells

$$\xi = \frac{\sum_{i=1}^{M} \int_{y_{2i-2}}^{y_{2i-1}} |\Phi(y)|^2 dy}{\int_{-\infty}^{\infty} |\Phi(y)|^2 dy}. \tag{A.12.29}$$

The variation of peak value of the mode gain $g_{peak}(\omega) = \xi g(\omega)$ with injection current density J is shown in Fig. A.12.7. This figure indicates that multiple wells with large M are not suitable in reducing the current density.

Index

M. Yamada, *Theory of Semiconductor Lasers*,
Springer Series in Optical Sciences 185,
DOI: 10.1007/978-4-431-54889-8, © Springer Japan 2014

MIX
Papier aus verantwortungsvollen Quellen
Paper from responsible sources
FSC® C105338

If you have any concerns about our products,
you can contact us on
ProductSafety@springernature.com

In case Publisher is established outside the EU,
the EU authorized representative is:
Springer Nature Customer Service Center GmbH
Europaplatz 3, 69115 Heidelberg, Germany

Printed by Libri Plureos GmbH
in Hamburg, Germany